Technology-Enabled Resilience: Innovations in Critical Infrastructure Protection

John Soldatos · Habtamu Abie
Editors

Technology-Enabled Resilience: Innovations in Critical Infrastructure Protection

From AI-based Situational Awareness to Workforce Upskilling: A Holistic Approach to Critical Infrastructure Protection

Springer

Editors
John Soldatos
INNOV-ACTS Limited
Nicosia, Cyprus

Habtamu Abie
Norwegian Computing Center
Oslo, Norway

ISBN 978-3-032-25916-5 ISBN 978-3-032-25917-2 (eBook)
https://doi.org/10.1007/978-3-032-25917-2

This work was supported by European Commission through various H2020 and Horizon Europe research and innovation actions under grant agreements AI4CYBER (Grant Agreement No. 101070450), ATLANTIS (Grant Agreement No. 101073909), CyberSecDome (Grant Agreement No. 101120779), DYNABIC (Grant Agreement No. 101070455), eFORT (Grant Agreement No. 101075665), EU-CIP (Grant Agreement No. 101073878), PHOENI2X (Grant Agreement No. 101070586), R2D2 (Grant Agreement No. 101075714), Resilmesh (Grant Agreement No. 101119681), SUNRISE (Grant Agreement No. 101073821), NEMECYS (Grant Agreement No. 101094323), and IRIS (Grant Agreement No. 101021727), which are part of the ECSCI cluster. Several contributions of the book have been also supported financially from other EU projects, including SKILLAB (Grant Agreement No. 101132663), as well as various national projects, initiatives, and funding authorities as indicated in the acknowledgement sections of each chapter. The Chapters have been funded by beneficiaries of the abovelisted actions including FUNDACION TECNALIA RESEARCH & INNOVATION, Universidad Pontificia Comillas, Norsk Regnesentral STI (NRS), CaixaBank, Netcompany S.A, Carr Communications LtD, FUNDACION VICOMTECH, Maggioli S.p.A, Technical University of Crete, Siemens S.R.L, OTE Group of Companies, ETRA INVESTIGACIÓN Y DESARROLLO SA, Jožef Stefan Institute, ICS Ljubljana, IBM ISRAEL – SCIENCE AND TECHNOLOGY LTD (IBM), Centre for Research and Technology Hellas, UNIVERSITY OF PATRAS, Engineering Ingegneria Informatica S.p.A., INNOV-ACTS LIMITED and Technological University of the Shannon.

This Springer imprint is published by the registered company Springer Nature Switzerland AG
The registered company address is: Gewerbestrasse 11, 6330 Cham, Switzerland

Preface

In recent years, we have witnessed the emergence of new challenges for critical infrastructure resilience. Nowadays, critical infrastructure operators need to operate their infrastructures in the scope of a rapidly evolving landscape, which is marked by an unprecedented number of disruptive and asymmetric incidents that take place in increased frequency. Such incidents range from global pandemics (e.g. the COVID19 pandemic outbreak) to armed conflicts (e.g. the war in Ukraine) and other hybrid threats. Such events threaten to destabilize essential services such as energy, transport, health, and digital infrastructure. At the same time, they highlight the complex interdependencies and vulnerabilities of critical assets in modern societies.

In this landscape, the accelerated digitalization of critical infrastructures—such as the increased deployment, integration and use of artificial intelligence (AI), Internet of things (IoT), and interconnected IT/OT systems—offers new opportunities for enhanced efficiency, improved situational awareness, and rapid incident response. However, these technological advancements also introduce novel risks, notably risks in the domain of cybersecurity. Cybersecurity attacks have grown in both sophistication and scale, as threats such as ransomware, supply chain attacks, and advanced persistent threats routinely challenge traditional defence mechanisms. This is because adversarial actors tend to exploit the expanding attack surface that is created by digital transformation and the increased deployment of the aforementioned technologies.

In addition to developing effective responses to evolving and increasingly sophisticated threats, critical infrastructure operators must also comply with a complex web of regulatory mandates such as the Critical Entities Resilience (CER) Directive, the second Network and Information Security Directive (NIS2), the General Data Protection Regulation (GDPR), and the AI Act. These regulations demand continuous improvement in cybersecurity risk management, transparency, incident reporting, and data protection. Moreover, the need for harmonized, cross-sectoral approaches highlights another critical challenge: critical infrastructures operators must bridge legacy gaps and align governance, processes, and standards in a highly volatile environment where technologies and regulations are constantly evolving.

The emergence of new tools and regulatory requirements also call for a significant shift in workforce capabilities. Successful management of digital risks and regulatory compliance depends on substantial investments in upskilling and reskilling towards developing worker and skills profiles that did not exist a few years ago. Soft skills, cross-sector collaboration, and the ability to adopt and operate modern technologies are now as important as technical and security know-how. Currently, critical infrastructure operators face a skills gap, which is among their primary concerns. The urgent need to fill-in this gap makes human capital and cultural adaptation vital "complementary assets" that go hand in hand with the development of technological capabilities for critical infrastructure resilience.

As previously outlined, modern technological solutions—such as knowledge modelling, advanced AI and machine learning, IoT-enabled real-time data analytics, the convergence of IT and OT domains—enable the development of highly effective systems for critical infrastructure protection and the resilience of critical entities. Specifically, these technologies empower critical entities to develop and deploy the following capabilities:

- Maintain real-time situational awareness and rapid threat detection through digital twins, knowledge graphs, and AI-driven monitoring systems.
- Implement automated and coordinated responses to emerging incidents towards reducing reaction times and improving incident mitigation.
- Perform predictive resilience planning based on dynamically updated, cross-domain datasets.
- Facilitate secure and compliant information sharing to support collective cyber–physical threat intelligence, while ensuring regulatory conformity.

These advancements require investments that go beyond hardware and software systems, notably investments in complementary assets like process reengineering, workforce development, and a culture that embraces cyber–physical resilience and continuous learning.

The urgency of these needs is recognized at the European level, where dozens of EU-funded projects on critical infrastructure resilience are at the forefront of researching, developing, and validating integrated solutions. Most of these projects bring together technology innovation, regulatory insight, and skills development towards developing and demonstrating real-world solutions to enhancing the resilience, adaptability, and trustworthiness of European critical infrastructures. Furthermore, several of the aforementioned projects have recognized the importance of sharing knowledge, insights, and best practices towards driving the next generation of critical infrastructure resilience that can effectively respond to modern challenges. In this direction, several EU-funded projects on critical infrastructure resilience have formed the European Cluster for Securing Critical Infrastructures (ECSCI) cluster, which is aimed at creating synergies and fostering emerging disruptive solutions to security issues through cross-projects collaboration and innovation. ECSCI (https://www.ecsci.eu/) has been founded back in 2019 and has brought together over 60 different projects that have collaborated on various knowledge sharing and innovation development activities for critical infrastructure resilience. Since 2022, ECSCI has also evolved into a knowledge network on Critical Infrastructure Protection, which provides experts insights into the present and future of critical entities' resilience. These insights are shared with the European Critical Infrastructure Protection community through a rich set of joint white papers, scientific papers, workshops, conferences, and policy briefs.

In this context, this book represents a joint effort of various EU-funded projects of the ECSCI cluster. It is dedicated to introducing and presenting of novel solutions for critical infrastructure resilience, notably solutions that have been developed and validated in selected projects of the cluster. Specifically, the book explores various strategies, challenges, and innovations involved in protecting and enhancing the resilience of critical infrastructures in light of rapidly evolving cyber and physical threats. It examines the shift from traditional, reliability-focused security models to emerging holistic, resilience-oriented frameworks that integrate risk-aware planning, cyber–physical security, advanced technology adoption, and regulatory compliance. Based on a wide range of practical case studies and stakeholder-driven analysis, the book sheds light on the intersection of policy, technology, and organizational preparedness required to safeguard infrastructures in essential sectors such as energy, telecommunications, healthcare, finance, and transportation.

Some of the key themes of the book include the increasing sophistication and interdependence of threats, ranging from cyberattacks on IoT and industrial control systems to cascading failures triggered by extreme weather events. The different chapters of the book present methods and solutions for developing adaptive and anticipatory responses to these threats. Emphasis is paid on real-world scenarios for protecting critical assets, such as the integration of AI for anomaly detection, dynamic risk assessment, stress-testing frameworks for regulatory alignment, and the use of digital twins and federated learning to support situational awareness and incident management. Furthermore, some of the chapters of the book address complementary assets such as skills profiles and learning paths for critical infrastructure protection roles and professionals, as a means to developing a workforce that can confront the modern challenges faced by critical entities and critical infrastructure operators.

The book comprises 17 chapters that present innovative systems, solutions and services for critical infrastructure resilience. The chapters describe work carried out in various EU-funded projects of the ECSCI cluster, including the HORIZON AI4CYBER (Grant Agreement No. 101070450), HORIZON ATLANTIS (Grant Agreement No. 101073909), the HORIZON CyberSecDome (Grant Agreement No. 101120779), the HORIZON DYNABIC (Grant Agreement No. 101070455), the eFORT (Grant Agreement No. 101075665), the HORIZON EU-CIP (Grant Agreement No. 101073878), the HORIZON PHOENI2X (Grant Agreement No. 101070586), the HORIZON R2D2 (Grant Agreement No. 101075714), the HORIZON Resilmesh (Grant Agreement No. 101119681), the HORIZON SUNRISE (Grant Agreement No. 101073821), the NEMECYS (Grant Agreement No. 101094323), and the H2020 IRIS (Grant Agreement No. 101021727) projects. Note that the EU-CIP project has assumed the role of coordinating the editing of the book, as part of its coordination role in the ECSCI ecosystem.

The book is structured into the following three parts:

Effective Risk Management and Impact Assessment of Complex Incidents
This part presents solutions for risk management, risk mitigation, and impact assessment of sophisticated incidents in critical infrastructures. The chapters of this part of the book address a wide range of risks—such as security, safety, privacy, business continuity risks—across various sectors (e.g. energy and power systems), and resilience settings (e.g. cyberattacks, pandemics, and healthcare crises).

Enhancing Resilience Based on Artificial Intelligence
This part of the book sheds light on the proliferating deployment and use of data-driven systems and AI-based system for protecting critical infrastructures and enhancing their resilience. It presents various AI-based techniques across different use cases such as AI-assisted cyber-resilience and AI-based orchestration of different systems and services towards increased resilience.

Knowledge Sharing, Situation Awareness and Skills in Critical Infrastructure Resilience
This part of the book complements the previous two parts by introducing novel complementary assets spanning various training and skills development resources, as well as tools and techniques for enhanced situational awareness and knowledge sharing.

Overall, the book provides a comprehensive overview of modern technology-enabled approaches to critical infrastructure resilience, along with complementary assets and resources that foster improved protection for critical entities. The book is published as an open access publication, which ensures that researchers and practitioners will have unrestricted access to it. In essence, this book represents a contribution of the ECSCI cluster and the EU-CIP knowledge network to the European Critical Infrastructures community. We hope that researchers, practitioners, integrators of resilience solutions, and critical infrastructure operators will find it useful and engaging.

Nicosia, Cyprus
Oslo, Norway
July 2025

John Soldatos
Habtamu Abie

Acknowledgments

This book has received funding from the European Union's H2020 and Horizon Europe research and innovation programmes under grant agreements AI4CYBER (Grant Agreement No. 101070450), ATLANTIS (Grant Agreement No. 101073909), CyberSecDome (Grant Agreement No. 101120779), DYNABIC (Grant Agreement No. 101070455), eFORT (Grant Agreement No. 101075665), EU-CIP (Grant Agreement No. 101073878), PHOENI2X (Grant Agreement No. 101070586), R2D2 (Grant Agreement No. 101075714), Resilmesh (Grant Agreement No. 101119681), SUNRISE (Grant Agreement No. 101073821), NEMECYS (Grant Agreement No. 101094323), and IRIS (Grant Agreement No. 101021727), which are part of the ECSCI cluster. Several contributions of the book have been also supported financially from other EU projects, including SKILLAB (Grant Agreement No. 101132663), as well as various national projects, initiatives, and funding authorities as indicated in the acknowledgement sections of each chapter. The editor and the chapter co-authors acknowledge valuable support from the partners of the above-listed EU projects. Views and opinions expressed are those of the authors only and do not necessarily reflect those of the European Union or the granting authorities. Neither the European Union nor the granting authorities can be held responsible for them.

Socioeconomic [illegible] Program Based on [illegible] Leveraging [illegible] and Data [illegible] and Awareness [illegible]

[illegible]

Abstract. The socioeconomic [illegible] cybersecurity training [illegible] nations [illegible] increasing cyber threats, [illegible]

e-mail: [illegible]

P. Papagiannis
e-mail: [illegible]

I. [illegible]

Contents

Effective Risk Management and Impact Assessment of Complex Incidents

Editors and Contributors

About the Editors

John Soldatos (http://gr.linkedin.com/in/johnsoldatos) holds a Ph.D. in Electrical and Computer Engineering from the National Technical University of Athens (2000) and is currently an honorary research fellow at the University of Glasgow, UK (2014–present). He was an associate professor and the head of the Internet of Things (IoT) Group at the Athens Information Technology (AIT), Greece (2006–2019), and an adjunct professor at the Carnegie Mellon University, Pittsburgh, PA (2007–2010). He has significant experience in working closely with large multinational industries (IBM Hellas, INTRACOM S.A, INTRASOFT International S.A, and Netcompany-Intrasoft S.A) as an R&D consultant and delivery specialist, while being scientific advisor to high-tech start-up enterprises. Dr. Soldatos is an expert in Internet of things (IoT) and artificial intelligence (AI) technologies and applications, including applications in smart cities, finance, and industry (Industry 4.0). Dr. Soldatos has played a leading role in the successful delivery of more than 80 (commercial-industrial, research, and business consulting) projects, for both private and public sector organizations, including complex integrated projects. He is a co-founder of the open-source platform OpenIoT (https://github.com/OpenIotOrg/openiot). He has published more than 220 articles in international journals, books, and conference proceedings. He has also significant academic teaching experience, along with experience in executive education and corporate training. Dr. Soldatos is a regular speaker in international conferences, as well as contributor to various international magazines and blogs, on topics related to artificial intelligence, IoT, Industry 4.0, and cybersecurity. Moreover, he has contributed to various standardization working groups and expert groups, while he has also been appointed at various boards. He has recently co-edited and coauthored ten (10) edited volumes (books) on artificial intelligence, Internet of things, and cybersecurity-related themes.

Habtamu Abie (https://home.nr.no/~abie/, https://www.linkedin.com/in/habtamu-abie-158746), Ph.D., Chief Research Scientist and the Digital Security Focus Area Leader at the Norwegian Computing Center (Norsk Regnesentral—NR). He received his B.Sc., M.Sc., and Ph.D. in Computer Science from the University of Oslo. Dr. Abie has previously held positions as a scientific associate and fellow at CERN (European Organization for Nuclear Research) in Geneva, Switzerland, a researcher at ABB Corporate Research, a software development engineer at Nera-AS in Billingstad, Norway, and at Alcatel Telecom Norway AS in Oslo, Norway. He also served as a senior engineer and research scientist at Telenor Research and Development. Dr. Abie has extensive experience in computing, working as both a practitioner and researcher. He has many years of experience in building large consortia and leading technical R&D work in both industry-driven and multinational projects. His expertise in leading interdisciplinary teams and projects at both national and international levels highlights his ability to foster collaborative environments aimed at enhancing security and resilience in critical infrastructures. Currently, he holds the position of IT-OT Integration Focus Area Leader at SFI NORCICS and is the originator and leader of the Norwegian Ecosystem for Secure IT-OT Integration (NESIOT). He served as the project manager and principal investigator for the ASSET project (2012–2016) and co-leads the CybAlliance project (2023–2028), funded by the Research Council of Norway. He has also played a key role in coordinating the European Cluster for Securing Critical Infrastructures (ECSCI), a significant cluster comprising 46 EU-funded projects dedicated to securing vital assets. Dr. Abie has co-edited nine books and proceedings, with two additional proceedings scheduled for publication in 2025. He has participated in several EU and national projects including GEMOM (2008–2010), uTRUSTit (2010–2013), IoTSec (2015–2020), ArsForensica (2015–2022), FINSEC (2018–2021), FAME (2023–2025), ENFIELD (2023–2025), EU-CIP (2022–2025), and NORCICS (2020–2028). He is the organizing co-chair of the CPS4CIP (2019–2025) and SecASure (2022–2025) in conjunction with ESORICS, SecIndustry (2023–2025) in conjunction with ARES, SUNRISE (2023–2025) in conjunction with NIKT, and CyberHunt 2024 in conjunction with IEEE Big Data. Dr. Abie has a solid and extensive background in the design and development of real-time systems for controlling and monitoring distributed and complex environments. His current research interests include adaptive security, adaptive AI for security, critical infrastructure protection and resilience, cybersecurity, adaptive threat hunting and intelligence, trust and risk management, 5G and beyond, Internet of things, and digital twins for enhancing cybersecurity and resilience. He is a member of the European Academy of Sciences (EURASC), a senior member of IEEE, and a fellow of the Asia-Pacific Artificial Intelligence Association (AAIA) and the International Artificial Intelligence Industry Alliance (AIIA). He has been recognized as a top scholar by ScholarGPS, ranking in the top 0.5% of all scholars worldwide and as the lifetime #16 top scholar in eHealth.

Contributors

Habtamu Abie Norwegian Computing Center, Blindern, Oslo, Norway

Emiliano Altobelli FS Technology SpA, Rome, Italy

Guy Amit IBM Research, IBM Israel—Science and Technology Ltd (IBM), Haifa, Israel

Maya Anderson IBM Research, IBM Israel—Science and Technology Ltd (IBM), Haifa, Israel

Constantinos Andreou National CSIRT-CY Digital Security Authority, Nicosia, Cyprus

Anna Andreoudi National Cybersecurity Authority, Athens, Greece

Athanasios Arvanitis Public Power Corporation S.A., Athens, Greece

Nikolaos Bakalis Public Power Corporation S.A., Athens, Greece

Jorge Bernal University of Murcia, Murcia, Spain

Denis Čaleta Institute for Corporative Security Studies, Ljubljana, Slovenia

Gianfranco Caputo Links Foundation, Torino, TO, Italy

Marianna Charalambous INNOV-ACTS Limited, Nicosia, Cyprus

Konstantinos Chisiridis Research and Innovation Development Department, Netcompany BELUX, Athens, Greece

Chrysanthos Chrysanthou Sphynx Analytics Limited, Nicosia, Cyprus

Grigorios Chrysos Technical University of Crete, Crete, Greece

Elias Dakos Athens Technology Center (ATC), Athens, Greece

Petros Daras CERTH, Thessaloniki, Greece

Ramon Martin de Pozuelo CaixaBank S.A. Sabino Arana, Barcelona, Spain

Ionut Vladut Dinu Siemens S.R.L., Bucharest, Romania

Kostas Drakonakis Technical University of Crete, Crete, Greece

Nicola Durante Engineering—Ingegneria Informatica Spa (ENG), Roma, Italy

Nicola Gregorio Durante Engineering Ingegneria Informatica SpA, Rome, Italy

Armend Duzha R&D Maggioli, Santarcangelo di Romagna, Italy

Lorenzo Eccher Engineering Ingegneria Informatica SpA, Rome, Italy

Abla Mimi Edjossan-Sossou Sixense Engineering, Nanterre, France

Martí Fabregat CaixaBank S.A. Sabino Arana, Barcelona, Spain

Ariel Farkash IBM Research, IBM Israel—Science and Technology Ltd (IBM), Haifa, Israel

Nicolas Ferry I3S/INRIA Kairos, Université Côte d'Azur, Nice, France

Evangelos Floros Department of Management Science and Technology, Hellenic Mediterranean University, Agios Nikolaos, Greece;
IT Department, University General Hospital of Heraklion, Heraklion, Greece

Kostantinos Fysarakis Department of Informatics, University of Oslo, Oslo, Norway

Marko Gerbec Jožef Stefan Institute, Ljubljana, Slovenia

Amaia Gil Vicomtech Foundation, Basque Research and Technology Alliance (BRTA), Donostia-San Sebastián, Spain

Gabriele Giunta Engineering Ingegneria Informatica SpA, Rome, Italy;
Engineering Ingegneria Informatica S.p.a., Palermo, Italy

Christos Gizelis Hellenic Telecommunications Organization (OTE), Athens, Greece

Abigail Goldsteen IBM Research, IBM Israel—Science and Technology Ltd (IBM), Haifa, Israel

Athanasios Grigoriadis Ministry of National Defence, Athens, Greece

Emilia Gugliandolo Engineering Ingegneria Informatica S.p.A, Roma, RM, Italy

Aleksandar Jovanović Steinbeis EU-VRi—European Risk & Resilience Institute, Stuttgart, Germany

Efi Kafali CERTH, Thessaloniki, Greece

Ioannis Kapatsoulias Hellenic Telecommunications Organization (OTE), Athens, Greece

Noam Koren IBM Research, IBM Israel—Science and Technology Ltd (IBM), Haifa, Israel

Petar Krstevski Ss. Cyril and Methodius University in Skopje, Faculty of Electrical Engineering and Information Technologies, Skopje, North Macedonia

Michael Kühr Computer Engineering, Technical University of Munich, Munich, Germany

Maxime Vinh Hoa La Montimage sarl., Paris, France

Kostantinos Lampropoulos Department of Electrical and Computer Engineering, University of Patras, Patras, Greece

Brian Lee Software Research Institute, Technological University of the Shannon, Athlone, Ireland

Gregorio López López Institute for Research in Technology, Comillas Pontifical University, Madrid, Spain

Konstantinos Magkos National Cybersecurity Authority, Athens, Greece

Jason Xabier Mansell TECNALIA, Basque Research and Technology Alliance (BRTA), Astondo Bidea, Derio, Biscay, Spain

Javier Matanza Domingo Institute for Research in Technology, Comillas Pontifical University, Madrid, Spain

Aleksandra Krkoleva Mateska Ss. Cyril and Methodius University in Skopje, Faculty of Electrical Engineering and Information Technologies, Skopje, North Macedonia

Vasileios Mavroeidis Department of Informatics, University of Oslo, Oslo, Norway

Dimitris Merkouris Public Power Corporation S.A., Athens, Greece

Jolanda Modic Institute for Corporative Security Studies, Ljubljana, Slovenia

Haralambos Mouratidis Security Labs Consulting, Cork, Ireland

Panagiotis Mpempis Hellenic Telecommunications Organization (OTE), Athens, Greece

Ekam Puri Nieto University of Murcia, Murcia, Spain

Nikolaos Nikoloudakis Public Power Corporation S.A., Athens, Greece

Dario Ogrizović Faculty of Maritime Studies, University of Rijeka, Rijeka, Croatia

Marc-Oliver Pahl IMT Atlantique, Rennes, France

Petros Panagiotidis Hellenic Telecommunications Organization (OTE), Athens, Greece

Mathaios Panteli Department of Electrical and Computer Engineering, University of Cyprus, Nicosia, Cyprus

Dimitris Papanikas Hellenic Telecommunications Organization (OTE), Athens, Greece

Spyridon Papastergiou R&D Maggioli, Athens, Greece

Aljosa Pasic Advanced Computing, Bull, Madrid, Spain

Charalampos Z. Patrikakis Department of Electrical and Electronics Engineering, University of West Attica, Egaleo, Athens, Greece

Sandeep Pirbhulal Norwegian Computing Center, Blindern, Oslo, Norway

Rauno Pirinen Laurea University of Applied Sciences, Vantaa, Finland

George Potamos Open University of Cyprus, Nicosia, Cyprus

George Pouraimis Ministry of National Defence, Athens, Greece

Natalia Razinkov IBM Research, IBM Israel—Science and Technology Ltd (IBM), Haifa, Israel

Angel Rego TECNALIA, Basque Research and Technology Alliance (BRTA), Astondo Bidea, Derio, Biscay, Spain

Néstor Rodríguez-Pérez Institute for Research in Technology, Comillas Pontifical University, Madrid, Spain

Jose Luis Rueda Torres Faculty of EEMCS, Delft University of Technology, Delft, The Netherlands

Lander Segurola Vicomtech Foundation, Basque Research and Technology Alliance (BRTA), Donostia-San Sebastián, Spain

Theodoros Semertzidis CERTH, Thessaloniki, Greece

Shlomit Shachor-Ifergan IBM Research, IBM Israel—Science and Technology Ltd (IBM), Haifa, Israel

Lukas Sigrist Institute for Research in Technology, Comillas Pontifical University, Madrid, Spain

Dimitrios Skias Research and Innovation Development Department, Netcompany BELUX, Athens, Greece

Michail Smyrlis Sphynx Technology Solutions AG, Nicosia, Cyprus

John Soldatos Innov-Acts LTD, Nicosia, Cyprus

George Spanoudakis Sphynx Technology Solutions AG, Zug, Switzerland; Sphynx Technology Solutions AG, Nicosia, Cyprus

Christoforos N. Spartalis CERTH, Thessaloniki, Greece

Eliana Stavrou Open University of Cyprus, Nicosia, Cyprus

Stavros Stavrou Open University of Cyprus, Nicosia, Cyprus

Ugo Stecchi New Technologies Department, ETRA Investigación y Desarrollo, Valencia, Spain

Branka Stojanovic Joanneum Research Digital, Graz, Austria

Stelios Trikos National CSIRT-CY Digital Security Authority, Nicosia, Cyprus

Konstantina Tripodi JRC Capital Management, Berlin, Germany

Sofia Tsekeridou Research and Innovation Development Department, Netcompany BELUX, Athens, Greece

Rita Ugarelli SINTEF Community, Blindern, Oslo, Norway

Balaji Venkatasubramanian Department of Electrical and Computer Engineering, University of Cyprus, Nicosia, Cyprus

Cristian Raul Vintila Siemens S.R.L., Bucharest, Romania

Iordanis Xanthopoulos Sphynx Technology Solutions AG, Zug, Switzerland

Tengfei Yang Software Research Institute, Technological University of the Shannon, Athlone, Ireland

Susana Gonzalez Zarzosa Advanced Computing, Bull, Valladolid, Spain

[illegible]

4 Advanced Cyber Range Features

4.1 [illegible] management

[illegible] organization's [illegible] and procedures [illegible] Meanwhile, the [illegible] personnel is [illegible] attacks in real-time [illegible]

The [illegible] improving the security processes.

4.2 Threat Hunting and Incident Response Training

[illegible] to detect, respond and mitigate [illegible] tion systems [illegible] simulating attack and response [illegible] and enhancing situational [illegible] real-world IT environments [illegible]

Abbreviations

AAA	Algorithmic Accountability Act
ACF	Awareness and Comprehension Framework
ADFS	Active Directory Federation Services
ADMIE	Independent Power Transmission Operator (Greece)
AI	Artificial Intelligence
AI/ML	Artificial Intelligence/Machine Learning
AI4CTI	AI for Cyber Threat Intelligence
AI4CYBER	AI for Cybersecurity (specific EU project)
AI4FIX	AI for Fixing (software/weaknesses correction)
AI4SOAR	AI for Security Orchestration, Automation, and Response
AI4TRIAGE	AI for Incident Triage
AIDA	Artificial Intelligence and Data Act
AI-IPS	AI-based Intrusion Prediction System
AIRE	Atos Incident Report Engine
AIS	Automatic Identification System
AMI	Advanced Metering Infrastructure
API	Application Programming Interface
APT	Advanced Persistent Threat
AR	Augmented Reality
ARERA	Regulatory Authority for Energy, Networks and Environment (Italy)
ARIMA	Autoregressive Integrated Moving Average
ART	The Adversarial Robustness Toolbox
ASR	Attack Success Rate
ATLANTIS	Improved resilience of Critical Infrastructures Against Large scale transNational and sysTemic rISks
ATLAS	Adversarial Threat Landscape for Artificial-Intelligence Systems
AutoML	Automated Machine Learning
AWS	Amazon Web Services
BCI	Business Continuity Institute
BERT	Bidirectional Encoder Representations from Transformers

BESS	Battery Energy Storage System
BPMN	Business Process Model and Notation
CACAO	Collaborative Automated Course of Action Operations
CASM	Cyber Attack Surface Management
CCI-SAAM	Cross-CI Sharing Assessment Analysis Mitigation
CCPA	California Consumer Protection Act
CDM	Cyber Defense Matrix
CE	Conformité Européenne
CEP	Complex Event Processing
CER	Critical Entities Resilience (Directive)
CERT	Computer Emergency Response Team
CGM	Continuous Glucose Monitoring
CI	Critical Infrastructure
CIA	Confidentiality, Integrity, and Availability
CIDR	Classless Inter-Domain Routing
CIP	Critical Infrastructure Protection
CIR	Critical Infrastructure Resilience
CIRIX	Composite Indicator for Resilience Index
CISA	Cybersecurity and Infrastructure Security Agency
CIWIN	Critical Infrastructure Warning Information Network
CK	Composite Knowledge
CoA	Course of Action
CoSAI	Coalition for Secure AI
CPH	Critical Physical Infrastructure.
CPPA	Consumer Protection Privacy Act
CPS	Cyber-Physical Systems
CRA	Cyber Resilience Act
CSA	Critical Service Awareness
CSA	Cybersecurity Act
CSIM	Converged Security and Information Management
CSIRT	Computer Security Incident Response Team
CTF	Capture The Flag
CTI	Cyber Threat Intelligence
CVEs	Common Vulnerabilities and Exposures
CXB	CaixaBank
CySA	Cyber Situational Awareness
DAG	Directed Acyclic Graph
DAIR	Dynamic and Adaptive Incident Response
DAO	Digital Artifact Ontology
DARS	Družba za avtoceste v Republiki Sloveniji (Slovenian Highways Operator)
DB	Database
DBMS	Database Management System
DDoS	Distributed Denial of Service
DERs	Distributed Energy Resources

DEXi	Decision EXpert for Intelligent systems
DIKW	Data, Information, Knowledge, Wisdom
DIN	Deutsches Institut für Normung
DL	Deep Learning
DoS	Denial of Service
DPO	Data Protection Officer
DRA	Dynamic Risk Assessment
DSN	Deep Stacked Network
DSS	Decision Support System
DT	Digital Twin
DYNABIC	Enhancing Cyber Resilience of Critical Infrastructures through Digital Twin Technology for Business Continuity
EasyEASM	Easy External Attack Surface Management
EBA	European Banking Authority
EBSI	European Blockchain Services Infrastructure
ECI	European Critical Infrastructure
ECSCI	European Cluster for Securing Critical Infrastructures
EDA	Event-Driven Architecture
EDR	Endpoint Detection and Response
EHDS	European Health Data Space Regulation
EHR	Electronic Health Record
EM-DAT	Emergency Events Database (CRED/OFDA International Disaster Database)
EME	Enabling Multi-layered and Executable resilience
ENISA	European Union Agency for Cybersecurity
ENSREG	European Nuclear Safety Regulators Group
ENTSO-E	European Network of Transmission System Operators for Electricity
EO	Earth Observation
EPFL	École Polytechnique Fédérale de Lausanne
EU	European Union
EU-CIP	European Knowledge Hub and Policy Testbed for Critical Infrastructure Protection
EUI	European University Institute
FAIR	Findability, Accessibility, Interoperability, and Reuse
FPGA	Field-Programmable Gate Array
FPR	False Positive Rate
FST	Ferrovie dello Stato Italiane (Italian Railways)
FTC	US Federal Trade Commission
GDPR	General Data Protection Regulation
GNN	Graph Neural Network
GNSS	Global Navigation Satellite System
GPS	Global Positioning System
GPT	Generative Pre-trained Transformer
GUI	Graphical User Interface

HIPAA	US Health Insurance Portability and Accountability Act
HiVIC	Humans in Vicinity
HMI	Human Machine Interface
HW-IDS	Hardware-based Intrusion Detection System
I2I	Image-to-Image (generative models)
IAM	Identity and Access Management
ICT	Information and Communication Technology
IDMEF	Intrusion Detection Message Exchange Format
IDPS	Intrusion Detection and Prediction System
IDS	Intrusion Detection System
IDS/IPS	Intrusion Detection and Prevention System
IoMT	Internet of Medical Things
IoT	Internet of Things
IP	Internet Protocol
IPS	Intrusion Prediction System
IR	Incident Response
IRIS	Increased Resilience and Efficiency through Information Sharing
IRP	Integration Reference Point
ISAC	Information Sharing and Analysis Center
ISIM	Infrastructure and Service Information Model
ISO	International Organization for Standardization
JRC	Joint Research Centre (European Commission)
JSON	JavaScript Object Notation
KG	Knowledge Graph
KPI	Key Performance Indicator
LEED	Leadership in Energy and Environmental Design
LGPD	Brazilian General Data Protection Law
LLava	Large Language and Vision Assistant
LLM	Large Language Model
LSP	Large Scale Pilot
LSTM	Long Short-Term Memory
LUK	Luka Koper (Port of Koper)
LUR	Luka Rijeka (Port of Rijeka)
MAIP	Montimage AI Platform
MCDA	Multi-Criteria Decision Analysis
MCSA	Maritime Cyber Situational Awareness
MDCG	Medical Device Coordination Group
MDI	Ministry of Interior, Railway State Police, Italy
MDMS	Meter Data Management System
MDR	Medical Devices Regulation
MFA	Multi-Factor Authentication
MI	Model Inversion
MIA	Membership Inference Attack
MISP	Malware Information Sharing Platform
MIT	Massachusetts Institute of Technology

ML	Machine Learning
MM	Mitigation Manager
MS	Member State
MTD	Moving Target Defense
MU	Machine Unlearning
MZI	Ministrstvo za Infrastrukturo (Ministry for Infrastructure, Slovenia)
NDR	Network Detection and Response
NERC	North American Electric Reliability Corporation
NESO	National Energy System Operator (UK)
NFT	Neural Fourier Transform
NHS	National Health Service
NIC	National Infrastructure Commission (UK)
NIS2	Network and Information Security Directive (version 2)
NIST	National Institute of Standards and Technology
NLP	Natural Language Processing
Nmap	Network Mapper
NSE	Network Situation Evaluator
NTK	Neural Tangent Kernel
NYIF	New York Institute of Finance
OASIS	Organization for the Advancement of Structured Information Standards
OECD	Organisation for Economic Co-operation and Development
OES	Operators of Essential Services
OOD	Out-of-Distribution
OODA	Observe-Orient-Decide-Act
OT	Operational Technology
OUC	Open University of Cyprus
OWASP	Open Web Application Security Project
PACS	Picture Archiving and Communication System
PAM	Privileged Access Management
PCTI	Physical/Cyber Threat Intelligence
PDC	Phasor Data Concentrator
PDCA	Plan-Do-Check-Act
PET	Petrol (Slovenian Oil Products Distributor)
PET	Privacy-Enhancing Technology
PHI	Protected Health Information
PII	Personally Identifiable Information
PIPL	Personal Information Protection Law
PMU	Phasor Measurement Unit
PoE	Pool of Experts
PPE	Public-Private Ecosystem
PPP	Public-Private Partnership
PT	Playbooks Tool
PV	Photovoltaic
QC	Quantum Computing

QKD	Quantum Key Distribution
R2D2	Reliability, Resilience and Defense technology for the grid (project)
RA	Regulatory Authority
RAE	Regulatory Authority for Energy (Greece)
RAG	Retrieval Augmented Generation
RCA	Root Cause Analysis
REST	Representational State Transfer
RF	Radio Frequency
RIIO-T3	Revenue = Incentives + Innovation + Outputs (Transmission price control, UK)
RL	Reinforcement Learning
RMF	Resilience Methodological Framework
RMF	Risk Management Framework
ROAR	Resilience Orchestration, Automation and Response
ROC	Receiver Operating Characteristic
RP	Resilience Playbook
RRIM	Risk Reduction and Incident Mitigation
RSE	Ricerca sul Sistema Energetico (Italian research center)
RTBF	Right to Be Forgotten
RTE	Réseau de Transport d'Électricité (French TSO)
SAAM	Situational Awareness and Mitigation Platform
SAFER	Situation Awareness for Enhancing Resilience
SAFER	Situational Awareness, Forecasting and Early Response
SCADA	Supervisory Control and Data Acquisition (referenced in relation to cyber exercises)
SCR	Sphynx Cyber Range
SCRM	Supply Chain Risk Management
SDIS73	Fire and Rescue Service Providers on the French side of the tunnel
SDLC	Software Development Life Cycle
SDO	Standards Development Organization
SHAP	SHapley Additive exPlanations
SIEM	Security Information and Event Management
SINA-SITAF	Societa Italiana Traforo Autostradale del Frejus Spa
SIR	Security Incident Response
SIS	Safety Instrumented Systems
SISA	Sharding, Isolation, Slicing, and Aggregation
SKILLAB	Skills LABoratory
SME	Store Medial Estimation
SMIR	Smart Mandatory Incident Reporting
SNIFFER	Air Quality Sensor Technology
SOAR	Security Orchestration, Automation, and Response
SOC	Security Operations Center
SP	Situational Picture
SPA	Security and Privacy Assurance

SPA	Sphynx Platform Architecture
SSU	Stable Sequential Unlearning
ST	Stress-Testing (frequent contextual use)
STIX	Structured Threat Information Expression
STREST	Harmonized Approach to Stress Tests for Critical Infrastructures against Natural Hazards
SUNRISE	Strategies and Technologies for United and Resilient Critical Infrastructures and Vital Services in Pandemic-Stricken Europe
SZ	Slovenske Železnice (Slovenian Railways)
TAC	Threat Actor Context
TAXII	Trusted Automated Exchange of Intelligence Information
TCN	Temporal Convolutional Network
TEN-T	Trans-European Transport Network
TF-IDF	Term Frequency Inverse Document Frequency
TIBER-EU	Threat Intelligence-based Ethical Red Teaming (European Central Bank framework)
TINTED	Threat Intelligence Node inTEgrated with Data enrichment services
TLS	Transport Layer Security
TPR	True Positive Rate
TS	Telekom Slovenije (Slovenian Telecommunication Services Provider)
TS	Time-Series
TSO	Transmission System Operator
TTPs	Tactics, Techniques, and Procedures
TUGraz	Technische Universität Graz.
UAS	Unmanned Aerial Systems
UC	University of California
UEBA	User and Entity Behavior Analytics
UI	User Interface
UK	United Kingdom
URI	Uniform Resource Identifier
URSIV	Government Information Security Office, Slovenia
UT	University of Texas
VISOR	Video Surveillance for Operations Room
VR	Virtual Reality
WAMS	Wide Area Measurement Systems.
WO	Workflow Orchestrator
XAI	Explainable Artificial Intelligence
XDR	Extended Detection and Response
XGBoost	eXtreme Gradient Boosting
XR	Extended Reality
XSS	Cross-Site Scripting
XTs	Extreme Threats

Effective Risk Management and Impact Assessment of Complex Incidents

Cybersecurity Incident Management during Slow and Rapid-Onset Adverse Events: Solutions and Challenges

Aljosa Pasic and Susana Gonzalez Zarzosa

Abstract The COVID-19 pandemic was an example of an adverse event that created temporary operational conditions for critical infrastructure (CI) operators. From remote work patterns to cybersecurity workforce absenteeism, cybersecurity incident management had to consider different risk indicators and adapt technology (e.g., detection of anomaly patterns) or processes (e.g., prioritization or ingestion of external threat intelligence). The use and sharing of cyber threat intelligence (CTI) proved to be valuable to stay up to date, but challenges related to trust and confidence emerged. This paper describes the design of a dynamic solution to be used by cybersecurity operators, which was validated in the SUNRISE project pilots with several CI operators.

Keywords Cybersecurity · Resilience · Threat intelligence · Risk management · Critical infrastructure · Incident management · Anomaly detection

1 Introduction

Resilience is a concept that spans different domains (physical, information, cognitive, or social) and has different capacities, abilities, and principles[1] in the face of different adverse events. These events can be categorized as slow-onset or rapid-onset based on how quickly they manifest and escalate. Climate change, geopolitical tensions, or rising unemployment can be considered slow-onset events. Natural disasters, stock market crashes, or power grid failures are examples of rapid-onset events.

[1] Definitions differ in literature but usually span some kind of division into prepare/absorb/recover/adapt.

A. Pasic (✉)
Advanced Computing, Bull, Madrid, Spain
e-mail: aljosa.pasic@bull.com

S. G. Zarzosa
Advanced Computing, Bull, Valladolid, Spain
e-mail: susana.gzarzosa@bull.com

J. Soldatos and H. Abie (eds.), *Technology-Enabled Resilience: Innovations in Critical Infrastructure Protection*, https://doi.org/10.1007/978-3-032-25917-2_1

An event such as a pandemic falls into both categories. It might initially spread slowly before reaching global impact, but during a critical period, situations might have characteristics of a rapid-onset event. Healthcare systems, or critical infrastructures in general, can experience near-instant service degradation due to a surge in cases, absenteeism, supply chain disruptions, and other risks, causing a rapid shift from containment to crisis management. Understanding this dual nature is therefore crucial for effective pandemic preparedness and response strategies.

Co-funded by the European Commission, Strategies and Technologies for United and Resilient Critical Infrastructures and Vital Services in Pandemic-Stricken Europe (SUNRISE) is a project that considers the lessons learned from the COVID-19 pandemic and the related challenges for critical infrastructure (CI) operators and comes up with solutions to help improve the latter's resilience. The project's approach is user-driven, with national workshops (in Italy, Slovenia, and Spain) as well as pan-EU workshops conducted in a systematic manner to investigate the challenges and requirements for the solutions' design.

The SUNRISE strategy consists of distinctive missions. One of them is awareness of the dynamic threat landscape related to and implied by pandemics. In this context, we consider cyber threats as a "multiplier", meaning a factor that amplifies the severity, scope, or impact of an existing risk or challenge (e.g., human risks and disruption of supply chain) derived from the main adverse event (pandemics). Therefore, "threat multiplier" can be any event that does not create the initial threat but can make the consequences worse. During a pandemic, for example, cyberattacks targeting healthcare systems (e.g., ransomware attacks on hospitals) can disrupt patient care and vaccine distribution, make certificate validation unavailable, or simply erode trust. Social engineering attacks (e.g., phishing) exploit fear or uncertainty, while other attacks are more likely to succeed when organizations are under-resourced or when defenses are lower. They also exploit the urgency, fear, and increased reliance on new or untested technology during pandemics to carry out malicious activities. Cyberattacks can also trigger cascading failures across sectors. For example, an attack on an energy grid or water management can paralyze other sectors.

As another example, we can mention Ryuk and Maze ransomware that were two of the most prominent cyber threats targeting hospitals and healthcare systems during the COVID-19 pandemic. Although they were identified before COVID-19, it was during the pandemic that they specifically targeted hospitals since hospitals were under immense pressure and very likely to pay ransoms quickly to restore operations. In October 2020 in the United States, the Department of Homeland Security (DHS) and the Cybersecurity and Infrastructure Security Agency (CISA) issued an alert about Ryuk ransomware attacks against hospitals [1], while Fresenius, Europe's largest private hospital operator, was targeted by Maze ransomware [2], disrupting operations at multiple facilities.

In this chapter, we present one of the key results of the SUNRISE project, which is an integral cyber–physical resilience (CPR) solution. We first discuss cyber resilience and its characteristics, as well as its relation to overall CI resilience during pandemics. Then we address challenges and gaps, before covering solutions for observation of risk indicators during pandemics when temporary operational conditions and usage patterns change. Subsequently, the incident reporting module is presented, followed by the validation in the pilot environment and conclusions.

2 Cyber Resilience and the Proposed Approach

Following the general definition of resilience, cyber resilience is focused on an organization's ability to prepare for, respond to, and recover from cyberattacks. As with the scope, which is limited to digital systems and data integrity with eventual consequences on the services that depend on them (e.g., disrupted logistics), the nature and speed of cyber resilience's response is also different from that of physical infrastructure. Technical countermeasures, such as attack containment, reconfiguration or patching vulnerabilities, do not depend as much on physical properties or resource management. However, there is a strong dependency on cyber experts and human knowledge, which in the case of pandemics might not be always available.

While successful cyberattacks often demand near real-time responses, not all cyber incidents are treated as cyberattacks. A cyber incident could be any event that poses a potential risk to an organization's systems, data, or processes. Unauthorized access detected in a system or a phishing attempt, where no damage occurs, are incidents but also indicators of a possible ongoing cyberattack.

The amount of time available to react during incident management, without experiencing any impact, depends on multiple factors. These factors include the nature of the incident, its detection speed, the organization's preparedness, and the type of systems or data involved. Based on predefined rules, automated defenses, such as firewalls and intrusion prevention systems, act immediately to block attacks like a DDoS or brute force attempts. For certain attacks, however, manual intervention combined with automated tools is needed. The so-called triage process is often the initial step in incident handling and response. It involves evaluating, prioritizing, and categorizing security alerts or incidents to determine their relevance, severity, and required actions. The goal is to focus resources on genuine threats while minimizing false positives.

If an incident is detected early and contained before escalation (e.g., identifying and isolating a phishing attack target), the reaction time window is longer. We might even have days to react to slower-developing threats (e.g., advanced persistent threats slowly exfiltrating data). In other words, not all attacks are the same. However, while the duration of a specific reaction window varies, organizations with proactive measures and a predefined response or course of mitigation actions are generally considered more cyber resilient than others.

Automated or semi-automated risk assessment can significantly enhance the efficiency and accuracy of the triage process. This was one of the starting points for SUNRISE's approach to the CPR tool design. Furthermore, it can also improve prioritization and ensure timely responses to threats. Automated systems can assign a risk score to alerts or incidents based on not only the risk model, severity of the threat, or contextual data (business impact such as asset criticality or operational conditions such as availability of workforce), but also based on external data such as indicators of compromise (IoCs) received from threat intelligence feeds. Enriched data can provide analysts with actionable insights and speed up triage, so the more complete the IOCs are, the stronger the cyber resilience.

One could argue that cyberattacks, like pandemic events, show the dual nature of both a slow and rapid-onset adverse event, and that the window of time to react depends on the detection of early symptoms in a pandemic and early warning signs of a cyberattack, such as network or access anomalies. Observation of "early warning" signs or events, and their use as risk indicators, was the second important consideration for the design of the SUNRISE CPR solution.

The SUNRISE project also defined a strategy process for operators of critical entities in general and critical infrastructures in particular, as well as decision-makers on a regional or national political level. It consists of five major steps (Fig. 1) that include the identification of pandemic-specific critical entities and services, the characterization of the pathogen causing the pandemic, the assessment of consequences according to different criteria (health, economic, societal, etc.), and the evaluation and selection of protection and mitigation measures. The SUNRISE strategy process is built on the cornerstones of the plan–do–check–act (PDCA) cycle and strongly relates to the generic risk management process defined in the ISO 31000 [3] and ISO 27005 [4].

Among the mitigation measures are non-pharmaceutical interventions (NPI) such as lockdowns or restrictions of movement that result in "derived risks" for cyber resilience management. For example, the introduction of NPI provoked more remote work, which resulted in unusual connectivity patterns, rendering existing AI-based anomaly detectors useless (so-called false positives were detected when remote users connected at unusual times to a server). In some situations, temporary configurations with adjusted parameters occurred (e.g., adaptive session timeouts can extend transaction validity during network congestion). The value of a digital asset changed (e.g., the vaccination appointment app) and new risks emerged from the introduction of new technologies (e.g., contact tracing) and changes in the digital supply chain. All of this was captured in the "temporary condition" model, the third pillar of the SUNRISE CPR tool design, which assures better situational awareness and adaptivity at operational level.

This approach resulted in the CPR solution's architecture being composed of the following (Fig. 2):

Log files from different pilot CI systems and components are analyzed by an AI-based anomaly detector (AD) module based on a preexisting tool called LOMOS [5] and open-source tool called Wazuh.

Alerts are sent from this module to a semi-automated risk assessment engine, extended to include physical risk indicators and models that correspond to pandemic-specific temporary conditions, which is linked to the strategies of CI operators under pandemics. It is based on a preexisting tool called the CybEr Risk Calculator (CERCA), which was used to bridge the gap between decision-makers and operational staff.

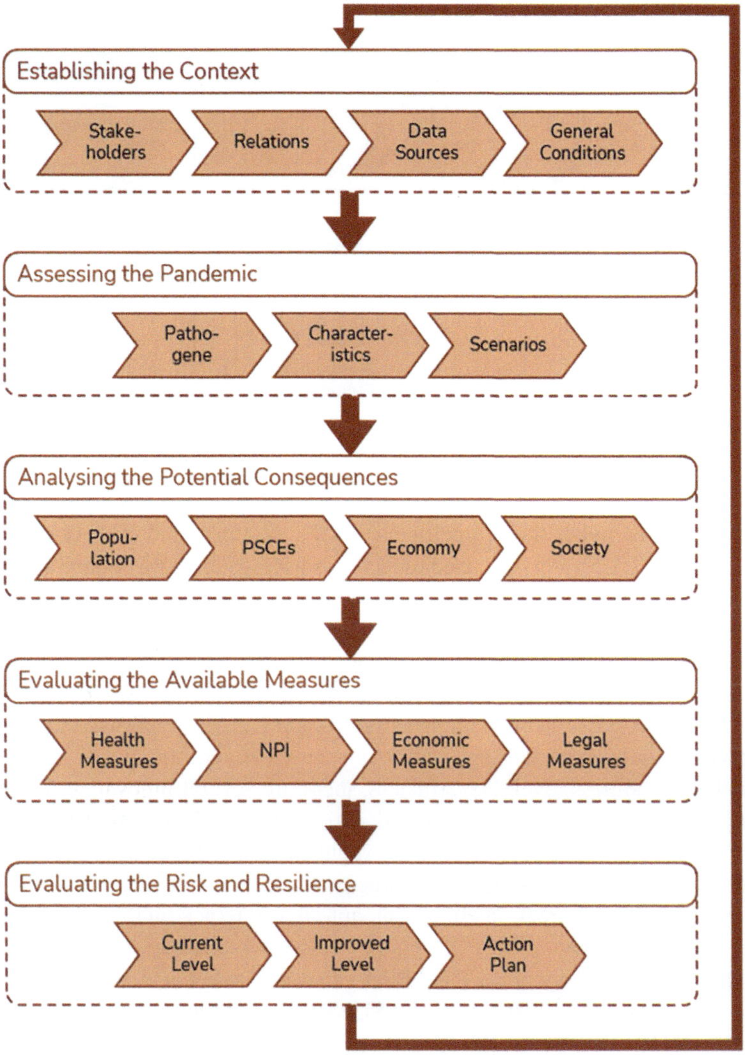

Fig. 1 SUNRISE strategy process

Alerts are also sent to the incident response and reporting management module that takes inputs from other tools, including legacy tools, and because of the input from the risk assessment engine (e.g., what is a "significant incident"), it considers reporting thresholds and workflows that are compliant with EU and national legislation for CI. This module builds on the preexisting tool Atos Incident Report Engine (AIRE), as well as the following open-source components: The Hive [6], Keycloak [7], and Cortex [8].

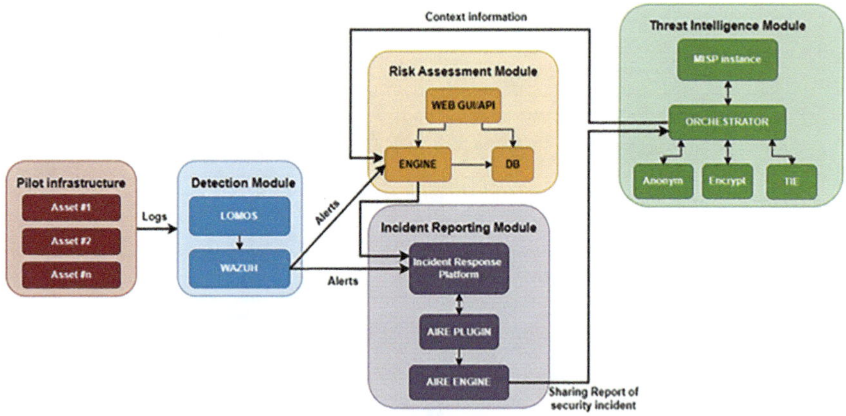

Fig. 2 Architecture of the cyber–physical resilience tool

An enhanced cyber threat intelligence (CTI) platform and sharing scheme and services for (CI) operators are used to feed risk assessment with context information about external threats, as well as to receive internal incident data that could be shared with outside stakeholders (e.g., other CI). It is built on top of the open-source platform MISP [9] and a tool called Threat Intelligence Node inTEgrated with Data enrichment services (TINTED) developed in Python.

Since SUNRISE is an innovation action (IA) project, the focus was on integration, prototyping, demonstrating, and validating the CPR solution and reaching a high technology readiness level. Nevertheless, there are several innovative features that will be described in more detail. The risk assessment module is placed at the core of the CPR solution, and it ingests all observable data that can serve as a risk indicator. Another decision was to reuse data sources already available in each SUNRISE use case with the new sources, such as a module that interacts with an external threat intelligence platform.

Enhancements designed in the CPR tool enable cyber incident response in CI organizations to align with the overall SUNRISE strategy process for resilience in CI [10]:

The step "establishing the context" includes collaboration procedures and impacts cyber threat intelligence sharing, since relevant stakeholders, relations, and available data sources might change temporary configurations (e.g., changes in supply chains due to pandemics). For example, the CPR tool contains a module to dynamically evaluate trust in cyber threat intelligence sources, such as a new provider or partner.

Analysis of the consequences of pandemics has a direct link to the temporary model, such as the value of digital assets. The role and importance of different digital assets and the impact on the provision of critical services changed during the last pandemic (e.g., healthcare applications such as vaccination appointment scheduling, the COVID-19 certificate repository, applications for contact tracing, epidemic modeling, COVID-19 dashboards, production, storage, and distribution of drugs, etc.).

Evaluating the available risk mitigation actions includes technical but also legal (e.g., compliance reassessment), economic and organizational measures (e.g., subcontracting workforce).

Comprehensive situational awareness at strategic level, "what-if" analysis, and the simulation of cascading effects can trigger re-analysis and recalculation of the cyber risk level or change the values of the variables in the "temporary condition" model. In this way, it is possible to optimize cyber risk mitigation measures for a specific scenario (e.g., what would happen in a hospital under a cyberattack if there was a lockdown for a certain class of employees who work on support and maintenance of digital assets that can only be accessed onsite, such as electricity substations).

3 Challenges and Related Work

As described in the previous section, both pandemics and cyberthreat contexts are very dynamic and are examples of adverse events where operational conditions can evolve rapidly. In [11], we explained that the PDCA approach, while flexible in its application across various sectors and industries, has a structured nature that can make it challenging to implement rapid changes. Its methodical approach involves predefined steps that may limit its adaptability in rapid-onset adverse events.

At the same time, the availability of real-time risk indicators, such as anomaly detection events, shared threat intelligence or security alerts in the case of cybersecurity, enabling faster reactions to adverse events. Unlike PDCA, the so-called observe, orient, decide, act (OODA) approach [12–14] focuses on the dynamicity and contextualization of the available information. It is a particularly suitable approach for a volatile, uncertain, complex, and often ambiguous inflow of data.

The OODA approach has already been used several times in the context of cybersecurity. For example, it was used for the design of the security operation center (SOC) tools [15], the simulation of the adversarial interaction between an intruder and a system administrator that intended to demonstrate the feasibility of implementing a rationally reconstructed OODA (OODA-RR) model [16], and even as a foundational model for incident management [17]. The strategic level PDCA approach in SUNRISE can be combined with the operational level OODA perspective and propose to combine the SUNRISE CPR tool sensing ("observation") with a cognitive process of "orientation" to make optimal decisions during temporary operational conditions. In the design of our cyber–physical resilience (CPR) tool, we used different observation sources (the anomaly detection tool, the security information and event management tool, and threat intelligence sharing) and integrated these with automated risk assessment and incident management tools, in order to better deal with uncertainties or partial information. We adjusted the probabilities of an attack or impact based not only on this data, but also "orientation" data, which includes temporary condition variables and parameters such as workforce availability, priority, and timeliness.

Implementing PDCA, which is a key approach of the SUNRISE strategy, involves significant resources for detailed planning, extensive data collection, and thorough analysis. The iterative nature of PDCA also means that resources are continually allocated to evaluate and refine processes through multiple cycles, often involving detailed documentation and rigorous oversight. Conversely, the OODA approach emphasizes speed. While it still demands resources at operational level for effective observation and orientation, including real-time data collection and efficient collaboration, many of these tasks can be automated through specialized technologies. We propose to combine the strategic level PDCA approach with the rapid cycle of OODA with near real-time data analysis, risk assessment, and decision-making, in order to improve the resilience of critical infrastructure during rapidly evolving temporary conditions.

Another challenge is system complexity and the inability to determine system-level properties solely from the properties of individual components. Here, we include issues such as compounding, cascading, and feedback. Some of these challenges are partially addressed in [18] that investigates the combination of a cascading effect simulation with a resilience methodological framework (RMF) to understand how a system of CIs reacts to threats, depending on the resilience of the individual components. The authors of this paper use the PRECINT RMF to determine how to quantify the service provided by CI, how to identify the parts of CI that are relevant for the services they provide, and how to quantify resilience directly by simulation and indirectly by using indicators [19].

Challenges related to finding the optimal number of risk indicators, given the resilience problem tackled, are described in the paper by Jovanović et al. [20]. In this paper, the authors mention that having learned lessons from COVID-19, resilience assessment should include several new elements, including emerging risks, transformation, and adaptation. The authors distinguish a "normal" operational timeframe from "scenario" time, which we call temporary conditions, and where different parameters and values apply. In their indicator-based approach, challenges related to finding the optimal number of indicators given the resilience problem tackled, as well as allowing "drill[ing] down" in cases when an indicator needs further explanation or breakdown.

The framework in Granadillo et al.'s paper [21] is broken down into three stages: (i) situation awareness (exploiting attack modeling), (ii) risk assessment (exploiting risk modeling), and (iii) response assistance (exploiting response modeling). In this framework, dynamic risk assessment evaluates possible attack scenarios before they originate in the system. Vulnerabilities and mitigation actions are modeled and analyzed through simulation or projection of data, making it possible to identify the best response set of actions that would be triggered after an attack is detected and the reactive mode is activated.

In Kamissoko et al.'s paper, [22] the authors propose a methodology for a continuous, real-time assessment of resilience, based on two phases: design and runtime. During the design phase, they focus on concept design, types of evaluation/assessments, determining types of indicators, selection of data sources, and evaluation frequency. During runtime, they continuously process and transform data and indicators into functionality models as time advances. The main differences in our approach are the contextualization of information that has been observed or shared, such as cyber intelligence, and how this is used to recalculate risk probability. Kamissoko et al.'s framework makes assumptions regarding quality control, which results in uncertainties in the trustworthiness of the external data sources. By contrast, we eliminate these uncertainties with a module dedicated to source confidence calculation. We also use a threat scoring module [23] that calculates the timeliness or relevance of indicators of compromise (IoC). Mapping of indicators of compromise (IoC) to tactics, threats, and procedures (TTP) [24] also helps with prediction and elimination of uncertainties.

4 Observation during Adverse Events

There are different types of data that can be observed or inferred in order to use them as risk indicators of a cybersecurity attack.

Indicators of compromise (IOCs) are specific, measurable artifacts or signs that indicate a potential security breach or compromise. These can include things like IP addresses, file hashes, domain names, URLs, registry keys, or patterns of behavior that suggest malicious activity. When IOCs are collected, such as through a cyber threat intelligence platform, they can be cross-referenced with past incidents or other sources to determine the probability of an attack. The main challenge is that they are often not specific to an organization, leading to potential noise. They might also be obsolete and rely on the organization's ability to integrate and act on the threat intelligence in a timely manner.

On the other hand, anomaly detector (AD) events come from internal tools that can detect previously unknown or emerging threats (Fig. 3). They detect unusual behaviors that might signify a compromise (e.g., lateral movement and data exfiltration), but they have a drawback of a high false positive rate that can overwhelm analysts, leading to potential oversight.

Security tools like security information and event management (SIEMs) and endpoint detection and response (EDRs) can generate alerts that increase the likelihood of an attack. An advantage is that they correlate multiple data points, helping to identify suspicious patterns. However, the quality often depends on the fine-tuning of detection rules, and these rules are not always available (e.g., in the case of new attack patterns).

Other sources of data and potential risk indicators include vulnerability scanners or other tools that provide proactive identification of security weaknesses (e.g., outdated software, misconfigurations, and attack surface management). The SUNRISE CPR tool has a layered approach and uses all these risk indicators for operational risk assessment (CERCA tool), while some of them, which are confirmed to be incidents,

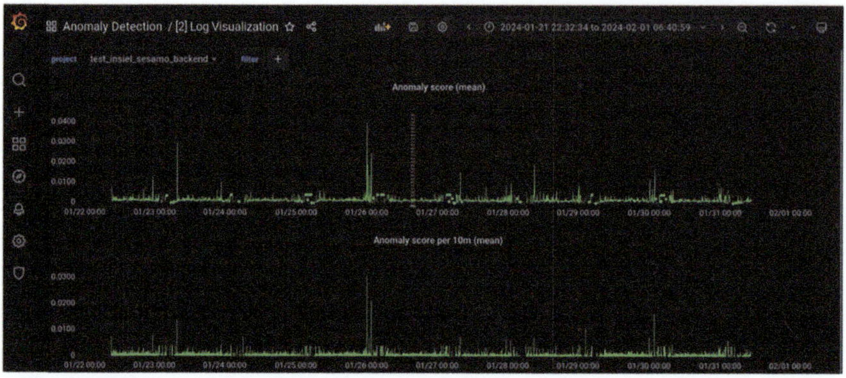

Fig. 3 Overview of detected anomalies in the last 10 days

also feed the incident response module (Fig. 2) to improve the organization's ability to act swiftly and decisively based on the information provided. However, incident management also needs to consider the dynamic adaptation of operational conditions or what can be classified as "orientation" data, which feeds the risk assessment module and can change the results of assessment dynamically.

Another important improvement to the risk assessment module in the CPR tool is the inclusion of not only physical security indicators, such as those related to unauthorized physical access or the detection of a malicious or unknown hardware component, but also human risk indicators such as those related to training and awareness, or psychological and behavioral risks. One example is the human ability to identify a phishing site. This can be included as a risk indicator [25]. Personal assessment (including subjective risk perception) can also change due to fear or stress, so the manual input of temporary condition parameters that describe observable behavior is also envisaged.

5 Orientation during Adverse Events

The SUNRISE CPR tool considers static and dynamic (near real time) inputs to perform operational cybersecurity risk assessment with multiple levels of granularity. The temporary conditions model also includes configurations or events that support "orientation" in risk assessment by modifying weight in risk models and supporting decision-making. Examples of this are as follows:

> The virtual private network (VPN) server that is used to enable secure remote working must accept a sudden increase in connections. SIEM and other security tools might also become overloaded with this peak in temporary teleworking conditions.

Equipment used for remote working, such as laptops, tablets, and phones, comes with different configurations and can have dual use.

Changes to workforce temporary conditions such as psychological resilience, the workforce's ability to perform, availability (absenteeism due to lockdowns or travel restrictions), motivation, compliance with policies, and team configuration.

Organizational asset model changes, adaptations to working conditions (e.g., access control and social distancing), priorities, collaboration procedural changes (authorities and law enforcement), and variation in security policies.

In some cases, these can be inserted manually into the risk module through the CERCA tool dashboard (Fig. 4). Some temporary conditions can be sudden and non-predictable, so prediction parameters can be used, such as in the case of workforce absenteeism. This feature tries to leverage the collaboration between CI operators, trend analytics, and, in the future, outputs from the tool for human resource demand prediction. See below for a graph of health-related trends including influenza, fever, and sick leave (Fig. 5). This is an example of sources that can be used for absenteeism prediction.

The probability (likelihood) associated with a cyberattack depends on many parameters: a phase or step of an attack, or a possible path the attacker may take. In the estimation of probability, it is therefore essential to identify tactics, techniques, and procedures (TTPs) that refer to the behavior and methods employed by cyber adversaries during an attack. These describe how attackers operate at different stages of their campaigns. An approximate equivalent in the SUNRISE strategy process that focuses on pandemics is "pathogen's characteristics", which describes the possible

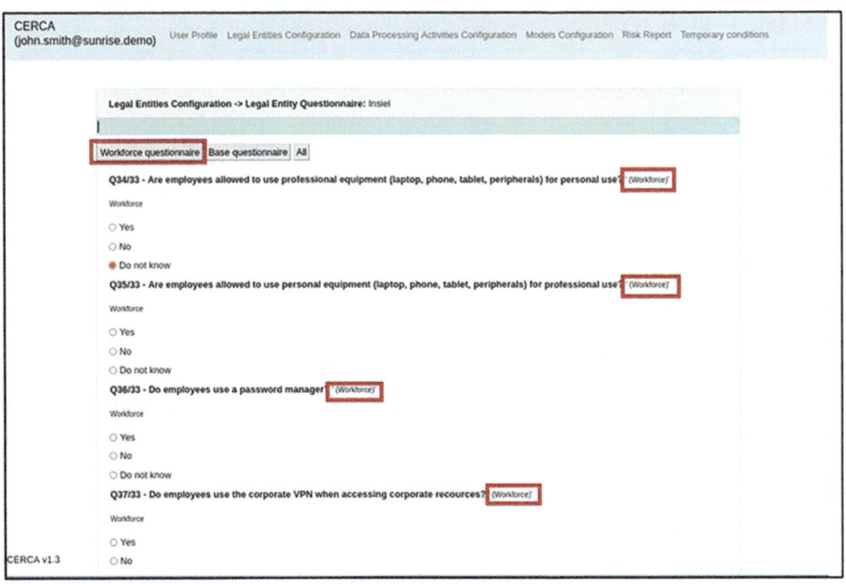

Fig. 4 Temporary condition input through workforce questionnaire

Fig. 5 Health-related trends used for absenteeism prediction

effects a pandemic might have on the population. Similarly, TTPs contribute to having a comprehensive understanding of cyberattacks' characteristics, helping defenders align their mental model of the situation with real-world conditions.

TTPs also make it easier to detect anomalies. For example, if an attacker attempts lateral movement using remote desktop protocol (RDP), the presence of such a technique can be compared to known organizational behaviors. If RDP is not typically used internally, it is an anomaly and therefore raises the probability of an ongoing attack. In other words, it helps to orient pattern detection or scarce resources toward the latest attack methodologies and trends. The prediction and probability of the future evolution of an attack and anticipation of attack stages is important to reduce reaction time. In addition, a team can prioritize response efforts based on the current stage of the attack. This ensures that the orientation is realistic and actionable, reducing misinterpretations or overreactions.

The ability to enrich the IoC with techniques from the MITRE ATT&CK matrix has been integrated in the CPR tool [24]. For each new event that is received in the MISP instance, this feature can deliver techniques associated with event attributes. This way, whenever CI operators pull limited information from generic intelligence feeds, they can get more insights (Fig. 6). We use VirusTotal API [26] to support better contextual relationships (e.g., linking malware hashes to domains and IPs) as well as give JSON-based responses, making them easier to parse and use in automated systems.

Although threat intelligence sharing is very useful, it can also be a source of vulnerability, since during temporary conditions rogue sources or fake events might appear. The new functionality "Source Confidence Calculation" (Fig. 7) has been implemented, introducing the following main capabilities:

Fig. 6 IoCs after enrichment with MITRE ATT&CK techniques

Dynamic Adjustment of Source Confidence: Allows dynamic updates to the confidence level of threat intelligence sources, ensuring users have access to the most up-to-date data reliability assessment.

NATO Admiralty Tag Assignment: Assigns a specific tag using the Admiralty taxonomy to each event associated with an intelligence source, facilitating identification and categorization within the system, and playing a crucial role in calculating the decay of IoCs in MISP.

Regarding the heuristics involved in the source confidence score, we use the following criteria based on the scoring model for IoCs [23]:

Extensiveness: Measures how much additional information an intelligence feed provides with each IoC, assigning higher confidence to feeds with more context. Context properties include timestamps, country, service, threat descriptions, etc. The system assumes all IoCs within a feed have the same context properties.

Timeliness: Determines how quickly an intelligence feed shares IoCs compared to others, using timestamps to assess recentness.

Completeness: Evaluates the quantity of IoCs contributed by an intelligence feed, prioritizing quantity over quality.

Whitelist Overlap Score: Calculates the extent to which IoCs in an intelligence feed overlap with trusted whitelists, adjusting for known sources like Cisco Umbrella domains and cloud provider IP ranges.

Finally, the source confidence score combines these factors into a weighted average to determine the confidence level of intelligence feeds as an additional input to risk-based decision-making (orientation phase).

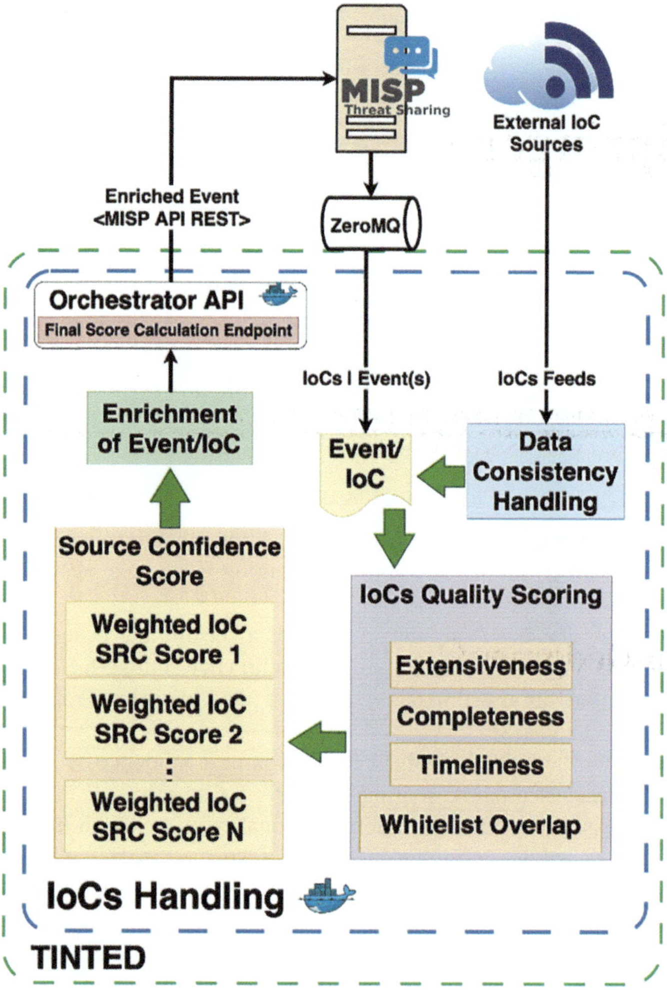

Fig. 7 Source confidence score architecture

6 Incident Response and Reporting

Determining the attack stage will lead to one set of response actions (immediate containment, assessment of impact, stop exfiltration, etc.). In addition, in the case of SUNRISE, there are some specificities related to the context of critical infrastructures, such as priority assignment during mitigation action (e.g., how the availability of a digital asset can be linked to physical safety).

The NIS2 directive [27] requires essential and important entities to report significant incidents to the competent authority within 72 h. A significant incident is typically defined as an event that has a substantial impact on the continuity of essential services, and this is again where the risk assessment module of CPR helps to define and filter the "significance" or "importance" of incidents in different scenarios and contexts.

Different workflows are implemented for reporting to different authorities, including the draft NIS2 workflow (Fig. 8). The incident reporting process includes:

Consideration of the risk assessment performed by the CPR cyber–physical risk assessment module, its usage for incident classification, and its notification to the service beneficiaries.
An early warning in case the incident is classified as significant.
A vulnerability notification procedure that could happen at any time during the collection and analysis of the incident data.
Consideration that after the early warning, CERTs can send some countermeasures related to the incident and associated information that can be added to the report.

During an investigation, several observations are generated that may or may not be related to the incident in question. These observations must be recorded for further analysis. If these observables are evidence of an attack or an attempted attack, they are designated as indicators of compromise (IOCs). The NIS2 directive also stipulates that incident notifications and final reports must include the IOCs obtained during the incident investigation. The CPR solution uses the incident reporting module based on the AIRE module that addresses this requirement by automating the aggregation of IOCs in the reporting process. Observables can be registered via web form (Fig. 9).

Each incident contains a list of observables (Fig. 10), which provides an overall view of the information collected on the incident.

To provide a proof of concept for this functionality, and in the absence of the final NIS2 templates[2,3] the pilots in SUNRISE provided the templates of the reports they currently use to report incidents to the authorities or internally (Fig. 11).

7 Pilots and Tool Validation

The CPR solution is demonstrated and validated by following a cluster-oriented piloting strategy with Critical Infrastructures in Italy and Slovenia, covering different sectors (the health, digital, energy, transport and water sectors). During the first phase of validation, the goal was to evaluate, at module level, the fulfillment of the different requirements defined for separated CPR modules, the integration of real operational data sources (in particular, the ones provided by the leading partner of the Italian CI cluster), and the validation of the main CPR workflows. Final validation of the tool is done through the execution of a selected set of realistic use cases on different pilot

---

[2] At the time of writing of this paper, a limited number of EU member states transposed the NIS2 directive into the national legislation.

[3] NIS2 page: https://digital-strategy.ec.europa.eu/en/policies/nis2-directive

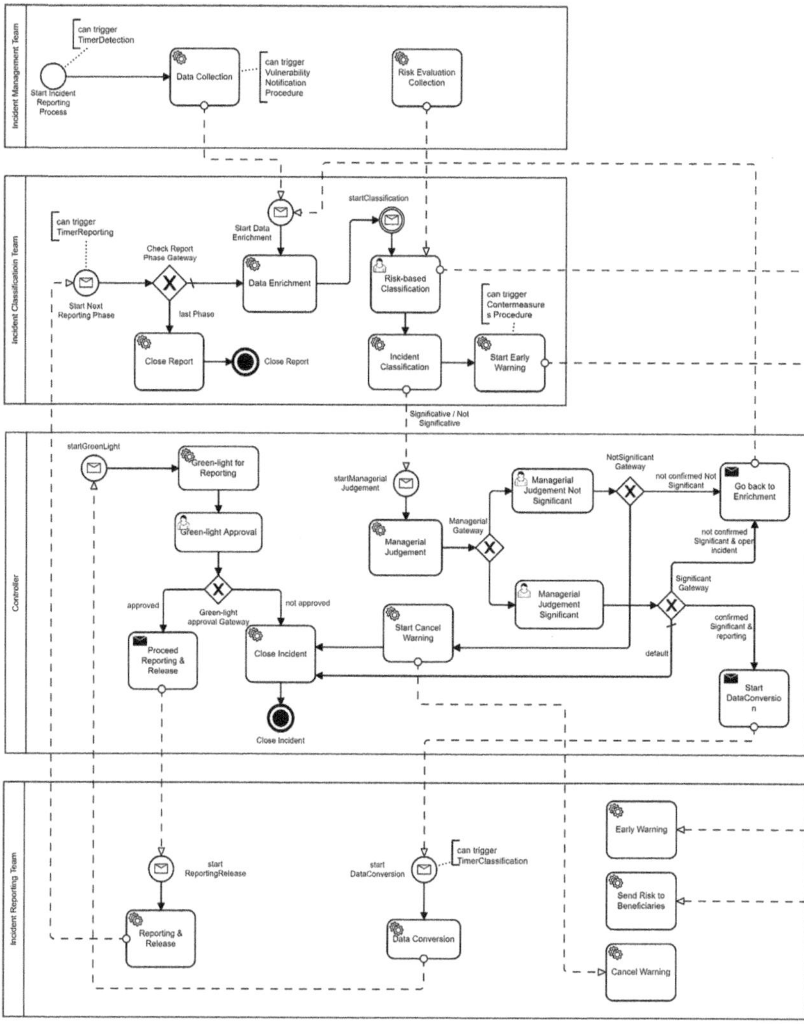

Fig. 8 Incident reporting in the CPR tool

sites to demonstrate the integrated version of the CPR solution within the operational environments.

The final evaluation and validation of the CPR solution involved:

Definition of realistic and credible use cases identified for each sector by the Critical Infrastructure operators involved in the project.

Selection of threats and risk models where temporary conditions may have a significant impact, based on the use cases defined.

Selection of available observable sources and data from the CIs to feed the risk assessment component of the CPR solution.

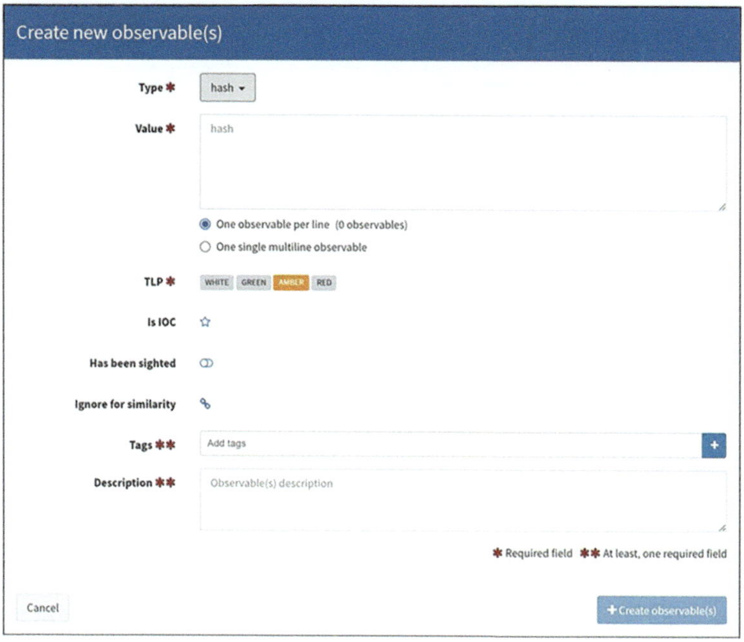

Fig. 9 Form to create new observables

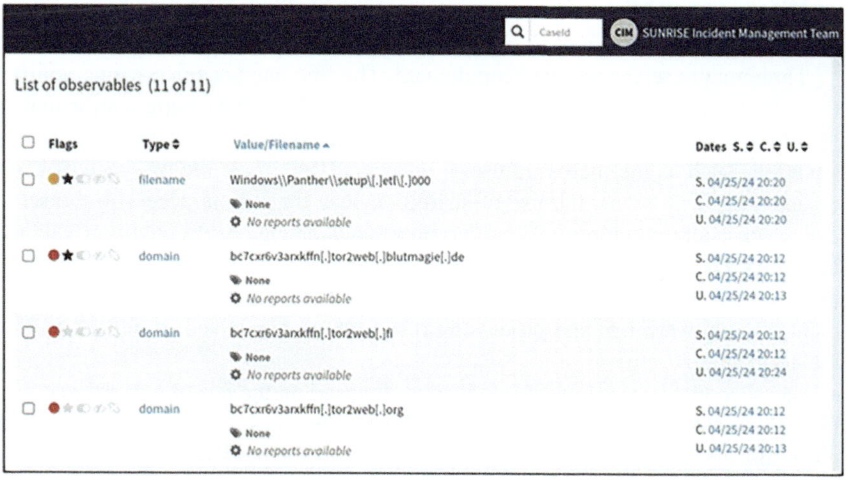

Fig. 10 List of observables, IOCs, and non-IOCs (star flag)

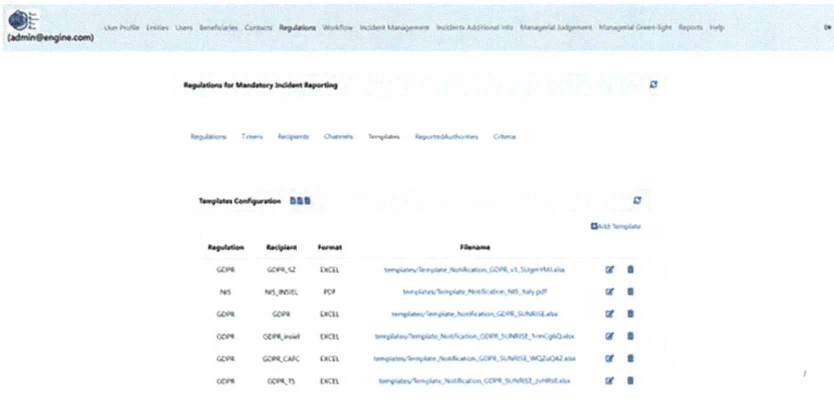

Fig. 11 List of available templates

- Selection of the observed or inferred temporary conditions and calculation of probabilities, based on strategic priorities and observations.
- Selection of the criteria for orientation in decision-making related to response, reporting or other actions that are relevant for resilience in the context of the different use cases.
- Execution of the demonstrations for end users including the simulation of attacks, sharing of relevant IoC and incident reporting process.

Three main use cases have been defined. The first one considers vulnerability in the software used for VPN remote access or in a healthcare web application that can lead to a session hijacking attack. It considers temporary conditions derived from situations such as the increased use of unsecured networks, the presence of simultaneous active sessions, the use of shared devices for remote access to the server, and inconsistent patching or configuration management. Additionally, it validated how the CPR solution also supports the CI operator to comply with the mandatory incident reporting process according to the new NIS2 directive, including management of early warnings and processing of information received from the CSIRT to the notification.

The second use case considered is more focused on the use of cyber threat intelligence, particularly those related to phishing campaigns. Phishing was one of the most observed initial infection vectors for ransomware incidents targeting CI during the COVID-19 pandemic. The goal of this use case is to validate the use of threat intelligence data sharing and enrichment to reduce incident response time, while also considering temporary conditions, such as changes in the circle of trust and data sharing policies.

The third use case involves the use of a mobile wallet. In this use case, the increase in online transactions due to specific emergency circumstances could lead to changes of temporary conditions at operational level, such as changes in configuration of service due to a network overload or the generation of false positives of denial-of-service attacks. Adaptive session timeouts can extend transaction validity during network congestion, while some wallets use progressive backoff, where retry attempts increase in duration (e.g., 5, 10, 20 s) to avoid overwhelming servers. Another option is to extend session durations (for login authentication), although longer sessions increase exposure to security threats like session hijacking. The SUNRISE CPR solution was evaluated in this context, through the adaptivity of different mitigation actions, to allow the users to continue using the mobile wallet (which could be any other mobile app, e.g., vaccination certificate checking).

At the time of writing this chapter, the final evaluation and validation was still in progress, but some early conclusions can already be reported. While the COVID-19 pandemic is an example of a temporary situation when critical infrastructure (CI) had to operate with continuously changing conditions, it is not easy to have a "one size fits all" model. The need for dynamicity in cyber–physical risk assessment, including real-time and external risk indicators as well as the dynamic change of values of temporary condition parameters, differs depending on the sector.

The move to remote working, combined with changes in the configuration of VPN and/or remote equipment, proved to be the most significant risk derived and the source of temporary operational conditions in pandemic scenarios. It had an impact on probabilities; events such as anomalies can indicate an attack. Human actors were also considered differently as they have different levels of expertise or awareness, different behaviors, or bias. Most previous approaches to cyber risk or resilience for critical infrastructure did not consider the essential role of specialists, also termed "essential workers", and mitigation measures should consider the availability of these workers.

Regarding the impact of timely data sharing, we considered the period between vulnerability or threat disclosure and the sharing of it. The perceived benefit is related to the determination of probability that a threat will be fully detected, and that the incident will be reported on time for security staff to act upon before the attack actually causes damage.

Finally, the potential effect of situations, such as network overload and congestion and its link to temporary conditions and derived risk indicators, proved to have an impact on false alarms, therefore leading to additional cost and fatigue of cybersecurity analysts. By leveraging adaptivity based on temporary conditions, it is possible to adjust the mitigation strategies to be applied in order to reduce the number of false positives and ensure the continuity of services such as the mobile wallet.

Other conclusions from the validation session included benefits such as:

- Increased awareness about dynamic changes and temporary conditions.
- Enhanced risk assessment by having more risk indicators and more frequent impact and probability updates.
- Informed decision-making to allocate scarce resources in incident detection more effectively.

Adaptive mitigation capabilities (e.g., decreased time to fine-tune the order of mitigation actions).

8 Conclusions

While sudden events may cause severe and immediate failure, events like the COVID-19 pandemic form slowly, and the operational conditions that impact cyber risk and resilience management change in several ways over time. The advantage of dealing with a slow crisis instead of a sudden disaster is that there is potentially time to fine-tune and adapt solutions, procedures, and strategies. However, our approach shows that these adaptations (e.g., a different course of action for mitigation measures) need to be considered and enabled in advance, such as, for example, considering the links between temporary conditions and risk indicators. Collaboration through tools, such as the threat intelligence module or semi-automated risk assessments, are essential for these adjustments. Strategic level decisions, such as the non-pharmaceutical interventions decided by the government (e.g., lockdown, restrictions on staff movements) or local context (e.g., supply chain changes), also create derived risks and temporary conditions that overwrite default values. Therefore, a new cyber risk assessment might be needed.

We started the SUNRISE project by investigating pandemic-related cybersecurity operational challenges. We then moved to mapping different static and dynamic risk indicators from internal "observation" sources that are relevant for cyber–physical risk models, usually from existing cyber tools. Situational awareness was also enhanced with information that can be statically provided by the CI operator about temporary conditions (e.g., absenteeism) and external information, such as threat intelligence. These events were also enhanced with enriched data about source trustworthiness or strategic information about adversary tactics, techniques, and procedures in what we called the "orientation" phase.

The threat scoring model, which measures threat extensiveness or timeliness, is needed to fine-tune the probability that a particular threat is relevant or important to CI operators. Timeliness is especially useful information since threat information such as indicators of compromise (IoC) could be outdated. Considering how long it takes to share its IoCs, the model will assign less source confidence to it. Mapping operational level indicators, such as IoC, to tactical and strategic insights (e.g., TTP) enables the anticipation of incidents before they escalate and have a significant impact. Regarding this approach, we think that new research is needed in the areas related to dynamic and adaptable threat intelligence sharing across different levels and sectors, likely as a part of a different research project.

In most cases, a pandemic, which changes temporary operational conditions, might not directly affect organizations or critical entities in the same way as it affects people. Essential employees and operators of digital assets of CIs were affected physically and mentally during COVID-19, both directly and indirectly (e.g., through family links). This resulted in increased absenteeism. Non-pharmaceutical interventions (NPI) implemented to contain pandemic spread (e.g., lockdown, movement restrictions, and closure of schools) sometimes destabilized normal maintenance and support of cyber infrastructure, as well as incident response teams. This information, considered in the CPR solution as "physical" risk indicators, is modeled and used in risk assessment. Both risk and resilience assessments calculate impact and/or performance loss, but resilience also considers recovery function and therefore has an important temporal dimension. In this sense, our approach of continuous risk assessment re-triggered with the new risk indicator values follows the OODA approach, therefore addressing "threat multipliers" and resilience capacities that are relevant during a prolonged adversary event with a different pace of evolution (slow and rapid onset) and different dynamics of risk indicator changes (e.g., human behavior or attitude toward potential phishing messages).

The remaining work and future direction are to optimize the balance between the system dynamics, organizational priorities, and mitigation aspects. Currently, we need reassessment every time there is a change of risk indicators or inputs, or dependencies between them, which could be cumbersome and inefficient. The development of risk models is also costly, while the gap between the strategic/tactical level decisions and the operational temporary conditions is not fully addressed (we used a survey to collect information about the relevant risk indicators). These challenges include "assessment fatigue" or congestion in decision-making. Fine-tuning might be needed to find optimal trade-offs between dealing with dynamicity and tool practicality.

Another line of work is considering the automation of mitigation actions and its optimization for temporary conditions. To respond effectively and timely to cybersecurity events and incidents, CI operators might create and utilize security playbooks, also known as course of action (CoA) playbooks. Mapping external threat intelligence that is shared among CI operators and using it to recalculate the probability of risks could help to adjust CoA at operational level or perform synchronized reassessment done by interconnected CI operators. This could build upon the previous work related to selecting optimal countermeasures for risk reduction, as described by Gonzalez-Granadillo et al. [28], where decisions were supported only by return on investment (ROI) criteria, which is not optimal for pandemic scenarios.

Acknowledgments This work has been supported by the European Commission through the SUNRISE project (grant no. 101073821) under the Horizon Europe research programme.

References

1. Newman, L. H. (2020). *Ransomware Hits Dozens of Hospitals in an Unprecedented Wave*. Wired. Retrieved April 24, 2025, from https://www.wired.com/story/ransomware-hospitals-ryuk-trickbot/

2. *Fresenius health group, victim of Snake ransomware.* Incibe. Retrieved April 24, 2025, from https://www.incibe.es/en/incibe-cert/publications/cybersecurity-highlights/fresenius-health-group-victim-snake-ransomware
3. International Organization for Standardization. (2018). *Risk Management—Principles and Guidelines* (ISO/IEC No. 31000:2018). https://www.iso.org/iso-31000-risk-management.html
4. International Organization for Standardization. (2022). *Information technology—Security techniques—Information security risk management* (ISO/IEC No. 27005:2022). http://www.iso.org/cms/render/live/en/sites/isoorg/contents/data/standard/07/52/75281.html
5. Cankar, M., Petrovic, N., Costa, J. P., Cernivec, A., Antic, J., Martincic, T., & Štepec, Dejan. (2023). Security in DevSecOps: Applying Tools and Machine Learning to Verification and Monitoring Steps. *Companion of the 2023 ACM/SPEC International Conference on Performance Engineering (ICPE '23 Companion). Association for Computing Machinery, New York, NY, USA*, 201–205. https://doi.org/10.1145/3578245.3584943
6. TheHive: a Scalable, Open Source and Free Security Incident Response Platform. Retrieved April 25, 2025, from https://github.com/TheHive-Project/TheHive
7. Keycloak: Open Source Identity and Access Management For Modern Applications and Services. Retrieved April 25, 2025, from https://github.com/keycloak/keycloak
8. Cortex: a Powerful Observable Analysis and Active Response Engine. Retrieved April 25, 2025, from https://github.com/TheHive-Project/Cortex
9. MISP (core software) Open Source Threat Intelligence and Sharing Platform. Retrieved April 25, 2025, from https://github.com/MISP/MISP
10. Schaeur, S., et al. (2024). D2.2 Strategy for awareness and resilience of CIs V1. Retrieved April 25, 2025, from https://sunrise-europe.eu/wp-content/uploads/2025/01/D2.2-Strategy-for-awareness-and-resilience-of-CIs-V1_v1.0.pdf
11. De Juan Fidalgo, P., Pasic, A., Del Álamo, J. M., Tourís, R., Álvarez, A (2023). TERME: a cyber-physical resilience toolset for risk assessment. *JNIC Cybersecurity Conference (JNIC), Vigo, Spain*, 2023, 1-6. https://doi.org/10.23919/JNIC58574.2023.10205687
12. *OODA loop*. Wikipedia. Retrieved April 24, 2025, from https://en.wikipedia.org/wiki/OODA_loop
13. Boyd, J. (1995). The Essence of Winning and Losing. Retrieved April 24, 2025, from https://www.coljohnboyd.com/static/documents/1995-06-28__Boyd_John_R__The_Essence_of_Winning_and_Losing__PPT-PDF.pdf
14. Boyd, J. (1976). Destruction and Creation. Retrieved April 24, 2025, from https://www.coljohnboyd.com/static/documents/1976-09-03__Boyd_John_R__Destruction_and_Creation.pdf
15. DeCusatis, C., Cannistra, R., Labouseur, A., Johnson, M. (2019). Design and Implementation of a Research and Education Cybersecurity Operations Center. *Cybersecurity and Secure Information Systems*. https://doi.org/10.1007/978-3-030-16837-7_13
16. Grant, T. J., Venter, H. S., Eloff, J. H. P. (2007) Simulating adversarial interactions between intruders and system administrators using OODA-RR. Proceedings of the 2007 annual research conference of the South African institute of computer scientists and information technologists on IT research in developing countries (SAICSIT '07). *Association for Computing Machinery, New York, NY, USA*, 46–55. https://doi.org/10.1145/1292491.1292497
17. Sager, T. (2020). The Cyber OODA Loop: How Your Attacker Should Help You Design Your Defense. Retrieved April 24, 2025, from https://csrc.nist.gov/CSRC/media/Presentations/The-Cyber-OODA-Loop-How-Your-Attacker-Should-Help/images-media/day3_security-automation_930-1020.pdf
18. König, S., Connolly, L., Schauer, S., O'Connor, A., Carroll, P., & McCrum, D. (2023). Combining Cascading Effects Simulation and Resilience Management for Protecting CIs from Cyber-Physical Threats. *Proceedings of the 32nd European Safety and Reliability Conference (ESREL 2022)*, 2741–2748. https://www.rpsonline.com.sg/proceedings/esrel2022/pdf/S20-03-343.pdf
19. Connolly, L., Konig, S., Egger, M., Schauer, S., O'Connor, A., Sheils, E. (2023). Critical Infrastructure Resilience in the PRECINCT Project. *14th International Conference on Applications of Statistics and Probability in Civil Engineering (ICASP14), Dublin, Ireland, 2023.*

20. Jovanović, A., Klimek, P., Renn, O., Schneider, R., Øien, K., Brown, J., DiGennaro, M., Liu, Y., Pfau, V., Jelić, M., Rosen, T., Caillard, B., Chakravarty, S., Chhantyal, P. (2020) Assessing resilience of healthcare infrastructure exposed to COVID-19: emerging risks, resilience indicators, interdependencies and international standards. *Environ Syst Decis*, 40(2), 252-286. https://doi.org/10.1007/s10669-020-09779-8
21. Granadillo, G., Dubus, S. Motzek, A., Garcia-Alfaro, J., Alvarez, E., Merialdo, M., Papillon, S., Debar, H. (2017). A dynamic risk management response system to handle threats in SCADA systems. *Future Generation Computer Systems*, 83, 535-552. https://doi.org/10.1016/j.future.2017.05.043
22. Kamissoko, D., Nastov, B., Allon, M. (2023) Improved model for continuous, real-time assessment and monitoring of the resilience of systems based on multiple data sources and stakeholders. *Structure and Infrastructure Engineering*, 19(8), 1122-1137. https://doi.org/10.1080/15732479.2021.2009883
23. Ermerins, J., Van Noort, N. (2020.) *Scoring model for IoCs by combining open intelligence feeds to reduce false positives.* [Doctoral dissertation, University of Amsterdam]. https://rp.os3.nl/2019-2020/p55/report.pdf#:~:text=The%20scoring%20model%20uses%20multiple%20open%20threat%20intelligence,of%20overlapping%20IoCs%2C%20and%20independence%20between%20the%20feeds
24. MITRE ATT&CK. Retrieved April 25, 2025, from https://attack.mitre.org/
25. Purkait, S., Kumar, S., Damodar, S. (2014) An empirical investigation of the factors that influence Internet user's ability to correctly identify a phishing website. *Information Management & Computer Security*. https://doi.org/10.1108/IMCS-05-2013-0032
26. *VirusTotal API v3 Overview*. (2023). Virus Total. https://docs.virustotal.com/reference/overview
27. Directive (EU) 2022/2555. Directive (EU) 2022/2555 of the European Parliament and of the Council of 14 December 2022 on measures for a high common level of cybersecurity across the Union. https://eur-lex.europa.eu/eli/dir/2022/2555/2022-12-27/eng
28. Gonzalez-Granadillo, G., Garcia-Alfaro, J., Alvarez, E., El-Barbori, M., Debar, H. (2015). Selecting optimal countermeasures for attacks against critical systems using the attack volume model and the RORI index. *Computers & Electrical Engineering*, 47, 13-34. doi:https://doi.org/10.1016/j.compeleceng.2015.07.023

Enhancing Power and Energy System Resilience during Extreme Events within the Current European Regulatory Framework

Aleksandra Krkoleva Mateska, Mathaios Panteli, Sina Hashemi, Balaji Venkatasubramanian, Petar Krstevski, and Ugo Stecchi

Abstract This chapter examines the resilience of European power and energy systems amid the increasing frequency of extreme weather events and cyberthreats. It highlights the need of making the shift from the traditional reliability-oriented planning and operation standards to resilience, risk-aware approaches for safeguarding the resilience of the European power systems. In this context, the chapter provides a holistic review of the evolving European regulatory landscape, including, for example, the Clean Energy Package, NIS2 Directive, and CER directive, emphasizing the need for integrated approaches to physical and cyber-resilience. Best practices and state-of-the-art initiatives across EU countries and beyond are analysed, and critical gaps in regulation, operation, and cybersecurity are identified. In order to bridge these gaps, the chapter concludes with the introduction of novel, state-of-the-art tools developed by the "Reliability, Resilience and Defense technology for the grid" (R^2D^2) Horizon Europe project, specifically designed to address the physical and cyber-resilience of current and future power systems.

Keywords Power Systems · Regulation · Resilience · Policy · Risk Assessment · Cybersecurity

List of Abbreviations and Terms

AI Artificial Intelligence

A. K. Mateska · P. Krstevski
Ss. Cyril and Methodius University in Skopje, Faculty of Electrical Engineering and Information Technologies, Skopje, North Macedonia
e-mail: krkoleva@feit.ukim.edu.mk

M. Panteli · S. Hashemi · B. Venkatasubramanian
Department of Electrical and Computer Engineering, University of Cyprus, Nicosia, Cyprus

U. Stecchi (✉)
New Technologies Department, ETRA Investigación y Desarrollo, Valencia, Spain

J. Soldatos and H. Abie (eds.), *Technology-Enabled Resilience: Innovations in Critical Infrastructure Protection*, https://doi.org/10.1007/978-3-032-25917-2_2

APT	Advanced Persistent Threat
CE	Critical Entity
CER	Critical Entity Resilience
CI	Critical Infrastructure
CIP	Critical Infrastructure Protection
CSIRT	Computer Security Incident Response Team
CTI	Cyber Threat Intelligence
DER	Distributed Energy Resources
DSO	Distribution System Operator
ECI	European Critical Infrastructure
ENISA	European Network and Information Security Agency
ENTSO-E	European Network of Transmission System Operators for Electricity
EPES	Energy and Power Electricity System
EPSS	Exploit Prediction Scoring System
EU	European Union
HILF	High Impact Low Frequency
HW	Hardware
IEC	International Electrotechnical Committee
IEEE	Institute of Electrical and Electronics Engineers
ICT	Information and Communication Technology
IoT	Internet of Things
ISO	International Organization for Standardization
IT	Information Technology
MISP	Malware Information Sharing Platform
ML	Mathine Learning
MTBF	Mean Time Between Failures
MTTF	Mean Time To Failure
NERC	CIP North American Electric Reliability Corporation Critical Infrastructure Protection
NIS	Network Information Security
NIST	National Institute of Standards and Technology
OT	Operational Technology
R^2D^2	"Reliability, Resilience and Defense technology for the grid" project
RA	Remedial Action
RES	Renewable Energy Sources
RSC	Regional Security Centre
SAIDI	System average interruption duration index
SAIFI	System Average Interruption Frequency Index
SW	Software
TEN-E	Trans-European Networks for Energy
TSO	Transmission System Operator

1 Introduction

In the twenty-first century, climate change has intensified the frequency and severity of natural hazards, posing unprecedented risks to global power systems. Between 2000 and 2023, weather-related disasters surged by 83% globally, with Europe experiencing a 65% increase in extreme events such as floods, wildfires, and storms [8, 64]. Notable events include the wildfires in Portugal [69] and the severe flooding in Germany and Spain [48], which resulted in a substantial loss of life and property. These events underscore the need for resilient and adaptive energy systems. In response, initiatives of the European Network of Transmission System Operators for Electricity (ENTSO-E), Germany's €30 billion resilience package [42], and the EU's Horizon Europe program have been introduced.

In addition to the extreme events, the electric industry and utilities in general are experiencing a significant digital transformation, as their operation can now be based on digital and automation technology, where real-time data analytics and artificial intelligence (AI)-based tools gain significance. The growing number of dispersed generators, the completion of smart meters roll-out, and the increased connectivity of the field devices are major factors driving the digitalization, enabling the energy transition. The side effect is that as an increasing number of systems become interlinked and dependent on digital technologies, the so-called attack surface (i.e. the potential entry points for malicious actors) is undergoing a fast and constant growth. Hence, increased connectivity and digitalization, along with the evolving geopolitical scenario, pose new security challenges. To ensure the security of the system in this new scenario, a comprehensive approach integrating advanced technologies, regulatory frameworks, and cross-sector collaboration to mitigate risks and enhance resilience is required.

Real-time data availability is a clear advantage to enhance the control, but it also presents risks that the integrity of the data is compromised. Manipulated data can lead to failures, blackouts, and economic losses. Moreover, the energy sector's reliance on external vendors, cloud services, and imported hardware has made the supply chain a key target for cyberattacks. Large energy firms may internally apply strict security protocols, but third parties and vendors often do not, originating vulnerabilities for their customers. Since 2020, supply chain attacks have increased, with attackers targeting the weaker links to access the chain [9], with major impacts on mission-critical systems too. To address this issue, it is necessary to revise procurement practices and compliance with evolving regulations as the EU's NIS2 Directive [35–37]. As global standards like North American Electric Reliability Corporation Critical Infrastructure protection (NERC CIP) and ISO/IEC 27001 [44] become more vital, continuous policy updates and cross-sector collaborations are essential to mitigate evolving cyber-risks.

This chapter is based on the findings of the "Reliability, Resilience and Defense technology for the grid" (R^2D^2) Horizon Europe project (Grant agreement N° 101,075,714 [43]). It is an innovation action project, aimed to enhance the resilience and reliability of electrical power and energy systems (EPES) in the face of increasing threats and vulnerabilities. It focuses on preventing and mitigating technical, cyber, and physical risks that could compromise the integrity of critical infrastructure, potentially causing harm to stakeholders and end users. The project is developed to encompass the electricity supply chain, taking in consideration a broad spectre of risks related to distribution, transmission, and regional security centres (RSC). The chapter explores the intersection of climate resilience and cybersecurity in the context of EPES in Europe. Section 2 reviews relevant EU legislation related to energy infrastructure, critical infrastructure protection, and cybersecurity. Section 3 highlights current best practices and state-of-the-art strategies for enhancing European EPES resilience. Section 4 identifies key gaps and barriers—both physical and cyber—and presents the contributions of the R^2D^2 project in addressing these challenges through innovative tools, coordinated frameworks, and recommendations for improvement of current practices of system operators and related legislative framework. Finally, Sect. 5 summarizes the chapter content and offers concluding reflections.

2 Resilience through the Perspective of EU Legislation

Since the Treaty on the Functioning of the EU [41] came into force on 1 December 2009, it has served as a legal foundation for the EU common energy policy. It defines energy as a shared competence among the EU and its Member States, covering electricity, energy systems, and networks. The key objectives of EU energy policy include ensuring market functionality, securing energy supply, promoting efficiency and renewable energy, and enhancing cross-border energy interconnections. The current EU energy legislation framework, known as the Clean Energy Package for all Europeans, facilitates the implementation of the above-mentioned objectives and the transition to a more sustainable and competitive energy system across the EU. This framework sets the environment for the secure operation of power systems, considering anticipated disturbances. However, as described in the introductory section, the increased frequency of occurrence of extreme events and incidents has motivated policymakers to focus on the introduction of the concept of resilience. This concept encompasses the timeline "before, during, and after" an unwanted event, thus covering the whole crisis management cycle [59]. While some aspects of resilience have been gradually included in the EU energy legislation, an additional set of legislative acts related to physical and cyber-resilience in critical economic sectors and infrastructure has become a necessity under current conditions. The following sections discuss in detail the above-mentioned aspects of EU legislation frameworks.

2.1 Aspects on Resilience in the Energy-Related Legislation

The implementation of the regulations and directives of the Clean Energy Package aims to empower consumers, enhance energy efficiency, increase the share of renewable energy, and strengthen the internal energy market while ensuring affordability, security of supply, and achievement of the climate targets. The acts of this package have been recently updated to foster higher EU targets for energy and climate. In addition, the regulatory canvas encompasses a set of network codes and guidelines dedicated to connections of demand and generation, power system operation, electricity markets, and cybersecurity.

The Clean Energy Package highlights parts of the broader perspective of resilience of EPES. The EU Energy Union Strategy and the Governance Regulation [31] of this package consider resilience of regional and national energy systems through the requirements that Member States develop plans to diversify energy supply, foster flexibility, and mitigate possible supply constraints. The amended Renewable Energy Directive III [38] highlights that self-consumption, such as local energy communities, contributes to increased resilience of the system while achieving the EU renewable energy targets. The broader concept of resilience is tackled in the Risk Preparedness Regulation [32–34], which is another act of the Clean Energy package focusing on security of supply. This Regulation addresses electricity incidents and crises beyond the operational challenges that can be solved by the implementation of the rules prescribed in System Operation Guidelines [23–25] and Network Code on Emergency and Restoration [23–25] and ensures that cyber-risks are identified among possible risks for security of supply. The Electricity Regulation [32–34] from the Clean Energy package introduces a clear reference to resilience through the requirement to assess the resilience of the system within the Ten-Year Network Development Plans. The System Operation Guidelines and the Network Code on Emergency and Restoration foresee the adoption of plans and measures that contribute to the resilience of power systems by outage planning coordination, ensuring system adequacy and security of supply and implementing coordinated restoration measures. The Trans-European Networks for Energy (TEN-E) Regulation [35–37] prescribes that infrastructure upgrades should contribute to increased reliability and resilience against natural or man-made incidents and adverse effects of climate change. It requires considering the impacts of such incidents as criteria for the evaluation of the projects of common interest in the process of upgrading the EU infrastructure.

2.2 Outlook of the EU Legislation for Critical Infrastructure

The domain of critical infrastructure (CI) and its protection is introduced in EU legislation by the European Programme for Critical Infrastructure Protection [21], followed by the adoption of the European Critical Infrastructure (ECI) Directive [5]. The objective of the ECI Directive was to provide a common approach to identify and designate ECIs in the energy and transport sectors and assess the need for their protection.

In its broader sense, the concept of resilience, especially of EPES, has been in the academic focus since the mid-2010s [47, 56–58]. With regards to legislation, the concept of resilience of CI has been introduced in the European Programme for Critical Infrastructure Protection [22], but it gained significance almost a decade later, with the adoption of the recommendation to strengthen the resilience of CI in 2022 and the Critical Entities Resilience (CER) Directive [35–37], which is in force since January 2023. The recommendation calls for completion of stress tests for CIs, which for the energy sector were finalized by the end of 2023. In 2024, the Council adopted the Recommendation for a Critical Infrastructure Blueprint, aiming to increase the EU Member States' response capabilities [6] and to provide a roadmap with measures to address major CI incidents affecting Member States, ensuring a targeted and proportionate response. The overview of relevant acts, key priorities, and implementation aspects are presented in Fig. 1.

The CER Directive [35–37] shifts the focus from protecting CI to enhancing the resilience of critical entities (CEs), which, in practice, are the operators of CI, by setting minimum rules across eleven key sectors. It defines resilience as a CE's ability "to prevent, protect against, respond to, resist, mitigate, absorb, accommodate, and recover from incidents", while CI refers to any asset or system that is critical for providing an essential service. The CER Directive requires Member States to identify CEs, support their resilience, establish implementation structures, ensure national and EU-level cooperation, adopt a CE resilience strategy, and conduct regular risk assessments.

The energy sector is among the eleven sectors encompassed by the CER Directive. Consequently, the list of CEs in the energy sector and electricity subsector includes transmission system operators (TSOs), distribution system operators (DSOs), producers, suppliers, nominated market operators, and providers of aggregation, demand response, or energy storage.

CEs are required to carry out risk assessment procedures at least every four years, accounting for relevant cross-sectoral and cross-border risks. Based on the outcomes of the risk assessment on Member State and CE level, the CEs are obligated to implement technical, security, and organizational measures related to prevention, protection, resistance, mitigation of consequences, recovery, and raising awareness for the identified risks. These measures should be included in the CE's resilience plan or equivalent documents.

2.3 *Outlook of the EU Legislation for Cybersecurity*

The integration of new digital technologies has profoundly transformed and digitized EPES, offering significant social and economic benefits. However, as discussed in the introductory section, it brings numerous cyberthreats, thus requiring the implementation of robust cybersecurity strategies to ensure the secure and reliable operation

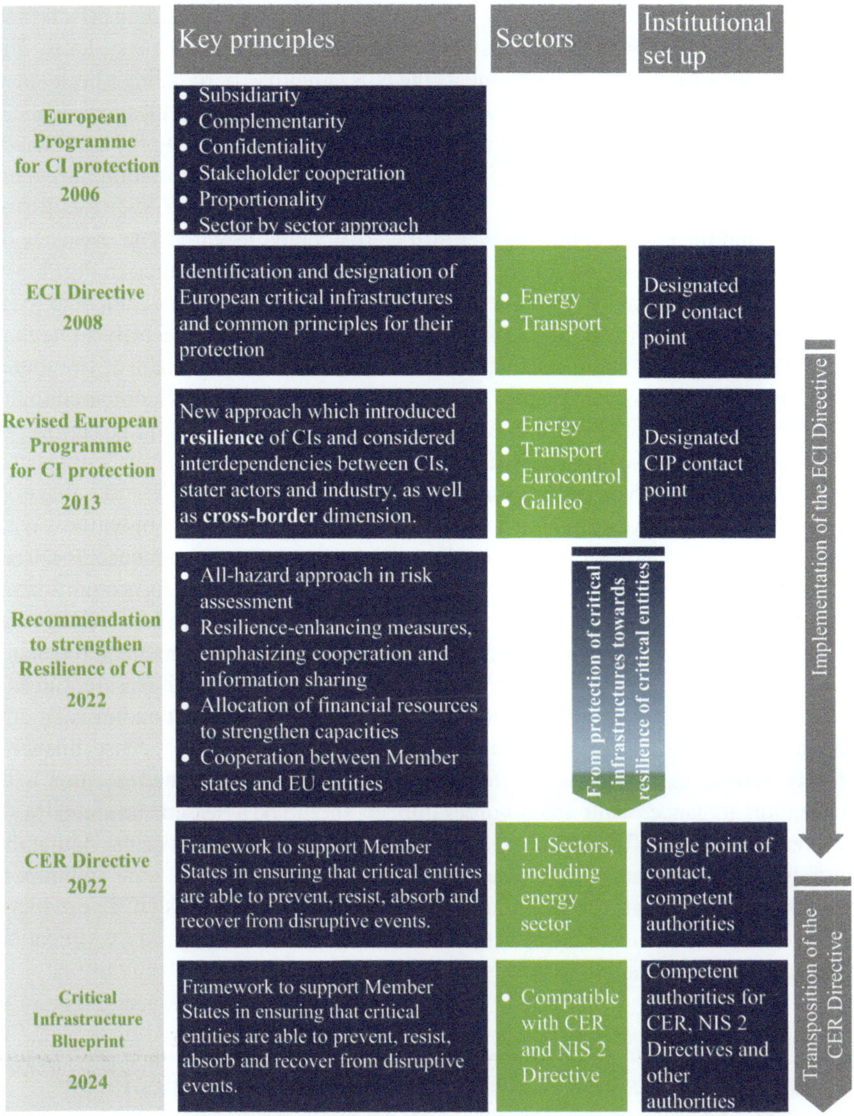

Fig. 1 Development of EU legislation related to protection and resilience of CIs

of power systems. To respond to these challenges, the NIS Directive [29, 30] was adopted in 2016. It aims to tackle cybersecurity in the sectors that are essential for the European Internal Market. In 2017, the Joint Communication on Cybersecurity [23–25] recognized the importance of sector-specific considerations at EU level, including the energy sector. In this context, the European Commission has adopted the Recommendation on cybersecurity in the energy sector [26], aiming to consider the real-time requirements of energy systems, the potential for cascading effects, and the coexistence of both legacy and state-of-the-art technologies in the systems. The Cybersecurity Act [32–34] complements the NIS Directive, introducing a permanent mandate for the EU Agency for Cybersecurity (ENISA) and EU rules for certification of products, processes, and services. These activities have led to the adoption of the **NIS2 Directive** [35–37], which now serves as the primary legislation for cybersecurity in the EU. Lastly, to complete the regulatory canvas, the Cyber Resilience Act [39] and the Cyber Solidarity Act [40] were adopted recently. The overview of relevant acts, key priorities, and implementation aspects are presented in Fig. 2.

The NIS2 Directive has a broader scope and expanded objectives compared to its predecessor. It requires Member States to develop national cybersecurity strategies, designate competent authorities, single points of contact, and establish Computer Security Incident Response Teams (CSIRTs). A novel requirement, compared to the NIS Directive, is to set up cyber crisis management authorities tasked to create large-scale incident response plans, which should be shared with the European Commission and the EU Cyber Crisis Liaison Organisation Network.

The NIS2 Directive distinguishes between essential and important entities, with the former corresponding to CEs from the energy sector according to the CER Directive. To ensure the cyber-resilience of essential entities, appropriate and proportionate technical, operational, and organizational measures should be taken to manage the security related to their network and information systems. Timely incident reporting is recognized as crucial for cybersecurity by the NIS2 Directive. Thus, essential and important entities are required to notify their CSIRT or competent authority of any incident significantly impacting their services. If disruptions may cause financial loss or serious harm to others, affected service recipients must be informed with mitigation guidance, while cross-border impacts should be assessed and shared with relevant authorities. To respond to the increased supply chain risks mentioned in the introductory section, the NIS2 Directive sets an obligation for essential and important entities to use Information and Communication Technologies (ICT) products, services, and processes that are certified under European cybersecurity certification schemes adopted pursuant to the Cybersecurity Act.

3 Best Practices and State of the Art in the Implementation of Resilience Strategies for EPES

Best practices in power system resilience encompass proven strategies and frameworks that have been successfully implemented to enhance the robustness, flexibility, and adaptability of power grids. These practices are typically endorsed by industry experts and backed by research, ensuring their effectiveness in real-world applications. They are developed in alignment with international standards, regulations, and policies, such as those established by IEEE, CIGRE, and ENTSO-E. Their reliability is further reinforced through validation in real-world power systems, where they are

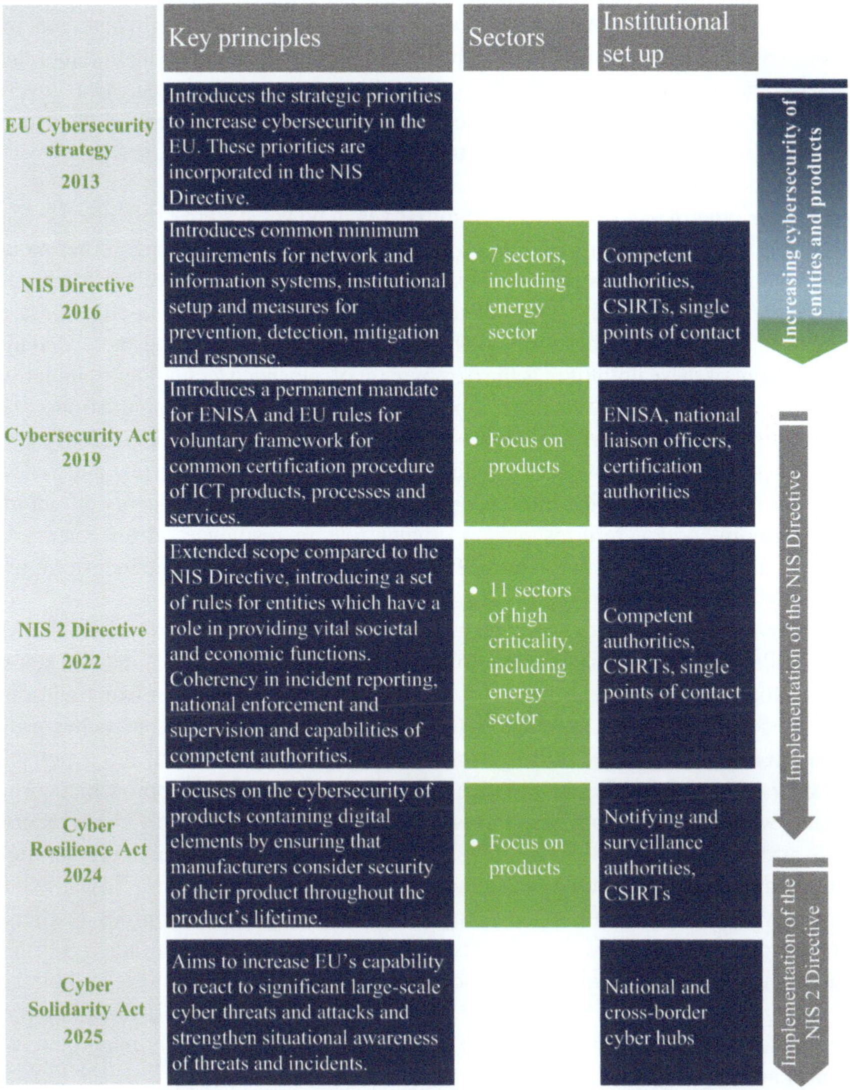

Fig. 2 Development of EU legislation for cybersecurity

tested under varying conditions. Moreover, these practices are continuously refined based on insights gained from past disruptions, including extreme weather events, cyberattacks, and large-scale blackouts, allowing for ongoing adaptation to emerging threats.

A key characteristic of best practices in power system resilience is their scalability and replicability, meaning that they should be adaptable to different grid environments and regulatory settings. Additionally, they are designed to enhance resilience across all phases of grid operation, covering anticipation, absorption, adaptation, and recovery from disruptions. Effective resilience strategies also require collaboration across multiple sectors and stakeholders, ensuring coordination between governments, regulatory authorities, utilities, TSOs, DSOs, and technology providers. This multi-stakeholder approach fosters a comprehensive and unified response to resilience challenges.

On the other hand, state-of-the-art initiatives in power system resilience focus on the latest advancements in technology, regulatory frameworks, and operational strategies to improve the grid's ability to withstand and recover from extreme disruptions. These initiatives shape the future of resilience by integrating cutting-edge research and pilot projects specifically aimed at enhancing power system robustness. A major component of these initiatives is the deployment of advanced technologies, including AI-driven predictive maintenance, digital twins for real-time grid simulations, self-healing grid architectures, and cyber–physical security frameworks. Additionally, they involve the integration of advanced energy storage, microgrids, and renewable energy sources, which improve grid flexibility and enhance energy autonomy. Furthermore, policy reforms play a crucial role in resilience advancements, as they establish new regulatory and operational benchmarks to ensure long-term grid security and stability.

State-of-the-art initiatives are characterized by their technology-driven innovation, utilizing AI, machine learning, Internet of Things (IoT) sensors, and blockchain technology to enable real-time grid monitoring and security. Unlike traditional resilience strategies that focus on post-event restoration, modern approaches prioritize proactive resilience, shifting from reactive recovery to anticipatory and self-healing grid solutions. These initiatives also emphasize a holistic approach, incorporating cyber-resilience, physical infrastructure hardening, and supply chain security into grid planning. Additionally, they rely on strong interdisciplinary collaboration, involving key players such as academia, private sector organizations, TSOs, DSOs, and government agencies to develop comprehensive and future-proof resilience frameworks. Such worldwide practices and initiatives are outlined in Table 1. Further, international professional bodies, including IEEE and CIGRE, have established working groups (e.g., CIGRE C4.47 "Power System Resilience") for addressing the topic of power systems resilience and how it can be seamlessly integrated into policy and regulatory standards.

4 Gaps and Barriers in Improving EPES Resilience

Despite the adoption of new legislative acts, the regulatory framework governing the energy sector is revealing significant shortcomings in addressing the direct and indirect damages from climate-related disruptions affecting the security of the infrastructure (intended as physical and operational security of the operational technology (OT)). While, from an ICT perspective, regardless of the introduction of new directives in Europe, challenges persist in ensuring the resilience of power systems.

Table 1 Summary of worldwide practices and initiatives towards energy resilience, with a focus on the EU

Stakeholder	Best practices	State-of-the-art initiatives
UK (National Infrastructure Commission—NIC, and RA: The Office of gas and Electricity Markets—Ofgem) [7, 50–53, 55, 70]	– Resilience standards and cross-sector planning – Cybersecurity integration – Long-term planning for climate adaptation and digitalization – Cross-sector approach to resilience, integrating energy, telecoms, transportation, and water networks – Policy framework embedding resilience into national energy policy and infrastructure planning – Emergency restoration upgrades	– Regulatory integration led by Ofgem and NESO – Advanced cyber and climate resilience frameworks – Strategic clean energy investments such as £4.3 billion eastern green link subsea project and £77.4 billion investment (2026–2031) in transmission expansion – Resilience-focused investment in the RIIO-T3 plan
Germany (TSOs: Amprion, TransnetBW, TenneT, and 50 Hz) [1, 20]	– Expansion of microgrids and self-sufficient energy hubs – Systematic deployment of distributed energy resources (DERs) and demand-side management – Development of peer-to-peer energy trading and blockchain-based decentralized control – Digitalization through integration of real-time digital monitoring systems and deployment of AI-powered self-healing grid technologies – Protection of automation and control systems from cyberattacks – Incident reporting and resilience assessment systems – Development of standardized resilience indicators – Embedding resilience into energy policy, regulatory frameworks, and market incentives	– Threat detection and automated mitigation for cyber-resilient grid management – Blockchain energy trading – Digitalized risk assessment models with anomaly detection and automated countermeasures – Self-healing grid technologies through autonomous detection, isolation, and rerouting of power flow during local failures to minimize disruptions – Integration of AI-driven digital twins with adaptive protection schemes for rapid response and recovery

(continued)

Table 1 (continued)

Stakeholder	Best practices	State-of-the-art initiatives
France (TSO: Réseau de transport d'Électricité—RTE) [49]	– Reinforcement of transmission infrastructure to storms and flooding – Expansion of underground transmission lines in high-risk areas – Expansion of demand-side response programs and battery storage solutions to improve operational adaptability during crises – Strengthening of cross-border interconnections with neighbouring countries (Spain, Italy, Germany, and UK) for enhanced regional stability – Embedding resilience in national energy security planning – Structured crisis management and long-term preparedness	– Digital twin technology: AI-driven digital simulations to model grid vulnerabilities and optimize resilience strategies proactively – AI-powered predictive maintenance – Integration of machine learning-based fault detection systems to anticipate and prevent equipment failures
Italy (TSO: Terna) [15, 65, 67, 68]	– Prospective climate hazard forecasting using climate models to evaluate risk exposure – Asset vulnerability analysis to identify and prioritize reinforcements for critical infrastructure – Preventive actions: transmission line upgrades, substation reinforcement, underground cabling in risk-prone areas, and anti-torsional devices – Fast restoration: mobile generators, automated self-healing systems, and streamlined emergency protocols – Expansion of interconnections with Greece, France, and Tunisia to enhance energy security and flexibility – Collaboration with national and European entities (ARERA, RSE, and ENTSO-E) to standardize resilience strategies	Digitalization for predictive resilience: AI-powered real-time grid monitoring and automated control systems for adaptive operation Satellite-based weather analytics for dynamic operational adaptation to climate conditions; demand response management through smart automation Advanced climate adaptation technologies: AI-driven predictive models and high-resolution meteorological simulations; real-time climate data to dynamically adjust grid behaviour during adverse conditions. €21 billion investment through the 2023 development plan focused on grid modernization and climate resilience

(continued)

Table 1 (continued)

Stakeholder	Best practices	State-of-the-art initiatives
Greece (RA: The regulatory Authority for Energy—RAE and TSO: ADMIE) [27]	– Upgrading and expanding 400 kV transmission corridors – Investments in digital grid infrastructure with real-time automation and predictive fault detection – Cross-border energy cooperation: development of 400 kV interconnections with Bulgaria, Italy, North Macedonia, Albania, and Turkey – Energy storage deployment: large-scale investment in pumped hydro and battery storage systems to balance renewables – Implementation of AI-driven analytics for real-time monitoring and predictive maintenance – Strengthened cybersecurity protocols to protect CI	– Supporting decarbonization and strengthening cross-continental energy security – EuroAsia and EuroAfrica interconnectors: pioneering subsea electricity link between Greece, Cyprus, and Israel; enhancing Mediterranean electricity trade and integrates isolated grids with Europe; subsea link between Greece and Egypt connecting north African renewables to the European grid – Cutting-edge integration PV/storage and wind/storage, especially in remote regions
ENTSO-E [13–17]	– Pan-European resilience framework – Developed standardized incident classification and risk assessment methodologies – System Defense and restoration plans, involving national and regional black start capabilities, islanding strategies, and cross-border emergency support – Grid flexibility enhancement through diverse flexibility resources, e.g. demand response, BESS Enhances system adaptability to renewable variability and load fluctuations – Improving interoperability between transmission and distribution networks and supporting cross-sector observability	– Vision 2050: long-term strategy to achieve a carbon-neutral, resilient, and affordable power system; incorporates system flexibility, adaptive markets, and strategic infrastructure planning – Cross-border interconnections: enhance regional energy flow and grid resilience through expanded interconnectivity – Climate-resilient grid planning: integrates climate adaptation and extreme weather modelling into grid development; geospatial analysis and AI-based weather forecasting to proactively strengthen the grid

4.1 Physical Systems Resilience

The relationship between higher RES integration and the increasing vulnerability in front of cascading failures has already been introduced along with the effectiveness of critical asset monitoring and maintenance as key elements of grid stability and reliability. Addressing these challenges is essential to ensure that modern power systems can withstand and recover from high-impact, low-frequency (HILF) events.

Past incidents, such as the Continental Europe separation [12], Balkan Incident [18], and the Rogowiec event [19], show how weather, environmental factors, and human errors can trigger widespread grid disturbances. These incidents emphasize the need for real-time situational awareness and coordinated emergency response. The growing frequency of extreme events and the expansion of distributed energy resources (DERs) reveal gaps in the ability to assess and manage cascading effects and highlight the need for improved operational strategies to mitigate vulnerabilities [66]. Enhanced system observability—through real-time data, external inputs, and predictive analytics—can help operators plan effective contingency scenarios and mitigate risks. Another key challenge is the lack of adaptive strategies to ensure maximum utilization of renewable energy sources (RES) during crises. While high-RES penetration increases grid instability due to reduced inertia, dispersed generation could provide system flexibility if properly integrated. Heatwaves and storms reduce RES availability while increasing demand, requiring advanced forecasting and demand response mechanisms. Microgrids also offer potential resilience benefits in disaster-prone regions, enabling autonomous operation and supporting critical loads. However, their integration is hindered by interoperability issues and regulatory barriers. Additionally, asset health monitoring remains underdeveloped, relying on reactive maintenance rather than predictive methods. Advanced sensor diagnostics and machine learning could enable proactive maintenance, improving system reliability and asset lifespan. In this sense, the adoption of new hardware (HW) (IoT, new sensors, automation, and robotic) and software (SW) (big data analysis, interoperability, and AI) technologies is crucial for moving from a manual to an automated outage management process, with tangible advantages such as: (i) early detection of incipient failures, (ii) optimization of the decision-making process, and (iii) reduction of response time and recovery after failure. The workflow approval process can be automated as well, with an additional reduction of the time for assessment and authorization of the request. These aspects are critical for a healthy and stable operation of the critical components, contributing to the improvement of the reliability indexes of the assets (mean time between failures (MTBF), mean time to failure (MTTF), etc.) and of the whole system (system average interruption duration index (SAIDI), system average interruption frequency index (SAIFI), etc.), with significant cost saving for the operator, improving the power quality and the service continuity.

Resilience in energy systems is not sufficiently addressed in current regulatory frameworks, especially considering the increasing risks from climate change and grid digitalization. Existing policies largely prioritize reliability, through the compliance with predefined performance standards, rather than fostering adaptive capabilities to confront emerging threats. Present resilience management frameworks often address discrete events, such as the failure of a single generating unit [46], neglecting the interconnected and complex failures typical of modern power systems, as demonstrated by incidents like the South Australian blackout [2]. The absence of comprehensive resilience metrics and standardized benchmarks leaves utilities without clear guidance for infrastructure hardening and operational improvements [4]. To address these gaps, the regulatory framework must integrate climate risk assessments into power system planning. Although the TEN-E Regulation [35–37] considers climate impacts when upgrading EU infrastructure, it insufficiently addresses real-time and cross-border operational aspects. Despite growing evidence of climate-induced threats, resilience planning is often secondary to cost-efficiency considerations, limiting the ability of EPES to shift from recovery-focused to preventive strategies. While the Risk Preparedness Regulation [32–34] encompasses rare and extreme risks, its main emphasis on supply security may overlook concurrent operational vulnerabilities. Effective resilience planning thus requires enhanced cooperation among TSOs, DSOs, regulators, and other stakeholders. Moreover, current market design also inadequately incentivizes investments in grid resilience, particularly for distributed energy resources and microgrid solutions [71]. Although the European Green Deal promotes climate resilience and sustainable infrastructure, its implementation is hindered by financial constraints, regulatory fragmentation, and technological barriers. Disparities in national frameworks further complicate the establishment of a unified European energy security strategy. Coordination between TSOs and DSOs also remains a challenge for outage management and system restoration. European blackout studies stress the need for harmonized frameworks ensuring real-time data exchange and cross-border collaboration [66]. Existing regulations, such as the System Operation Guidelines [23–25], primarily address normal operations, highlighting the necessity for dedicated crisis cooperation protocols. Finally, regulatory frameworks insufficiently emphasize critical asset monitoring and maintenance. Although infrastructure hardening measures are recognized, financial and administrative barriers often impede their implementation [46]. Furthermore, incentives for adopting advanced monitoring technologies, such as real-time diagnostics and predictive maintenance, are lacking, limiting utilities' ability to mitigate the impacts of extreme weather events and enhance overall system reliability.

4.2 Cyber Systems Resilience

The increasing digitization of energy systems has introduced significant cybersecurity challenges (as anticipated in previous sections), making them easy targets for cyberthreats such as ransomware, state-sponsored attacks, and supply chain vulnerabilities. The integration of DERs, smart grid technologies, and cross-border electricity flows has widened the attack surface for malicious actors, leading to the urgent need for more robust cybersecurity structures. However, numerous operational and regulatory challenges remain unresolved, limiting the ability of energy systems to withstand and recover from cyberthreats. The topic is incredibly wide and for what is concerning the work within R^2D^2 project the following aspects have been investigated: (i) prioritize the adoption of AI-driven anomaly detection, (ii) facilitate implementation of harmonized regulatory and normative framework for resilience, (iii) enhance information technology (IT)-OT security integration, and (iv) enforce stringent supply chain protections, and facilitate knowledge sharing.

Most defence mechanisms rely on a reactive strategy that cannot fully meet the need for resilience against advanced persistent threats and zero-day exploits. The integration of machine learning (ML) into resilience strategies offers solutions that rely on anomaly detection and automatic response to such threats. However, AI implementation remains hampered by concerns about data integrity, computational demands, and adversary threats that can trick a model. Although there is not yet a full exploitation in the electrical industry, the leap is even greater when it comes to the effective use of quantum computing in the near future.

The resilience and reliability of **critical assets and equipment** play a key role in providing grid stability amid cyberthreats. From the European perspective, the NIS2 Directive recognizes system operators as critical entities, thus introducing obligations to increase their cybersecurity. As the European energy market becomes increasingly integrated, cyberthreats may trespass national boundaries. To overcome this risk, the Network Code on Cybersecurity [28] underscores the necessity for coordinated security protocols and information-sharing mechanisms among EU Member States. Nevertheless, variations in national cybersecurity strategies and insufficient data-sharing practices among system operators can limit the effectiveness of such regulation, hindering the capacity to effectively address large-scale cyber incidents. Current standards, including IEC 62443 for industrial automation security and the IEEE 1686 standard for intelligent electronic devices, define security guidelines, but their adoption can be inconsistent due to varying industry-specific requirements and implementation complexity. Therefore, operators struggle with the resource-intensive nature of aligning their existing infrastructure with these frameworks, leading to partial or inconsistent compliance [3]. Without more structured regulatory mandates, the adoption of these standards remains fragmented, leaving critical infrastructure vulnerable to emerging cyberthreats [10]. Strengthening resilience requires enhanced asset monitoring, real-time intrusion detection systems, and built-in security measures to ensure the operation of critical components even under cyber duress.

Another significant gap is the inadequate **coordination between IT and OT security**. Fragmentation in cybersecurity standards also creates challenges for power system operators, since the two major standards (IEC 62443 and ISO 27001) present challenges in their applications. There are a number of difficulties when implementing IEC 62443 in electricity settings, but for an effective implementation, the gap between the OT and IT has to be closed, considering the teams have historically worked in silos. On the other hand, ISO 27001 provides general cybersecurity best practices, and its applicability to power systems remains limited, unless a strong custom adaptation is undertaken. Thus, conventional IT security protocols frequently present security limitations when implemented in OT environments, supervising industrial operations, and energy distribution networks. Many OT systems operate on legacy infrastructure that lacks built-in security features, making them particularly vulnerable to cyberattacks. The problem persists even when focusing on new devices connected to the grid as DERs, which are becoming ubiquitous, even though they are not considered mission-critical. Some studies [72] observe that there is no unified certification framework to validate the cybersecurity resilience of DER components, leading to inconsistencies in security measures across different manufacturers and vendors. The current regulatory framework for DERs shows significant operational and regulatory gaps that hinder effective cybersecurity measures. The NIS2 Directive introduces obligations for all electricity firms as defined in the Electricity Market Directive [32–34]. The implementation of proportionate measures only ensures cybersecurity of IT systems, leaving out some of the OT equipment.

Supply chain security is another critical factor in cyber-resilience, considering the risk of supply chain attacks has grown significantly since 2020, as mentioned in the introductory section. While legislation acts such as the NIS2 Directive and the EU Cyber Resilience Act provide guidance on securing supply chains, enforcement remains inconsistent, and many system operators lack the resources and experience to conduct comprehensive vendor risk assessments. Strengthening supply chain resilience requires more rigorous security evaluations, continuous monitoring of third-party software dependencies, and a compliance requirement for vendors supplying critical infrastructure components. Following this collaborative approach, knowledge sharing plays a central part in enhancing the resilience of organizations against cyber crises, particularly in the context of supply chain attacks. ENISA has highlighted that effective knowledge sharing can significantly improve the preparedness and response capabilities of Member States and critical entities in general [11].

4.3 R^2D^2 Contributions to EPES Resilience

The R^2D^2 project is dedicated to improving the resilience and reliability of EPES against a growing number of threats and vulnerabilities. The project develops four tools dedicated to the prevention, protection, and restoration of EPES, which can be used to prevent and mitigate cyber and physical risks in the electric infrastructures: C3PO—Multi-risk assessment framework for power system [60]; IRIS—Resilience suite for TSO and DSO [61]; PRECOG—Prevention Systems for Energy Infrastructures Security [62]; and EMMA—Enhanced Assets Maintenance and Management Toolkit [63].

4.3.1 Contributions and Recommendations for Physical Resilience

As mentioned above, amid the growing frequency of extreme climate-induced events, it is essential that the critical entities are able to assess and manage possible cascading effects that may have a multilayered impact on the economy and society. As HILF events cannot be anticipated, nor there are enough data to serve as benchmarking rules, critical entities would benefit from tools that will support them in the modelling and simulation of such events. Hence, the C3PO product includes a spatial and temporal event and cascading simulator tool, with the main objective to simulate extreme events and evaluate their impact on the networks. It is a holistic research effort including both wind and wildfire events modelling, cascading failure and ML models. ML models are extensively used for locating critical power lines in the network, as well as to predict the resulting load shedding due to the cascade propagation. Moreover, a security study is conducted to show the vulnerability of ML models deployed in power systems against malicious machine unlearning. C3PO product enables transmission and distribution system operators to derive a set of operational measures to enhance resilience and to implement cascading-aware investment planning. While this approach is not included in standard planning methodologies yet, the tool enables the operators to simulate extreme wind and wildfire events, assess the impact and the proposed measures for their mitigation, and use the results to reshape their planning methodologies. These actions should serve as a basis for the development of new resilience-oriented planning methodologies and the integration of resilience as a planning criterion in the energy-related legislation (distribution network codes and guidelines). In this way, the product contributes to bridging the gap identified before, related to the integration of climate risk assessments into distribution system planning.

Furthermore, C3PO introduces a novel approach based on networked microgrids formation that handles multiple line outages and limited DERs operation. It outlines a microgrid-based post-disruption operation and restoration strategy, demonstrating the potential benefits of microgrid formation in distribution system restoration in the face of extreme events. In addition, the product includes a tool that enhances energy resilience by integrating microgrids and mobile sources into the load restoration process of distribution networks. A defender–attacker–defender model optimizes the placement of mobile generators, mobile storage systems, and repair crews. AI-based control algorithms enhance rapid response during outages. The demonstration of these tools in the project strongly supports the need for establishing clear guidelines on microgrid operation under normal and extreme events and their integration in distribution systems. However, the regulatory authorities and operators are still lagging in taking concrete actions for microgrid integration.

The R^2D^2 project additionally contributes to the shift from recovery-focused to preventive strategies for strengthening the physical resilience of EPES by introducing a new solution called EMMA. This product has been developed to deal with the planned and unplanned outage management, it integrates novel autonomous and AI-based technologies in equipment monitoring and inspection with predictive maintenance capabilities. For planned outages, it improves the network reconfiguration and the planning of maintenance interventions through the automation and integration of ICT processes and novel optimization techniques. Both aspects contribute to establishing an effective predictive maintenance approach in the daily process of TSOs, DSOs, and RSCs. The product enables improved maintenance and intervention planning under complex scenarios of outages and implements optimal remedial actions.

Through the adoption of EMMA, system operators will improve their awareness and response before and after a critical event, thus contributing to the enhancement of the system's resiliency.

To address the challenges of TSO/DSO coordination under extreme events, the R^2D^2 project has developed the IRIS Resilience Suite product. It is designed to be utilized by TSOs, DSOs, and RSCs and provides tools and functionalities that facilitate their collaboration, synchronize their actions, and enable them to make swift, well-informed decisions in situations where the security of the grid is at risk. Several tools are dedicated to TSO/DSO coordination for enhanced distributed energy resource control, monitoring congestion and power quality in the process of provision of flexibility by distribution-connected resources, as well as validation and coordination of remedial actions to be applied in real-time during crisis situations, and improved observability of low-voltage networks. Another set of tools aims to respond to the challenge of potential instability in EPES, which may be related to high-RES penetration. The tools enable replacing the current practices of stability assessment through off-line simulations, enhancing the process of restoration during system split and introducing a novel remedial actions cost-sharing methodology. Based on the development and demonstration evidence of these tools, it becomes clear that real-time stability monitoring would significantly improve the system resilience, thus arising the need to introduce this requirement in existing network codes. In addition, a novel remedial actions cost-sharing methodology between TSOs is introduced by the project, as an alternative solution to existing cost-sharing methodologies.

4.3.2 Contribution and Recommendations for Cyber-Resilience

AI Integration for IT and OT traffic security. The integration of AI into cybersecurity systems for power infrastructure, as implemented in the PRECOG product of the R2D2 project, represents a major step forward in advancing early threat detection and cyber-resilience. Central to this innovation is the CARMEN tool, which utilizes AI-driven techniques for deep packet inspection, dynamic traffic characterization, anomaly detection, and advanced persistent threat (APT) identification. CARMEN can continuously monitor network activity, and through its ML capabilities, it identifies subtle deviations from expected behaviour, flags potential intrusions, and generates actionable alerts. One of the key advancements offered by this approach is the unified inspection and analysis of both OT and IT traffic. PRECOG addresses this well-known gap by enabling AI-powered traffic dissection across both domains, with support for industrial communication protocols. This holistic visibility not only improves threat detection accuracy and reduces false positives but also enhances situational awareness and incident response capabilities for cybersecurity teams. Moreover, it works in parallel with another tool able to assess the cyber-risk from IT traffic dynamically in real time [60], which also includes AI algorithms (exploit prediction scoring system (EPSS) and fuzzy cognitive maps). TSOs and DSOs are able to prioritize vulnerabilities based on actual exploitation of risk rather than a theoretical severity.

Vendors' attack and supply chain security. The supply chain assessment tool developed within the PRECOG suite introduces a significant innovation in addressing one of the most critical and often under-secured aspects of cybersecurity in the energy sector: the supply chain. This tool confronts this challenge head-on by providing a unified platform where both operators and vendors can engage in structured, evidence-based evaluations of cybersecurity posture. Through this tool, it is possible to assess the current practices against industry standards, identify gaps, and receive tailored recommendations for improvement. It allows stakeholders to evaluate cybersecurity maturity, compliance with best practices, and risk exposure across the supply chain. The tool generates detailed reports, benchmarking results against recognized standards (such as IEC270002 and National Institute of Standards and Technology (NIST)), and offers targeted recommendations for improvement, thereby turning assessments into actionable strategies.

Knowledge sharing. Within the context of the C3PO framework in the R^2D^2 project, the cyberthreat intelligence tool (CTI), and the shared knowledge sharing tool form an integrated knowledge sharing system to strengthen the resilience of the electricity system, addressing cyberthreats and operational risks related to physical and climate events, respectively. The CTI tool can monitor and share cyberthreats in real time, allowing the collection, analysis, and distribution of information from internal and external sources thanks to the integration with established platforms such as malware information sharing platform (MISP) and reference frameworks such as MITRE ATT&CK and EPSS. It provides system operators and IT security managers with access to indicators of compromise, emerging vulnerabilities and other critical data, significantly improving the ability to prevent and respond to attacks, and bridging information between organizations to facilitate a collective and proactive defence. The knowledge sharing tool is designed to securely collect, store, and share knowledge and data related to cascading events triggered by physical disturbances, such as storms or wildfires. The tool allows the integration of information from different resilience modules developed in the project, thus facilitating scenario development, impact analysis, and definition of operational strategies. It fosters a systematic sharing of best practices, improves the decision-making process of network operators (DSO/TSO), and increases collective preparedness through a structured, accessible, and continuously updated operational memory.

As mentioned in the Introduction1, the implementation of the requirements from the NIS2 Directive is a current challenge for system operators, as it introduced a number of new requirements that have to be met. Performing regular cyber-risk assessments is among these requirements, and therefore, the project envisaged the development of risk assessment tools within the C3PO product. It provides a comprehensive approach to the evaluation and mitigation of cyber-risks in critical infrastructures. It starts with defining the infrastructure and identifying assets and their criticality and analysing vulnerabilities utilizing well-established libraries. The tool facilitates a gap analysis by allowing the selection of various cybersecurity frameworks, such as NIST SP 800–53 [54] or ISO/IEC 27002 [45], to evaluate the maturity of existing security controls. It calculates vulnerability severity based on the gap analysis, enables the selection of predefined risk scenarios tied to notable incidents, and supports threat identification and likelihood assessment. After calculating risk levels, the tool generates risk treatment plans with recommended controls and provides extensive reporting capabilities through numeric reports, histograms, spider web charts, and exportable formats to track progress over time. A step further is the development of a dynamic risk assessment tool which addresses the challenge of a continuous evolution of threats, considering the organization's network topology and deployed security controls.

5 Conclusion

This chapter discusses the challenges of increasing the resilience of EPES in order to withstand the evolving and more frequent threats caused by extreme weather events or man-provoked incidents. As explained in the introductory section, the complex field of physical and cyberthreats to the EPES has drastically increased in the past few years, requiring new approaches and solutions to prevent, protect, resist, and mitigate them, thus increasing the level of resilience of EPES. Understanding the criticality of the problem, international standardization bodies are developing various sets of standards to cover this field, while governments and official entities are establishing regulatory frameworks to create an environment for the implementation of measures and policies in critical sectors. The example of the EU regulatory framework has been discussed in this chapter, confirming the importance of resilience of all critical sectors, although the emphasis of the discussion is on the energy sector. The regulatory framework sets the requirements to critical entities to implement resilience measures by applying adequate technical solutions. These solutions should be tailored to the needs of system operators and employ innovative approaches that can grasp the evolving cyber and physical threats of EPES. The innovative work of the R^2D^2 project has succeeded in filling some of the existing gaps in this area, as explained in this chapter.

Acknowledgments This research has been supported by the EU Horizon project R^2D^2, "Reliability, Resilience and Defense technology for the grid" (grant agreement 101075714). The paper reflects only the authors' views, and neither the Agency nor the Commission are responsible for any use that may be made of the information contained therein.

References

1. acatech—National Academy of Science and Engineering (2021) "The Resilience of Digitalised Energy Systems: Options for Reducing Blackout Risks", April 2021. Available from: https://en.acatech.de/publication/rde/.
2. AER—Australian Energy Regulator (2018) The Black System Event Compliance Report. Investigation into the Pre-event, System Restoration, and Market Suspension aspects

surrounding the 28 September 2016 event. Available from: https://www.aer.gov.au/system/files/Black%20System%20Event%20Compliance%20Report%20-%20Investigation%20into%20the%20Pre-event%20System%20Restoration%20and%20Market%20Suspension%20aspects%20surrounding%20the%2028%20September%202016%20event.pdf
3. Barrett, M. (2018) Framework for Improving Critical Infrastructure Cybersecurity Version 1.1, NIST Cybersecurity Framework, Available from: https://doi.org/10.6028/NIST.CSWP.04162018, https://www.nist.gov/cyberframework
4. Ciapessoni, E., Cirio, D., Pitto, A., Van Harte, M., Panteli, M. (2023) Power System Resilience: definition, features and properties. Cigré Science & Engineering Journal, 30.
5. Council of the European Union (2008) Council Directive 2008/114/EC of 8 December 2008 on the identification and designation of European critical infrastructures and the assessment of the need to improve their protection Available from: https://eur-lex.europa.eu/eli/dir/2008/114/oj/eng
6. Council of the European Union (2024) Council Recommendation of 25 June 2024 on a blueprint to coordinate a response at Union level to disruptions of critical infrastructure with significant cross-border relevance, OJ C, C/2024/4371, 2024. Available from: https://eur-lex.europa.eu/legal-content/EN/TXT/PDF/?uri=OJ:C_202404371
7. Eastern Green Link 1 (2024) About the Project. 2024. Available from: https://www.easterngreenlink1.co.uk/about-the-project
8. EM-DAT (2023) The Emergency Events Database. Trends in weather-related disasters globally and in Europe (2000–2023). 2023. Available from: https://www.emdat.be/
9. ENISA-European Union Agency for Cybersecurity (2021). Threat Landscape for Supply Chain Attacks. Public report. Available from: https://www.enisa.europa.eu/publications/threat-landscape-for-supply-chain-attacks
10. ENISA—European Union Agency for Cybersecurity (2022). Cyber Threat Landscape for the Energy Sector. Public. report. Available from: https://www.enisa.europa.eu/sites/default/files/publications/ENISA%20Threat%20Landscape%202022.pdf
11. ENISA—European Union Agency for Cybersecurity (2024). Best practices for Cyber Crisis Management. Public report. Available from: https://www.enisa.europa.eu/publications/enisa-threat-landscape-2024
12. ENTSO-E (2021), Continental Europe Synchronous Area Separation on 24 July 2021. Technical report, 15 November 2021. Available from: https://eepublicdownloads.entsoe.eu/clean-documents/SOC%20documents/SOC%20Reports/entso-e_CESysSep_210724_211112.pdf
13. ENTSO-E (2022) "ENTSO-E Vision 2050: A Power System for a Carbon-Neutral Europe", October 2022. Available from: https://eepublicdownloads.entsoe.eu/clean-documents/tyndp-documents/entso-e_Vision_2050_report_221006.pdf
14. ENTSO-E (2022) "System Defence Plan, Regional Group Continental Europe", February 2022. Available from: https://eepublicdownloads.entsoe.eu/clean-documents/SOC%20documents/Regional_Groups_Continental_Europe/2022/220215_RGCE_TOP_03.2_D.1_System%20Defence%20Plan_v8_final.pdf
15. ENTSO-E (2023) "Guideline for Cost Benefit Analysis (CBA) 4.0", April 2023. Available from: https://eepublicdownloads.blob.core.windows.net/public-cdn-container/tyndp-documents/CBA/CBA4/230424_for-opinion/CBA_4_Guideline_for_ACER_opinion.pdf
16. ENTSO-E (2024) "ENTSO-E Strategic Roadmap 2023–2025", February 2024. Available from: https://eepublicdownloads.blob.core.windows.net/public-cdn-container/clean-documents/Publications/ENTSO-E%20general%20publications/entso-e_strategic_roadmap_WEB_240215.pdf
17. ENTSO-E (2024) "Transversal Challenges in the Energy Transition", 2024. Available from: https://vision.entsoe.eu/chapters/transversal-challenges
18. ENTSO-E ICS Investigation Expert Panel (2025), Grid Incident in South-East Europe on 21 June 2024. Final Report, 25 February 2025. Available from: https://eepublicdownloads.blob.core.windows.net/public-cdn-container/clean-documents/Publications/2024/entso-e_incident_report_240621_250225_02.pdf

19. ENTSO-E, ICS Investigation Expert Panel (2022), Local Grid Incident in Rogowiec (Poland) substation on 17 May 2021. Technical report, 18 March 2022. Available from: https://www.entsoe.eu/news/2022/03/18/final-report-on-the-local-power-grid-incident-in-rogowiec-poland-substation-on-17-may-2021/
20. envelio GmbH (2023) "Digital Twin of Power Grid: Best Practices and First Steps", May 5 2023. Available from: https://envelio.com/insights/digital-twin-power-grid-best-practices
21. European Commission (2006) Communication from the Commission of 12 December 2006 on a European Programme for Critical Infrastructure Protection, Brussels, 2006. Available from: https://eur-lex.europa.eu/legal-content/EN/TXT/?uri=celex:52006DC0786
22. European Commission (2013) Commission Staff Working Document on a new approach to the European Programme for Critical Infrastructure Protection, Brussels, 2013. Available from: https://eur-lex.europa.eu/legal-content/EN/TXT/?uri=legissum:jl0013
23. European Commission (2017) "COMMUNICATION FROM THE COMMISSION TO THE EUROPEAN PARLIAMENT AND THE COUNCIL on Resilience, Deterrence and Defence: Building strong cybersecurity for the EU," COM (2017) 450 final, 13.09.2017. Available from: https://eur-lex.europa.eu/legal-content/EN/TXT/?uri=celex:52017JC0450
24. European Commission (2017) Commission Regulation (EU) 2017/1485 of 2 August 2017 establishing a guideline on electricity transmission system operation. Available from: https://eur-lex.europa.eu/eli/reg/2017/1485/oj/eng
25. European Commission (2017) Commission Regulation (EU) 2017/2196 of 24 November 2017 establishing a network code on electricity emergency and restoration. Available from: https://eur-lex.europa.eu/eli/reg/2017/2196/oj/eng
26. European Commission (2019) Commission Recommendation (EU) 2019/553 of 3 April 2019 on cybersecurity in the energy sector (notified under document C (2019) 2400) Available from: https://eur-lex.europa.eu/legal-content/EN/TXT/?uri=CELEX%3A32019H0553
27. European Commission (2022) "Risk-Preparedness Plan—Greece", September 2022. Available from: https://energy.ec.europa.eu/system/files/2022-09/EL_%20RPP%20electricity.pdf
28. European Commission (2024) Commission Delegated Regulation (EU) 2024/1366 of 11 March 2024 supplementing Regulation (EU) 2019/943 of the European Parliament and of the Council by establishing a network code on sector-specific rules for cybersecurity aspects of cross-border electricity flows Available from: https://eur-lex.europa.eu/eli/reg_del/2024/1366/oj/eng
29. European Parliament (2016) "Cyber Security Strategy for the Energy Sector, Study for the ITRE Committee," 2016. Available from: https://www.europarl.europa.eu/RegData/etudes/ATAG/2017/602019/IPOL_ATA(2017)602019_EN.pdf
30. European Parliament (2016) Council of the European Union, Directive (EU) 2016/1148 of the European Parliament and of the Council of 6 July 2016 concerning measures for a high common level of security of network and information systems across the Union Available from: https://eur-lex.europa.eu/eli/dir/2016/1148/oj/eng
31. European Parliament (2018) Council of the European Union, Regulation (EU) 2018/1999 of the European Parliament and of the Council of 11 December 2018 on the Governance of the Energy Union and Climate Action, amending Regulations (EC) No 663/2009 and (EC) No 715/2009 of the European Parliament and of the Council, Directives 94/22/EC, 98/70/EC, 2009/31/EC, 2009/73/EC, 2010/31/EU, 2012/27/EU and 2013/30/EU of the European Parliament and of the Council, Council Directives 2009/119/EC and (EU) 2015/652 and repealing Regulation (EU) No 525/2013 of the European Parliament and of the Council Available from: https://eur-lex.europa.eu/eli/reg/2018/1999/oj/eng
32. European Parliament (2019) Council of the European Union, Regulation (EU) 2019/881 of the European Parliament and of the Council of 17 April 2019 on ENISA and on information and communications technology cybersecurity certification and repealing Regulation (EU) No 526/2013 (Cybersecurity Act), Brussels: OJ L 151, 2019. Available from: https://eur-lex.europa.eu/eli/reg/2019/881/oj/eng
33. European Parliament (2019) Council of the European Union, Regulation (EU) 2019/943 of the European Parliament and of the Council of 5 June 2019 on the internal market for electricity (recast). Available from: https://eur-lex.europa.eu/eli/reg/2019/943/oj/eng

34. European Parliament (2019), Regulation (EU) 2019/941 of the European Parliament and of the Council of 5 June 2019 on risk-preparedness in the electricity sector and repealing Directive 2005/89/EC. Available from: https://eur-lex.europa.eu/eli/reg/2019/941/oj/eng
35. European Parliament (2022) Council of the European Union, Directive (EU) 2022/2555 of the European Parliament and of the Council of 14 December 2022 on measures for a high common level of cybersecurity across the Union, amending Regulation (EU) No 910/2014 and Directive (EU) 2018/1972, and repealing Directive (EU) 2016/1148 (NIS2 Directive) Available from: https://eur-lex.europa.eu/eli/dir/2022/2555/oj/eng
36. European Parliament (2022) Council of the European Union, Directive (EU) 2022/2557 of the European Parliament and of the Council of 14 December 2022 on the resilience of critical entities and repealing Council Directive 2008/114/EC, Brussels: OJ L 333, 2022, p. 164–198. Available from: https://eur-lex.europa.eu/eli/dir/2022/2557/oj/eng
37. European Parliament (2022) Council of the European Union, Regulation (EU) 2022/869 of the European Parliament and of the Council of 30 May 2022 on guidelines for trans-European energy infrastructure, amending Regulations (EC) No 715/2009, (EU) 2019/942 and (EU) 2019/943 and Directives 2009/73/EC and (EU) 2019/944, and repealing Regulation (EU) No 347/2013 Available from: https://eur-lex.europa.eu/eli/reg/2022/869/oj/eng
38. European Parliament (2023) Council of the European Union, Directive (EU) 2023/2413 of the European Parliament and of the Council of 18 October 2023 amending Directive (EU) 2018/2001, Regulation (EU) 2018/1999 and Directive 98/70/EC as regards the promotion of energy from renewable sources, and repealing Council Directive (EU) 2015/652. Available from: https://eur-lex.europa.eu/eli/dir/2023/2413/oj/eng
39. European Parliament (2024) Council of the European Union, Regulation (EU) 2024/2847 of the European Parliament and of the Council of 23 October 2024 on horizontal cybersecurity requirements for products with digital elements and amending Regulations (EU) No 168/2013 and (EU) 2019/1020 and Directive (EU) 2020/1828 (Cyber Resilience Act) Available from: https://eur-lex.europa.eu/eli/reg/2024/2847/oj/eng
40. European Parliament (2025) Council of the European Union, Regulation (EU) 2025/38 of the European Parliament and of the Council of 19 December 2024 laying down measures to strengthen solidarity and capacities in the Union to detect, prepare for and respond to cyber threats and incidents and amending Regulation (EU) 2021/694 (Cyber Solidarity Act) Available from: https://eur-lex.europa.eu/eli/reg/2025/38/oj/eng
41. European Union (2007) "Treaty of Lisbon amending the Treaty on European Union and the Treaty establishing the European Community", Official Journal of the European Union, C 306, 17 December 2007. Available from: https://eur-lex.europa.eu/legal-content/EN/TXT/?uri=celex:12007L/TXT
42. Federal Government of Germany (2021) "Flood Aid: Federal Cabinet sets "Aufbauhilfe 2021" in motion", August 2021. Available from: https://www.bundesregierung.de/breg-en/service/archive/flood-aid-1952030
43. Horizon Europe (2022) "Reliability, Resilience and Defense technology for the grid (R2D2)" Project Grant No. 101075714, 2022–2025, https://r2d2project.eu/
44. ISO—International Organization for Standardization/IEC International Electrotechnical Committee (2022a) Information security, cybersecurity and privacy protection—Information security management systems—Requirements, ISO/IEC 27001:2022
45. ISO—International Organization for Standardization/IEC International Electrotechnical Committee (2022b) Information security, cybersecurity and privacy protection — Information security controls, ISO/IEC 27002:2022.
46. Jacobs J., Kumar C., Sarmento H., Sharafi D., Srivastava V. (2020) Power system operational resilience—What it means and where we stand. Cigre Science & Engineering, N°18 June 2020
47. Lee, A.V.,Vargo J. and Seville, E., (2013) "Developing a tool to measure and compare organizations' resilience," Natural Hazards Review, vol. 14, no. 1, pp. 29-41, 2013. https://doi.org/10.1061/(ASCE)NH.1527-6996.0000075
48. Martin-Moreno JM, Garcia-Lopez E, Guerrero-Fernandez M, Alfonso-Sanchez JL, Barach P. Devastating (2025) "DANA" Floods in Valencia: Insights on Resilience, Challenges, and

Strategies Addressing Future Disasters. Public Health Reviews. 2025; 46:1608297. https://doi.org/10.3389/phrs.2025.1608297
49. Ministry of Ecological Transition, France (2022) "Risk Preparedness Plan in the Electricity Sector", February 2022. Available from: https://energy.ec.europa.eu/system/files/2022-03/FR_RPP_electricity.pdf.
50. National Grid Electricity Transmission (NGET) (2024) "RIIO-T3 Business Plan", December 2024. Available from: https://www.riiot3.nationalgrid.com/document/30069/download
51. National Infrastructure Commission (NIC) (2019) "Resilience Standards Report", September 2019. Available from: https://nic.org.uk/app/uploads/NIC-Resilience-Standards-Report-Final-190924.pdf
52. National Infrastructure Commission (NIC) (2020) "Anticipate, React, Recover: Resilient Infrastructure Systems", May 2020. Available from: https://nic.org.uk/app/uploads/Anticipate-React-Recover-28-May-2020.pdf
53. National Infrastructure Commission (NIC) (2023) "National Infrastructure Assessment 2", 2023. Available from: https://nic.org.uk/app/uploads/Final-NIA-2-Full-Document.pdf
54. NIST—National Institute of Standards and Technology (2020), "NIST Special Publication 800-53 Revision 5 Security and Privacy Controls for Information Systems and Organizations", November 2023, https://doi.org/10.6028/NIST.SP.800-53r5
55. Ofgem (2022) "NIS Guidance for Downstream Gas and Electricity Operators of Essential Services in GB", April 2022. Available from: https://www.ofgem.gov.uk/sites/default/files/2022-04/NIS%20Guidance%20for%20Downstream%20Gas%20and%20Electricity%20Operators%20of%20Essential%20Services%20in%20GB%20v2.0.pdf
56. Panteli, M. and Kirschen, D. S. (2015) "Situation awareness in power systems: Theory, challenges and applications," Electric Power Systems Research, vol. 122, pp. 140-15, 2015. https://doi.org/10.1016/j.epsr.2015.01.008
57. Panteli, M. and Mancarella, P. (2017) "Modelling and evaluating the resilience of critical electrical power infrastructure to extreme weather events" IEEE Systems Journal, vol. 11, no. 3, pp. 1733-1742, 2017. https://doi.org/10.1109/JSYST.2015.2389272
58. Petrakos N. and Kotzanikolaou, P. (2019) "Methodologies and Strategies for Critical Infrastructure Protection" In: Gritzalis, D., Theocharidou, M., Stergiopoulos, G. (eds) Critical Infrastructure Security and Resilience. Advanced Sciences and Technologies for Security Applications. Springer, Cham. https://doi.org/10.1007/978-3-030-00024-0_2
59. Pursiainen C. and Kytomaa, E. (2022) "From European critical infrastructure protection," Sustainable and Resilient Infrastructure, vol. 8, no. S1, pp. 85-101, 2022. https://doi.org/10.1080/23789689.2022.2128562
60. R^2D^2 Consortium (2023a). Deliverable 3.1 Design of the Multi-risk assessment framework for power system. Public deliverable, available at: https://r2d2project.eu/sdm_downloads/d3-1-design-of-the-multi-risk-assessment-framework-for-power-system/
61. R^2D^2 Consortium (2023b). Deliverable 4.1 Design of the resilience suite for TSO and DSO. Public deliverable, available at: https://r2d2project.eu/sdm_downloads/d4-1-design-of-the-resilience-suite-for-tso-and-dso/
62. R^2D^2 Consortium (2023c). Deliverable 5.1 Design of the Prevention Systems For Energy Infrastructures Security. Public deliverable, available at: https://r2d2project.eu/sdm_downloads/d5-1-design-of-the-prevention-systems-for-energy-infrastructures-security/
63. R^2D^2 Consortium (2023d). Deliverable 6.1 Design of the enhanced maintenance and asset management tool. Public deliverable, available at: https://r2d2project.eu/sdm_downloads/d6-1-design-of-the-enhanced-maintenance-and-asset-management-tool/
64. Ritchie H, Rosado P, Roser M. (2022) "Number of recorded natural disaster events", Natural Disasters. 2022. Available from: https://ourworldindata.org/grapher/number-of-natural-disaster-events
65. Silverio Casulli. (2023), "Terna's strategy to increase resilience of Italian Transmission Network", March 2023. Available from: https://renewables-grid.eu/fileadmin/user_upload/Files_RGI/Event_material/2023_03_Online_Conference_ENTSO-E/04._RGI_ENTSOE_Resilience_Conference_Best_Practice_Showcase_2_Terna_compressed.pdf

66. Stankovski, A., Gjorgiev, B., Locher, L., & Sansavini, G. (2023). Power blackouts in Europe: Analyses, key insights, and recommendations from empirical evidence. Joule, 7(11), 2468-2484. https://doi.org/10.1016/j.joule.2023.09.005
67. Terna (2023) 2023 Development Plan for the National Electricity Grid Presented. Available from: https://www.terna.it/en/media/press-releases/detail/2023-development-plan.
68. Terna, ARERA, and RSE (2021) "Tackling Climate Change: Terna, RSE and ARERA Take Action to Support the Resilience of the Electricity System", Available from: https://download.terna.it/terna/CS%20Terna-Arera-Rse%20per%20Resilienza%20rete%20-%20EN_8d90fd8ed4675e6.pdf
69. Turco, M., Jerez, S., Augusto, S. et al. (2019) Climate drivers of the 2017 devastating fires in Portugal. Sci Rep 9, 13886 (2019). https://doi.org/10.1038/s41598-019-50281-2
70. UK Government (2024), "Strategy and Policy Statement for Energy Policy in Great Britain", 2024. Available from: https://assets.publishing.service.gov.uk/media/6631ff75ed8a41eeaf58c0eb/strategy-and-policy-statement-for-energy-policy-in-great-britain.pdf
71. Zhou, Y. (2023). Climate change adaptation with energy resilience in energy districts—A state-of-the-art review. Energy and Buildings, 279, 112649. https://doi.org/10.1016/j.enbuild.2022.112649
72. Zografopoulos, I., Hatziargyriou, N. D., & Konstantinou, C. (2023) Distributed Energy Resources Cybersecurity Outlook: Vulnerabilities, Attacks, Impacts, and Mitigations. IEEE Systems Journal, 17(4), 6695-6709. https://doi.org/10.1109/JSYST.2023.3305757

Enhancing Cyber Resilience of Critical Infrastructures through Digital Twin Technology for Business Continuity

Eider Iturbe, Jason Xabier Mansell, Angel Rego, and Nicolas Ferry

Abstract Enhancing the resilience of critical infrastructures has become a strategic priority in the era of a rapidly evolving cyber-physical threat landscape. The DYNABIC framework addresses this imperative by developing an integrated approach that ensures the continuity and protection of essential services such as energy, transport, health, and digital infrastructure. This chapter presents DYNABIC's approach, which leverages the use of digital twin (DT) technology to enable real-time simulation, monitoring, and proactive intervention in physical systems. By embedding advanced cybersecurity controls within the DT, DYNABIC facilitates dynamic risk mitigation and adaptive response strategies. The framework also incorporates a diverse set of technologies, including AI-driven decision support, automated orchestration, and human-in-the-loop systems. It enhances resilience across four phases of the resilience cycle: Plan/Prepare, Absorb, Recover, and Adapt. A resilience matrix illustrates how each of DYNABIC's capabilities contributes to these phases. This holistic approach positions DYNABIC as a solution for safeguarding critical infrastructures against evolving threats.

Keywords Digital twin (DT) · Cyber resilience · Critical infrastructure protection · Adaptive cybersecurity · Situational awareness · Security orchestration automation and response (SOAR) · Resilience matrix

---

E. Iturbe (✉) · J. X. Mansell (✉) · A. Rego
TECNALIA, Basque Research and Technology Alliance (BRTA), Astondo Bidea, Derio, Biscay, Spain
e-mail: Eider.Iturbe@tecnalia.com

J. X. Mansell
e-mail: Jason.Mansell@tecnalia.com

N. Ferry
I3S/INRIA Kairos, Université Côte d'Azur, Nice, France

J. Soldatos and H. Abie (eds.), *Technology-Enabled Resilience: Innovations in Critical Infrastructure Protection*, https://doi.org/10.1007/978-3-032-25917-2_3

1 Introduction

In an era of increasing cyber-physical threats, enhancing the resilience and ensuring the continuous management of critical infrastructures has become a top priority for the European Union. The protection of essential services, such as energy, transport, health, and digital infrastructure, is vital for maintaining societal stability and economic continuity. This imperative is underscored by the EU's Critical Entities Resilience (CER) Directive[1], which establishes a framework for strengthening the resilience of critical entities against a wide range of risks, including natural hazards and cyber incidents.

In response to these challenges, the DYNABIC project emerges as a forward-thinking solution designed to support the objectives of the CER Directive. The primary goal of DYNABIC is to enhance the resilience and ensure the continuity of essential European services in the face of sophisticated cyber-physical threats.

The DYNABIC framework incorporates digital twin (DT) technology as a key enabler, supporting its approach to resilience and adaptive cybersecurity. This DT not only simulates and monitors the behaviour of physical systems in real time but also integrates advanced cybersecurity controls. Thanks to their bidirectional nature, DT in the DYNABIC framework can go beyond passive monitoring: it can proactively intervene in the physical twin when necessary, enabling dynamic risk mitigation and adaptive response strategies. This capability empowers critical infrastructure operators with real-time situational awareness, disruption prevention, and agile decision-making.

This document introduces the key concepts underpinning the DYNABIC framework and offers an overview of the integrated cybersecurity controls. It also illustrates how the DYNABIC solution addresses the different phases and system domains of the resilience management.

Chapter 2 introduces the foundational concepts of digital twins (DTs) and resilience. The section highlights key frameworks like NIST's CSF and Linkov's R4 model, emphasizing the importance of resilience metrics and adaptive strategies in cyber-physical systems. Chapter 3 presents the DYNABIC framework, a European initiative aimed at strengthening the resilience of critical infrastructures against cyber-physical threats. Chapter 4 presents the resilience matrix which maps DYNABIC technologies to resilience phases and domains, showing how tools like AI, digital twins, and human-in-the-loop systems enhance planning, response, and recovery, enabling adaptive, cross-domain cyber resilience in critical infrastructures. Finally in Chap. 5, the conclusions are drawn. The conclusions section concludes that DYNABIC enhances cyber resilience in critical infrastructures through digital twin-based monitoring, adaptive cybersecurity controls, and real-time decision support.

---

[1] Directive (EU) 2022/2557 of the European Parliament and of the Council of 14 December 2022 on the resilience of critical entities and repealing Council Directive 2008/114/EC (Text with EEA relevance), OJ L 333, 27.12.2022, p. 164–198. Available at: https://eur-lex.europa.eu/eli/dir/2022/2557/oj.

Validated in four sectors, the framework supports all resilience phases and will launch by end of 2025.

2 Background

To set the grounds for the authors contributions, in the upcoming sub sections, the foundational concepts of DTs and resilience are provided.

2.1 Digital Twin

A digital twin (DT) platform is a collection of software tools designed to support the creation and management of DT solutions. These tools address various aspects such as data collection, storage, modelling, and visualization. Different platforms may focus on specific needs and sectors. In manufacturing, DTs are extensively used, with platforms like Oracle IoT Production Monitoring Cloud Service and SAP Leonardo IoT services providing real-time insights, proactive maintenance, and efficient data management. The a priori digital manufacturing platform [6] aids in product design and decision-making by integrating CAD models for data-driven insights.

General-purpose DT platforms, while offering similar functionalities, are not tailored specifically for manufacturing. For instance, Eclipse Ditto, part of the Eclipse IoT programme, connects IoT devices with their DTs using interfaces like MQTT, HTTP, and Apache Kafka. It stores real-time data in a central database, updating DT properties accordingly. AugmentCity [8] focuses on simulating and visualizing data from smart cities, integrating with virtual reality tools for enhanced data presentation and interaction.

DTs are increasingly recognized as valuable tools for bolstering security in critical infrastructures and cyber-physical systems (CPS). These virtual models enable simulations that facilitate security assessments through non-invasive testing environments and real-time threat analysis, all without compromising the integrity of the actual systems.

The recent studies highlight the diverse applications of DTs in cybersecurity. For example, a proposed cyber digital twin framework expands traditional DT functionalities to tackle specific cybersecurity issues such as intrusion detection, security testing, and anomaly detection [4]. This framework utilizes the MiniCPS simulator to showcase its effectiveness in securing CPS, underscoring the necessity of ongoing security testing and adaptive threat response strategies to ensure system resilience against evolving threats.

As highlighted by the authors in [4], DTs contribute significantly to achieving a wide range of security goals:

- **Security Testing**: DTs provide controlled environments where system defences can be thoroughly tested without affecting live operations.
- **Cybersecurity Training**: They enable operators to engage with realistic simulations of system behaviour, allowing them to build experience while minimizing the risks of interacting with actual systems.
- **Threat Detection**: By comparing real-time and historical data with predefined behavioural models, DTs help in identifying potential threats at an early stage.
- **Selection of Security Controls**: DTs offer valuable insights into how new security measures might impact the system, aiding in more informed decision-making.
- **Cyber Threat Intelligence**: Simulated attack scenarios help uncover system vulnerabilities, supporting proactive cybersecurity strategies.
- **Diagnostics**: DTs assist in analysing historical data to investigate and determine the root causes of security incidents.

By leveraging intelligent systems, DTs support the development of flexible, smart, and responsive industrial operations [15].

In addition to real-time insights, DTs are instrumental in forecasting and analysing failures, as well as facilitating the recovery of their physical counterparts [15]. Within industrial environments, DT applications span a variety of areas, including enhanced cybersecurity, improved operational efficiency, remote monitoring and control, scenario-based risk evaluation, and data-driven decision-making support [3].

In summary, DTs are becoming increasingly versatile in cybersecurity, providing innovative methods to address security challenges and strengthen the resilience of critical infrastructures and cyber-physical systems. Their evolution is driven by rapid technological advancements, broadening their scope and enhancing their effectiveness in complex security scenarios.

2.2 *Resilience*

While having a protection strategy is vital to reduce the likelihood and impact of adverse events in critical systems, it is not always possible to prevent every incident, nor can all threats be reliably detected [29]. Therefore, when disruptions do occur, systems must demonstrate resilience and the capacity to respond effectively.

Understanding the concept of resilience is essential in this context. Although various definitions exist in academic literature, most highlight three core capabilities: the system's ability to (1) absorb shocks, (2) adapt to disruptions, and (3) recover from them [30, 7].

In the realm of critical infrastructure, Vugrin et al. describe resilience as comprising three capacities: absorptive, adaptive, and restorative [18]. Achieving overall resilience involves minimizing both the impact of disruptions and the resources needed for recovery.

The Critical Entities Resilience (CER) directive [13] defines resilience as 'a critical entity's ability to prevent, protect against, respond to, resist, mitigate, absorb, accommodate and recover from an incident.' In the cyber domain, the U.S. National Institute of Standards and Technology (NIST) defines cyber resilience as 'the ability to anticipate, withstand, recover from, and adapt to adverse conditions, stresses, attacks, or compromises on systems that use or are enabled by cyber resources' [3].

According to NIST, cyber resilience and cybersecurity are closely linked, as both aim to ensure systems remain secure and operational under adverse conditions. NIST has developed a cybersecurity framework to support the protection and resilience of critical infrastructure [14]. This framework, known as CSF version 1.1, is organized around five core functions: identify, protect, detect, respond, and recover; this provides a comprehensive approach to managing cybersecurity risks throughout their lifecycle.

Several studies have explored the relationship between cybersecurity and resilience. For instance, Caron examined how cybersecurity testing methods can be used to evaluate an organization's resilience by assessing the effectiveness of its security controls [12]. Malatji et al. introduced a cybersecurity framework for critical infrastructure that extends the CSF with an additional dimension, enabling the assessment of resilience against cyberattacks on industrial systems [25].

Cyber systems constantly face threats that can compromise their functionality. To assess their recovery capabilities, resilience metrics have been developed to measure how systems resist, adapt to, and recover from adverse events. These metrics apply across physical, informational, cognitive, and social domains. Their analysis helps identify vulnerabilities and improve incident preparedness.

The resilience approach goes beyond traditional security, incorporating robustness, redundancy, responsiveness, and rapidity. These dimensions allow for a more holistic system evaluation, considering both technical design and organizational capacity. Implementing these metrics supports resource prioritization and mitigation strategies. This strengthens operational continuity in the face of cyberattacks or critical failures.

According to Linkov [22], resilience metrics for cyber systems draw on frameworks like R4 (robustness, redundancy, resourcefulness, and rapidity), originally used in natural disaster contexts. These metrics quantify resilience at various system levels, supporting informed cybersecurity investment decisions. They also encourage adaptive, proactive responses to emerging threats, integrating both technical and human factors.

Linkov [21] contrasts two key paradigms in cyber system design: cyber efficiency, which focuses on optimizing performance and minimizing costs under normal conditions, and cyber resilience, which emphasizes the system's ability to adapt, recover, and continue functioning under adverse conditions. While efficiency-driven systems may perform well in stable environments, they often lack the flexibility and redundancy needed to withstand disruptions such as cyberattacks or system failures.

Linkov and colleagues also argue that over-prioritizing efficiency can inadvertently increase vulnerability, as streamlined systems tend to eliminate buffers and redundancies that are crucial for resilience. The authors advocate for a balanced approach that integrates resilience thinking into cyber infrastructure planning. This includes adopting frameworks like the resilience matrix, which evaluates system performance across different domains (physical, informational, cognitive, and social) and phases (plan, absorb, recover, adapt), enabling more robust and adaptive cyber systems.

Wood [26] proposed a technique to comprehensively evaluate organizational robustness by dividing it into smaller elements such as units, branches, and hierarchical tiers. This approach can be adopted to qualitatively analyse the strength-related capabilities expected to be implemented by existing cybersecurity frameworks.

3 The DYNABIC Ecosystem

The primary objective of the DYNABIC project is to strengthen the resilience and ensure the uninterrupted operation of critical European services in the face of advanced cyber-physical threats. At the core of the DYNABIC approach is DT technology, which is an advanced innovation that enables real-time monitoring and situational awareness of physical assets. This chapter presents the foundational concepts and key contributions of the DYNABIC ecosystem, which leverages DT technology to bolster cyber resilience. By integrating cybersecurity solutions that facilitate real-time risk assessment, the system helps maintain business continuity, prevent operational disruptions, improve situational awareness, and support adaptive response strategies.

3.1 Foundations of the DYNABIC Ecosystem

The DYNABIC ecosystem is built upon a sophisticated interplay of concepts that collectively define its approach to securing and managing cyber-physical systems as depicted in Fig. 1. At its core lies the notion of an Observable Cyber-Physical System (ISO23247-1, 2020), which refers to engineered systems whose physical and cyber components are both visible and measurable within operational contexts. These systems are susceptible to a range of disruptions, including both deliberate cyber-physical attacks and unintentional incidents, each capable of causing real-world consequences.

To effectively monitor and manage such systems, DYNABIC introduces the concept of a DT, which is a synchronized digital replica of the physical system. This virtual representation enables real-time monitoring and risk management, predictive analytics, and informed decision-making. The DT is a central element of the broader DYNABIC Cybersecurity Control, element in the DYNABIC framework that supports resilience across all phases of incident management: planning, absorption, recovery, and adaptation.

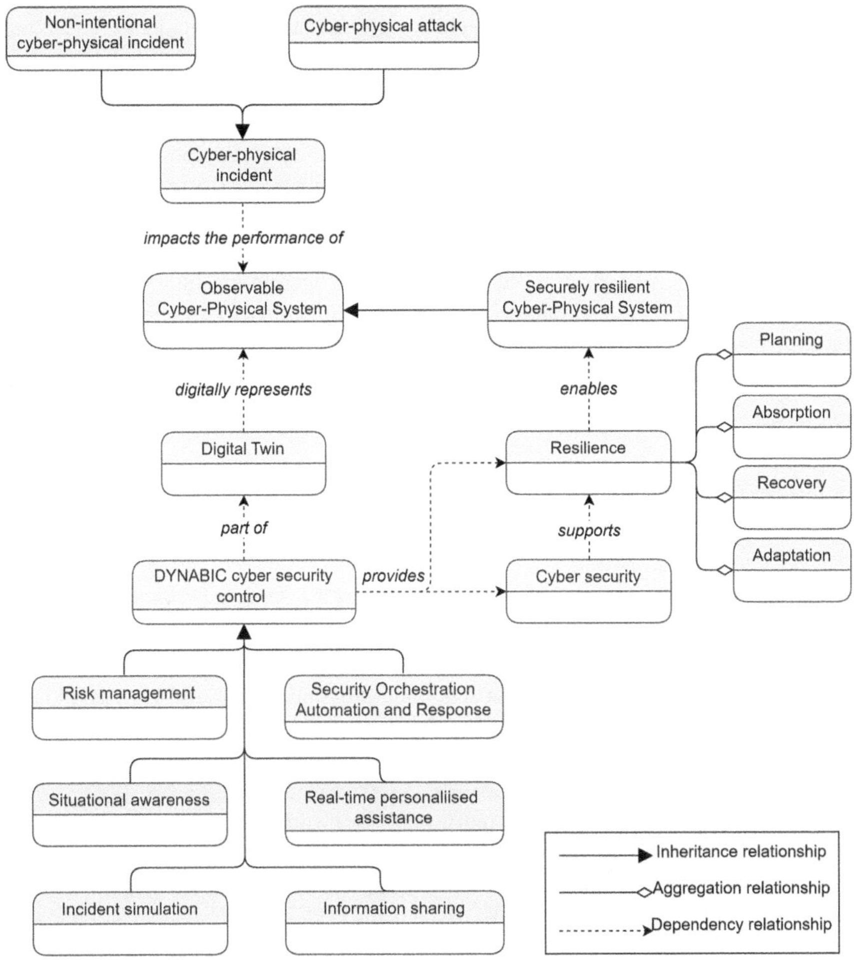

Fig. 1 Foundational concepts of the DYNABIC ecosystem

Cybersecurity within DYNABIC is not limited to traditional IT protection. It encompasses a comprehensive strategy that includes risk management, situational awareness, and security orchestration supported by personalized assistance. Risk management involves identifying and mitigating threats to operations, assets, and individuals, while situational awareness provides a real-time understanding of the system's security and resilience posture and threat landscape. Security orchestration, automation, and response (SOAR) technologies further enhance the system's ability to respond swiftly and effectively to incidents. This is complemented by real-time personalized assistance, often delivered through augmented reality, which supports human operators in making timely and informed decisions during crises.

A key feature of DYNABIC is its ability to simulate cyber incidents. These incident simulations replicate intentional disruptions, allowing stakeholders to evaluate system resilience and refine response strategies at the virtual space.

Another pillar of the DYNABIC framework is information sharing. By facilitating the exchange of threat intelligence and incident data among organizations, DYNABIC enhances collective cybersecurity capabilities and fosters a more resilient infrastructure.

Ultimately, the DYNABIC approach aims to create securely resilient cyber-physical systems, i.e. systems that can maintain operational integrity even under attack or failure. This resilience is grounded in the ability to anticipate, withstand, recover from, and adapt to adverse conditions, ensuring continuity and safety in an increasingly interconnected world.

3.2 DYNABIC Cybersecurity Controls

3.2.1 Risk Management

The recent research has focused on developing real-time cyber threat risk assessment and management methods to enhance cybersecurity awareness and decision-making. Several approaches have been proposed, including ontology-based systems [11] and multicriteria decision frameworks [5]. These methods aim to provide dynamic risk assessment by considering various factors such as threat agents, vulnerabilities, and potential consequences.

A significant advancement in real-time risk evaluation within the DYNABIC framework is the development of a specialized engine [16] designed to assess the vulnerability of critical infrastructures by modelling their components and interrelations through Bayesian Networks. By structuring systems as directed acyclic graphs, the engine can efficiently infer the likelihood of failures and trace their potential origins and ripple effects. It supports real-time simulations and dynamic updates without disrupting operations, enabling continuous risk evaluation. The model categorizes nodes to reflect different types of system elements, such as attacks, functions, assets and events, allowing for granular analysis and the integration of historical data to refine predictions. This engine contributes to the DT-driven cybersecurity ecosystem to provide a coordinated response framework where risk assessments inform and are informed by other tools, enhancing the overall resilience strategy.

Paduraru [10] uses DT technology to enhance process intelligence by transferring knowledge across similar business workflows. Modelled with BPMN 2.0, these digital representations enable the system to predict outcomes in unfamiliar scenarios by analysing execution data and applying machine learning. Through compact vector-based process embeddings, the framework identifies patterns and reuses insights to support decision-making in new but related contexts.

3.2.2 Situational Awareness

Situational awareness, in the context of cybersecurity, refers to an organization's ability to perceive and understand its current security status and threat landscape within a specific timeframe and operational environment. It involves not only recognizing ongoing events and vulnerabilities but also interpreting their significance in terms of risk and anticipating how they might evolve in future [24]. This dynamic understanding is essential for proactive decision-making and rapid response. As such, situational awareness forms a core capability of a Security Operations Centre (SOC), [30] by maintaining a continuous and coherent status of the organization's security and resilience posture.

Within DYNABIC, [19] introduces an integrated model for evaluating urban mobility systems, addressing a broad spectrum of dimensions including financial, societal, institutional, ecological, and technical factors. It leverages a semantic infrastructure based on RDF-star and a dedicated inference mechanism to enhance knowledge representation and reasoning. The study also presents a quantitative structure for interpreting performance metrics through sequences of observed data. Furthermore, it employs a multivariate methodology to characterize system states using diverse attributes. The main objective is to investigate how disruptions evolve and to gain insights into the interdependencies that lead to systemic breakdowns across various domains.

As another key contribution of DYNABIC, [33] presents an innovative evaluation approach that leverages change-impact analysis to measure the resilience of adaptive security mechanisms, with a focus on Moving Target Defence (MTD) strategies. It incorporates automated tools for generating and executing test scenarios, along with mechanisms for collecting relevant performance data. The framework is seamlessly embedded within the software engineering process, enabling continuous validation of system behaviour under evolving conditions. It not only examines internal system responses but also explores how disruptions may influence interconnected platforms. Additionally, the research explores the integration of advanced technologies like DTs to enhance the efficiency of test optimization processes, including selection, reduction, and prioritization of test cases.

3.2.3 Security Orchestration Automation and Response

Security orchestration automation and response (SOAR) refers to technologies that allow organizations to quickly and efficiently perform incident analysis, and design and execute incident response procedures [24]. SOAR mechanisms enhance cybersecurity resilience by integrating multiple security tools and automating processes. SOAR platforms streamline workflows, reduce workload, and improve response times in Security Operations Centres (SOCs) [23]. Integration of Artificial Intelligence (AI) in SOAR enhances threat detection, automation and situational awareness

[31]. In critical infrastructure like smart grid-based SCADA systems, SOAR implementation addresses security challenges by utilizing AI, machine learning and threat intelligence [9].

DYNABIC introduces a comprehensive cybersecurity framework tailored for IoT and cyber-physical systems, aiming to overcome the limitations of traditional SOAR solutions. By integrating DT technology with intelligent automation, the proposed system enables proactive threat detection and adaptive response mechanisms that maintain operational continuity during cyber incidents. Reinforcement learning (RL) is employed to support autonomous decision-making and continuous system improvement [27]. The framework is validated through experiments involving data manipulation and distributed denial-of-service scenarios, demonstrating its effectiveness in real-world conditions [28]. Additionally, the architecture supports self-healing capabilities, such as isolating compromised components and mitigating cascading failures, which are critical for resilient infrastructure.

As a key component of DYNABIC SOAR solution, [2] introduces Chat4XAI, a novel tool that enhances the transparency of RL-driven decisions within service-oriented systems. Unlike traditional visual explanation methods, Chat4XAI leverages conversational AI to provide intuitive, natural language justifications for the actions taken by deep reinforcement learning agents. This improves user trust and understanding, particularly in environments like critical infrastructure protection. By bridging the gap between complex AI behaviour and human interpretability, Chat4XAI strengthens the overall reliability and usability of the DYNABIC framework.

3.2.4 Incident Simulation

Incident simulation is designed to imitate cyber-physical incidents to evaluate the cyber security and resilience status of a cyber-physical system. This enables the prognosis and analysis of impact propagation estimation and a more reliable evaluation of the cyber security and resilience of the system.

As part of DYNABIC contribution, [17] introduces a sophisticated simulation platform designed to strengthen the cybersecurity posture of cyber-physical systems by enabling safe and controlled simulation of attack scenarios. The system employs a graph-based modelling approach to map potential threat pathways and assess their impact across interconnected components. By synthesizing realistic incident data through AI-driven techniques, it enhances the training and performance of anomaly-based intrusion detection mechanisms. Unlike traditional methods, this solution captures behavioural patterns across multiple system layers, including network, system and operational domains, offering a multi-dimensional view of incident dynamics. Furthermore, the use of attack graphs as a strategic planning tool supports informed configuration and optimization of detection systems, making the platform an asset for proactive defence and resilience testing without interrupting live operations.

3.2.5 Real-Time Personalized Assistance

Within DYNABIC, augmented reality and large language model (LLM)-based real-time personalized assistance are integrated to support the resilience of critical infrastructures by enabling timely, human-level responses. These technologies enhance operators' situational awareness and interaction with cyber-physical systems, empowering them to make informed decisions and execute response actions more effectively. Complementary module based on augmented reality enhance the DYNABIC ecosystem by providing human-in-the-loop support.

Furthermore, as part of DYNABIC smart assistance, [10] presents a forward-looking framework for integrating LLMs into SOCs, aiming to enhance cybersecurity workflows through modular and intelligent automation. The proposed architecture combines Agentic AI, Retrieval-Augmented Generation (RAG), and user-specific customization to streamline threat detection, analysis, and response. A key innovation is the use of a teacher model, specifically GPT-4o, to generate and refine a diverse set of cybersecurity questions and answers, forming a high-quality dataset that supports model training and evaluation. This process ensures the system is grounded in both foundational knowledge and current best practices.

The framework also emphasizes adaptability to user preferences and organizational norms. Through prompt engineering and fine-tuning, the system can tailor its tone, detail level, and communication style to match user expertise and cultural expectations. Access control mechanisms and safety filters, such as Llama Guard [20], are incorporated to enforce secure and respectful interactions. Finally, the fine-tuning process, demonstrated using Llama 3.1–8B-Instruct [1], aligns the model with cybersecurity-specific tasks and user profiles, ensuring both technical relevance and usability across diverse SOC environments.

3.2.6 Information Sharing

Information sharing involves the exchange of information about threats, attacks, and vulnerabilities among different organizations, to enhance their cyber security capabilities. Sharing incident information is usually advantageous as multiple organizations are often affected simultaneously by the same threats and attacks [32].

DYNABIC presents a next-generation Cyber Threat Intelligence (CTI) sharing platform that enhances collaborative defence through seamless integration with existing cybersecurity ecosystems. By incorporating DT technology, the platform enables virtual modelling and simulation of threat scenarios, supporting proactive analysis and response planning. A key innovation lies in its automated pipeline for transforming unstructured threat data into standardized, actionable intelligence, significantly reducing manual effort. The system features a dual-layer architecture: a powerful backend that handles real-time data ingestion, anonymization and processing, and an intuitive dashboard that offers tailored visualizations and interaction modes for different user roles. The platform also emphasizes operational efficiency by ensuring continuous synchronization with external tools and data sources.

Overall, it delivers a privacy-aware solution that empowers organizations to share and act on threat intelligence with greater speed and precision.

3.2.7 DYNABIC Resilience Matrix

This section illustrates how various DYNABIC technologies contribute to resilience across four system domains (Physical, Information, Cognitive, Social) and four resilience phases (Plan/Prepare, Absorb, Recover, Adapt), as defined by [22].

Based on the description of the DYNABIC cybersecurity controls provided in the previous section, the following list identifies the unique contributions specifically related to resilience. These are the elements incorporated into the resilience matrix in Table 1.

- AI-driven anomaly detection: AI-based detection of abnormal system behaviour for early threat identification.
- Automated playbook generation and orchestration logic: Automates predefined response actions to streamline incident handling.
- Bayesian network-based risk evaluation engine: Probabilistic risk modelling engine for threat prediction and cascading failure analysis.
- Change-impact analysis for MTD: Evaluates the effects of changes in Moving Target Defence strategies on system resilience.
- Collaborative threat intelligence with digital twin integration: Enables collaborative cyber threat intelligence sharing using DTs.
- Conversational explainable AI: Provides human-understandable explanations for AI-driven decisions to enhance trust.
- Coordinated response across systems to restore functionality: Synchronizes recovery actions across interconnected systems and teams.
- Cross-workflow knowledge transfer via process embeddings: Reuses learned insights across similar workflows to improve decision-making.
- Digital twin-based process intelligence framework (BPMN 2.0): Uses DTs to model and analyse business processes for resilience.
- Dynamic reconfiguration of security policies and workflows: Adjusts security configurations in real time to respond to evolving threats.
- Human-in-the-Loop augmented reality support: Enhances operator situational awareness and decision-making through AR interfaces.
- Isolation of compromised components: Contains threats by isolating affected system parts to prevent spread.
- LLM-based SOC assistant with conversational decision support: AI assistant that supports SOC analysts with natural language interaction and decision guidance.
- Organizational and cultural customization of LLM-based SOC assistant: Adapts AI tools to align with organizational norms and user expectations for better usability.
- Reinforcement learning for adaptive response: Learns optimal response strategies through trial-and-error to improve over time.

- Resilience evaluation model: Framework for assessing system resilience across domains and lifecycle phases.
- Self-healing capabilities in IoT/CPS: Automatically detects and corrects faults to maintain system functionality.
- Simulation platform for attack scenarios: Simulates cyberattacks to evaluate system vulnerabilities and response effectiveness.

Each cell identifies specific technologies, models, or tools developed within DYNABIC that enhance resilience capabilities in that context. The matrix underscores the integrative nature of DYNABIC's approach, which combines physical infrastructure protections, information system intelligence, cognitive decision support and social collaboration mechanisms. By mapping these capabilities to the phases of resilience, the table illustrates how DYNABIC supports a holistic and dynamic resilience strategy tailored for complex cyber-physical and socio-technical systems.

In complex systems like those addressed by DYNABIC, certain technologies and tools naturally contribute to multiple resilience phases and span across different system domains. For example:

- Self-healing capabilities are relevant to both the Absorb and Recover phases, and they impact both Physical and Information domains.
- LLM-based SOC assistant: Appears in cognitive and social domains across all phases, reflecting its central role in human-AI collaboration.
- Human-in-the-Loop AR support: Spans physical, cognitive, and social domains, by supporting both system and human resilience.

The matrix highlights how DYNABIC's innovations, ranging from AI-driven decision support and DTs to automated orchestration and human-in-the-loop systems, work together to anticipate, withstand, recover from, and adapt to disruptions.

4 Conclusions

In this chapter, we presented the DYNABIC ecosystem to enhance cyber resilience and increase business continuity of European critical infrastructures. The cybersecurity control-based framework developed in DYNABIC proposes a set of advanced and innovative tools for real-time monitoring, risk management, intelligent incident response and simulation, enabling informed decision-making and reaction. These tools revolve around a DT as a central element that provides a bi-directionally synchronized digital replica of an Observable Cyber-Physical System.

As indicated in our resilience matrix, the DYNABIC framework contributes to the four phases of incident management planning: preparation, absorption, recovery, and adaptation over the physical, information, cognitive and social domains.

This cybersecurity controls in DYNABIC framework are evaluated in the context of four use cases focusing on smart preparedness, prevention, and response to possible

Table 1 Mapping of DYNABIC controls to the resilience matrix by [22]

Domain/ Phase	Plan/Prepare	Absorb	Recover	Adapt
Physical	Resilience evaluation model	Self-healing capabilities in IoT/ CPS	Human-in-the-loop augmented reality support Isolation of compromised components Resilience evaluation model Self-healing capabilities in IoT/ CPS	Human-in-the-loop augmented reality support
Information	Automated playbook generation and orchestration logic Bayesian network-based risk evaluation engine Change-impact analysis for MTD Collaborative threat intelligence with digital twin integration Digital twin-based process intelligence framework (BPMN 2.0) Simulation platform for attack scenarios Resilience evaluation model	AI-driven anomaly detection Bayesian network-based risk evaluation engine	Change-impact analysis for MTD Coordinated response across systems to restore functionality Digital twin-based process intelligence framework (BPMN 2.0) Resilience evaluation model	Dynamic reconfiguration of security policies and workflows Reinforcement learning for adaptive response

(continued)

Table 1 (continued)

Domain/ Phase	Plan/Prepare	Absorb	Recover	Adapt
Cognitive	Conversational explainable AI LLM-based SOC assistant with conversational decision support	Human-in-the-loop augmented reality support LLM-based SOC assistant with conversational decision support	Conversational explainable AI Human-in-the-loop augmented reality support LLM-based SOC assistant with conversational decision support	Cross-workflow knowledge transfer via process embeddings LLM-based SOC assistant with conversational decision support Reinforcement learning for adaptive response
Social	Collaborative threat intelligence with digital twin integration Organizational and cultural customization of LLM-based SOC assistant	Collaborative threat intelligence with digital twin integration	Human-in-the-loop augmented reality support	Conversational explainable AI LLM-based SOC assistant with conversational decision support Organizational and cultural customization of LLM-based SOC assistant

cascading business disruption risks of four types of critical infrastructures: electric vehicle charging stations, critical transport services, telecommunication infrastructures, and hospital services. At the time of writing this chapter, the DYNABIC framework is under its final stage of validation and is expected to be released by the end of 2025.

Acknowledgments This work has received funding from the European Union's Horizon Europe research and innovation programme under grant agreement No 101070455 (DYNABIC).

References

1. A Grattafiori, A. D. (2024). The llama 3 herd of models. (pág. https://arxiv.org/abs/2407.21783.). arxiv.org.
2. A Metzger, J. B. (2023). An AI chatbot for explaining Deep Reinforcement Learning decisions of service-oriented systems. *International Conference on Service-Oriented Computing, 2023* (págs. 323-338). Switzerland: Springer.
3. A Pokhrel, V. K.-P. (2020). Digital twin for cybersecurity incident prediction: A multivocal literature review. *Proceedings of the IEEE/ACM 42nd International Conference on Software Engineering Workshops.*
4. A Somma, V. C. (2023). A Cyber Dgital Twin Framework to support Cyber Physical Sustems Security. *IEEE Smart World Congreee (SWC)*, (págs. 1-10). Portsmouth, UK.

5. AA Ganin, P. Q. (2020). Multicriteria decision framework for cybersecurity risk assessment and management. *Risk Analysis.* Wiley Online Library.
6. aPriori. (09 de 06 de 2025). *aPriori.* Obtenido de digital manufacturing platform: https://www.apriori.com/?utm_source=google&utm_medium=search&utm_campaign=fy26-organic-search&utm_term=organic-search&utm_content=web-content
7. AS Kahnamouei, T. B. (2017). The conceptual framework of resilience and its measurement approaches in electrical power systems. *IET International Conference on Resilience of Transmission and Distribution Networks (RTDN 2017).*
8. augmentcity. (09 de 06 de 2025). *https://augmentcity.no/.* Obtenido de https://augmentcity.no/
9. AW Mir, R. R. (2021). Implementation of security orchestration, automation and response (SOAR) in smart grid-based SCADA systems. *Sixth International Conference on Intelligent Computing and Applications* (págs. 157-169). Singapore: Springer.
10. C Paduraru, C. P. (2024). *CyberGuardian 2: Integrating LLMs and Agentic AI Assistants for Securing Distributed Networks.* Scitepress.org.
11. C Sánchez-Zas, V. V.-B.-N. (2023). Ontology-based approach to real-time risk management and cyber-situational awareness. *Future Generation Computer Systems.* Elsevier.
12. Caron, F. (2019). Obtaining reasonable assurance on cyber resilience., (págs. 193-217).
13. Commission, E. (13 de Nov de 2023). *EUR-Lex—32022L2557—EN—EUR-Lex,.* Obtenido de https://eur-lex.europa.eu/legal-content/EN/TXT/?uri=CELEX%3A32022L2557
14. Cybersecurity, C. I. (2018). *Framework for Improving Critical Infrastructure Cybersecurity Version 1.1.*
15. D Holmes, M. P. (2021). Digital twins and cyber security–solution or challenge? *6th South-East Europe Design Automation, Computer Engineering, Computer Networks and Social Media Conference.*
16. DYNABIC. (10 de 2024). *cordis.europa.eu.* Obtenido de Dynabic: https://ec.europa.eu/research/participants/documents/downloadPublic?documentIds=080166e51265e699&appId=PPGMS
17. E Iturbe, J. A. (2024). A Multi-layer Approach through Threat Modelling and Attack Simulation for Enhanced Cyber Security Assessment. *ARES '24: 19th International Conference on Availability, Reliability and Security* (págs. 1-8). ACM.
18. ED Vugrin, D. W. (2011). A resilience assessment framework for infrastructure and economic systems: Quantitative and qualitative resilience analysis of petrochemical supply chains to a hurricane. *Process Saf. Prog., vol. 30, (3),* 280-290.
19. G Rocher, J. T. (2024). A Framework Towards Assessing the Resilience of Urban Transport Systems. *19th International Conference on Availability, Reliability and Security* (págs. 1-10). ACM.
20. H Inan, K. U. (2023). Llama Guard: LLM-based Input-Output Safeguard for Human-AI Conversations.
21. I Linkov, A. L. (2023). Cyber Efficiency and Cyber Resilience. *ACM Communications.*
22. I Linkov, D. E. (2013). Resilience metrics for cyber systems. *Environment Systems and Decisions, vol. 33,* 471-476.
23. J Johnson, C. J.-M. (2023). Soar4der: security orchestration, automation, and response for distributed energy resources. *Springer International Publishing,* 387-411.
24. K Knerler, I. P. (2023). 11 Strategies of a World-Class Cybersecurity Operations Center. *MITRE.*
25. M Malatji, A. M. (2022). Cybersecurity capabilities for critical infrastructure resilience. *Information & Computer Security, vol 30 (2),* 255-279.
26. MD Wood, E. W. (2019). Quantifying and mapping resilience within large organizations. *OMEGA vol 87,* 117-126.
27. P Nguyen, R. D. (2023). Towards smarter security orchestration and automatic response for CPS and IoT. *2023 IEEE International Conference on Cloud Computing Technology and Science (CloudCom)* (págs. 298-302). ieeexplore.ieee.org.
28. PH Nguyen, A. R. (2024). Digital Twin-Based Security Orchestration, Automation and Response for IoT and CPS. *IFIP International Internet of Things Conference, 2024* (págs. 243-260). Switzerland: Springer.

29. R Ross, V. P. (2019). Developing cyber resilient systems: A systems security engineering approach. NIST special publication 800-160, volume 2, revision 1.
30. S Hosseini, K. B.-M. (2016). A review of definitions and measures of system resilience. *Reliability Engineering & System Safety 145* (págs. 47-61). Elsevier.
31. Sarker, I. (2014). AI-driven cybersecurity and threat intelligence: cyber automation, intelligent decision-making and explainability. *Springer Nature.*
32. V Valdés Ríos, F. Z. (2024). Towards the adoption of automated cyber threat intelligence information sharing with integrated risk assessment. *Proceedings of the 19th International Conference on Availability, Reliability, and Security (ARES '24)* (págs. 1-9). ACM.
33. VV Ríos, A. C. (2023). Testing techniques to assess impact and cascading effects. *2023 IEEE International Conference on Cloud Computing Technology and Science (CloudCom)* (pág. 269.272). IEEE.

Stress-Testing for Optimizing Resilience of Critical Infrastructures against Extreme Threats

Aleksandar Jovanović

Abstract This chapter focuses on stress-testing and its role in evaluating and strengthening the resilience of critical infrastructures (CIs) exposed or potentially exposed to extreme threats (XTs). We propose a framework where the main elements include (a) private–public partnership (PPP) solutions necessary for aligning stakeholders' opinions and acceptance of stress-testing resilience, (b) the use of resilience indicators, (c) the introduction of parametric insurance against XTs into reinsurance and insurance practices for CIs, and (d) further development and integration of stress-testing into international (ISO), EU, and national standards, such as DIN SPEC 91461. Examples from the EU, national, and industrial insurance projects are briefly described, and recommendations regarding regulation, particularly at the EU level, are provided in the conclusions.

Keywords Resilience · Stress-testing · Resilience indicators · Standards · EU-CER directive

1 Introduction—The Overall Stress-Testing Framework

In the world constantly becoming riskier (e.g., due to the catastrophic climate change-related risks), in which the global growth is slowing down, large macroeconomic regime shifts happen (e.g., the tariff regime), and the world order is getting more fragmented, new solutions for enhancing resilience against such extreme threats (XTs—as they are defined in more detail in [1]) need to be explored. In that sense, this chapter of the book examines the issue of stress-testing resilience in a broader context of the critical infrastructure (CI) resilience and protection, including:

1. Private–public partnership (PPP), e.g., the partnership with insurance and innovative insurance concepts when dealing with XTs.
2. Private sector policies and rules, e.g., in insurance.

A. Jovanović (✉)
Steinbeis EU-VRi—European Risk & Resilience Institute, Stuttgart, Germany
e-mail: jovanovic@risk-technologies.com

J. Soldatos and H. Abie (eds.), *Technology-Enabled Resilience: Innovations in Critical Infrastructure Protection*, https://doi.org/10.1007/978-3-032-25917-2_4

3. Public policies, e.g., the EU policies.
4. Standards (primarily the ISO and national standards).
5. Extreme threats (XTs, e.g., the nature-related or the man-made ones, for which the acceptability limits need to be defined).
6. Acceptability limits in resilience stress-testing (e.g., max. Loss of functionality).
7. Advanced technologies, methods, and tools (primarily looking at smart/soft tools and big data), including those for the stress-testing itself.
8. The stress-test performance (here on critical infrastructures subject to stress-testing (entities as defined by the EU CER Directive [2], including the cyber-CIs covered by the NIS2 Directive [47].
9. Results/outcomes of the stress-testing.
10. Implementation and communication of the results, broader implications.

The above elements are structured within the proposed stress-testing framework (Fig. 1) focused on the XTs, resilience indicators, and PPP, mainly within the insurance and the EU practices. The components of the framework are discussed in the sections below. The recommendations related to these elements are given in the conclusion section of the paper, within the context of current and planned trends in the private sector and governance, especially in the insurance industry and the EU.

The starting point of the framework is represented by the needs of the main PPP actors when dealing with the XTs. Its end-point and main outcome is the embedding of the stress-testing results into the policies. The framework establishes the feedback loops, both in the case of negative or partially negative results and in the case of positive results. In the case of negative results, the feedback can be, e.g., a request to redefine the acceptability limits or the indicators used; in the case of positive results, these can be suggestions to include, for instance, further aspects to be stress-tested. The framework helps structuring the stress-testing and XT-related activities better in all the phases of the resilience cycle, as they are defined in the SmartResilience project [3] and ISO TS 31050 [43], namely helping to:

- Better prepare for the XTs.
- Better protect the CIs against XTs.
- Better absorb the impact of the XTs.
- Better recover from the XT events, and.
- "Build-back-better" after the XTs and better adapt to the possible future XTs, integrating anticipation and adaptivity aspects as an important aspect of the long-term and sustainable resilience of CIs.

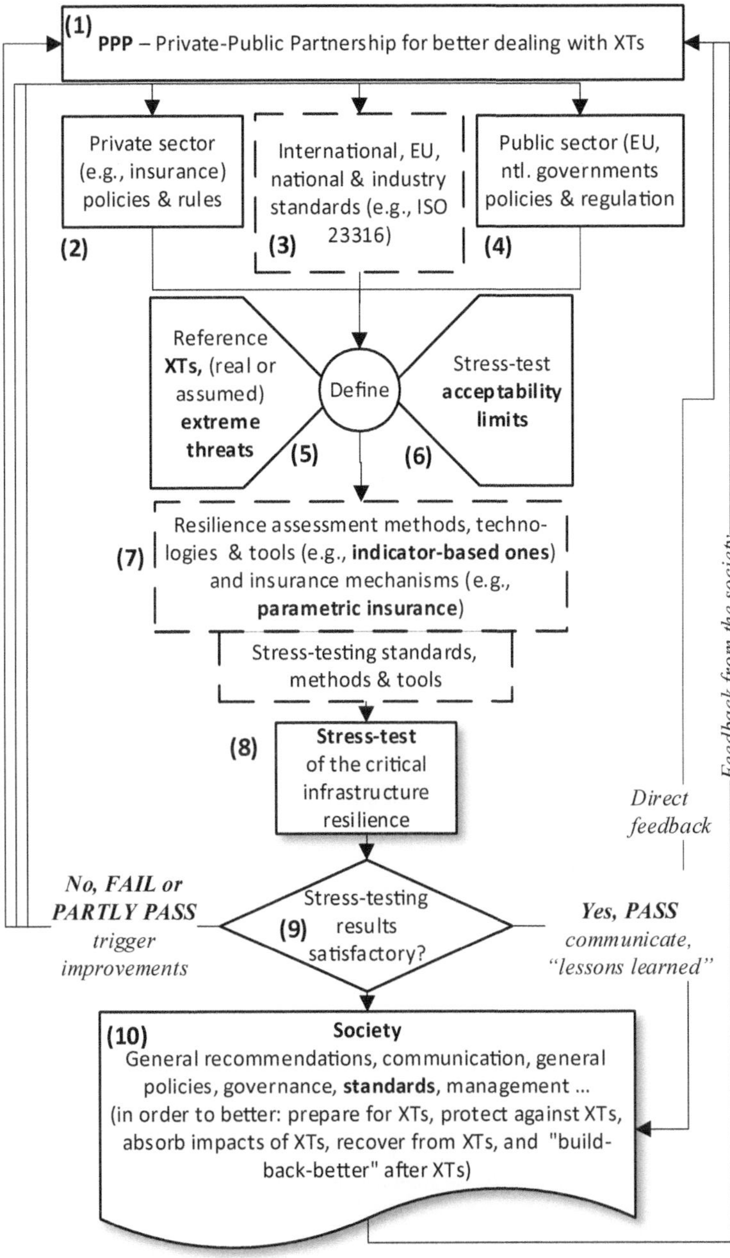

Fig. 1 The stress-testing framework

2 Extreme Threats—What Should One Stress-Test Against?

2.1 Extreme Threats

The concept of extreme threats, x-Threats (XTs), assumes these threats have some specific characteristics, requiring particular handling and management (see elements (1–5) of the framework). The example in Table 1 gives a hint at some of the main attributes of these threats. In the framework, we argue that the "extremeness" of XTs should be quantified, i.e., should be "measured" and made part of the overall resilience index (measuring the resilience against a given XT scenario or a portfolio of XT scenarios) and used in stress-testing.

Main XTs' characteristics can be summarized as follows:

- Lack of early warnings, surprise, e.g., "no advance warnings".
- Unprecedented intensity, e.g., "extreme intensity".
- Extreme spread and speed of spread, e.g., "the whole country", leading to "global shocks".
- Cascading/systemic character, interconnectedness, e.g., "follow-up events or aftershocks".
- Lack of knowledge, e.g., "unknown", no historical data, "black swans".
- More frequent than the extrapolation of past data might suggest.
- Impacts amplified by the context, e.g., "winter, night", "high infrastructure density".
- Impacts amplified by inadequate communication.

Table 1 Examples of the description of the main values for the scenarios of "significant", "major", and "extreme" intensity, using earthquake and pandemics hazards as examples (adapted from [30, 32])

Label	Earthquake	Disease outbreak
Significant	Magnitude approx. 5.5; intensity (EMS) VII (building damage) no of aftershocks damage radius 25 km radius of damage epicenter 5 km low infrastructure density; time: Afternoon	Individual regions in Switzerland affected appears first in surrounding countries (early warning) the causative agent is known low infection rate
Major	Magnitude approx. 6.5 intensity (EMS) IX (destructive); aftershocks occur damage radius 80 km radius of damage epicenter 25 km high infrastructure density; time: Late spring, weekdays in the morning	Whole of Switzerland affected appears first in surrounding countries (advance warning) the causative agent is known high rate of infection
Extreme	Magnitude approx. 7.0 intensity (EMS) XI (devastating); aftershocks occur damage radius 120 km radius of damage epicenter 40 km high infrastructure density time: Winter, at night	Whole of Switzerland affected occurs first in Switzerland (no advance warning) the causative agent is known or unknown high rate of infection

- Impacts amplified by perception biases, e.g., "risk aversion".

The difficulties in managing XTs, including their anticipation and adaptation to them, are considered in IRGC [4] and similarly elaborated in Table 2, [5].

To conclude, XTs are "here to stay"; the measures to overturn the natural trends are yet to be established: the technological ones are yet to be developed, and the political ones face serious difficulties as indicated by IPPC [34]. As for stress-testing, the XTs are the main events when choosing "what to stress-test against", as when answering

Table 2 Examples of factors possibly influencing the perception of XTs [5]

Factor	Short explanation
Misinterpretation	Possible misperception of opportunities or benefits; susceptibility to deliberate or other misinterpretation
Catastrophic potential	If fatalities occur in large numbers in a single event, instead of in small numbers dispersed over time, the perception of risk rises
Miscommunication	Communication of XTs can be a factor in the perception of the risk level; in extreme cases, the miscommunication can be a risk of its own
Lack of familiarity	Unfamiliar or novel risks make one worry more
Lack of understanding	If one believes that the way of how an activity or technology works is not well understood, his or her sense of risk goes up
Lack of personal control or voluntariness	If one feels the potential for harm is beyond one's control (like a passenger in an airplane) one worries more than if she or he feels that he or she is in control (e.g., as the driver of a car); if one cannot decide personally to engage the risk or not, the risk feels more threatening
Affected persons	Identifiable people who are affected by the consequences of risk, rather than statistical abstractions, make the sense of risk rise
Lack of trust	If the institutions involved are not trusted, risk perception rises
Lack of reversibility	If the effects of an outcome cannot be reversed, risk rises
Personal risk	If the risk involves oneself, one tends to perceive the risk as higher (unless we think we are in better control than the average person)
Cognitive biases	Biases such as the confirmation bias, and other cognitive biases, including information fatigue, can significantly increase risk perception
Protection or precaution	Possibility to envisage the implementation of protective or precautionary measures
Lack of possibility to control/mitigate	Perception of risk rises when the possibility of monitoring the risk is low, false signals appear, inherent and coherent risk management is missing, and control or planned mitigation mechanisms are poor
Susceptibility to ignorance	The complexity of the modern world often leads to a situation in which people are flooded with information about a phenomenon or an issue, but the basic understanding of the phenomenon is missing

the questions like "has the LA fire protection been well prepared for the 2025 fire disaster or not[1]" or "will be better prepared to do it in the future". The stress-testing can provide a quantified and evidence-based, not just an opinion-based, answer.

2.2 XTs: The Insurance/Reinsurance Side

XTs are gaining importance and cause an ever-increasing negative impact on society. This is well reflected in the insurance records and general statistics (e.g., the EMDAT database https://www.emdat.be). The losses increase in terms of numbers and frequency of catastrophic events, their intensity, their complexity, and their impact. For instance, the losses in disasters like the one in the Los Angeles area 2025 wildfires will lead to the record high insured losses of over b$ 40. The overall worldwide economic losses in 2024 are assessed by Sigma [6] to be b$ 328 (all numbers in US $ 2024 prices). These losses are caused by 336 registered catastrophic events, in which the world has experienced more than 14,000 victims. The losses are equal to 0.3% of global GDP. This is a significant increase from a year before (b$ 303 in 2023) and far above the 10-year average (b$ 254). The nature-related catastrophic losses prevail: b$ 318 versus b$ 10 for the man-made ones, but these numbers need to be considered in the context of the uninsured losses—the insured ones are far better investigated, assessed, and documented.

The term nature-related catastrophic events or natural catastrophe (often shortened NatCat—floods, storms, earthquakes, droughts, forest fires, wildfires, heat waves, cold waves, frost waves, hail, tsunamis, and other natural catastrophes) refers to an event caused by natural forces, generally resulting in multiple and severe losses. The scale of the losses, in most of the cases, is a combination of the intensity of the natural forces and the type of affected region (population density, type of buildings, disaster management, and other factors).

The man-made disasters, on the other hand, refer to major events associated with human activities, including technical disasters. They are usually linked to large objects/assets in a limited area. The insurance normally excludes the war-like or the war-related events and covers only losses from major fires and explosions, transportation disasters, mining accidents, collapse of buildings/bridges, and similar.

Most reinsurers also distinguish between primary and secondary perils, where the term primary is used for perils causing very large single event losses, whereas the term secondary covers all other perils, excluding the primary ones. The losses caused by the cumulative impact of secondary losses can also be significant, and, in fact, they have been growing over the years. Conditionally, one can say that the primary perils are of greater concern for reinsurers and the secondary perils for the insurers. The XTs considered in this chapter belong mainly to the category of primary perils. Insured losses from natural catastrophes grow (up to 7% annually), and with that

[1] *"Los Angeles 'not prepared' for size, growth of raging wildfires: Fire chief"* https://6abc.com/post/los-angeles-not-prepared-size-growth-raging-wildfires-fire-chief/15779557/.

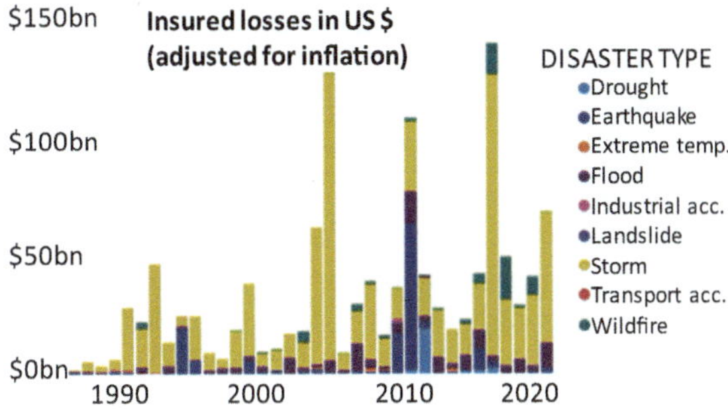

Fig. 2 The constantly increasing importance of XTs: EMDAT data processed by the author, showing the losses by the disaster cause

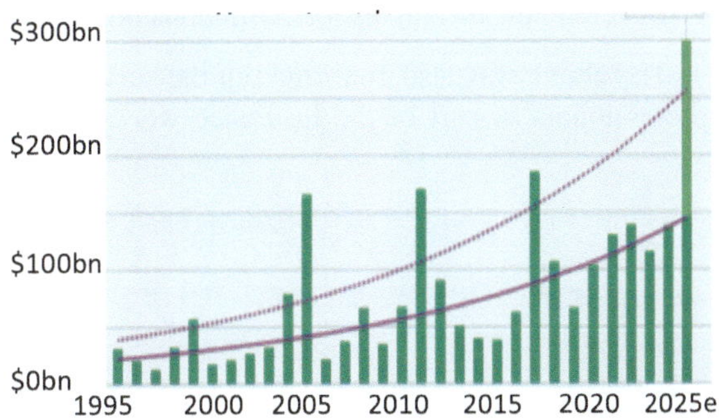

Fig. 3 The constantly increasing insured losses due to XTs (green), their trend (line) and +1 standard deviation (dotted line): *Source* Sigma (2025) data with the estimate for 2025

growth, the costs in 2025 are expected to be about b$ 145, exceeding the b$ 300 mark in 2025. This would make 2025 a "peak loss year" in insurance terms, similar to the years of the US hurricanes Harvey, Irma, and Maria. It is interesting to note that the frequency of these "peak years" also increases: defining a peak loss year as a year in which insured losses are more than one standard deviation above trend, there have been five such years in the last 30 years, three of them in the last decade. Although the particular causes of nature-related causes vary from year to year (Fig. 2), their aggregated impact clearly increases year on year (Fig. 3).

Single events can cascade toward unforeseen scenarios. Consequences are difficult, if not impossible, to assess beforehand[2]. Then, total risk is not the sum of single risks having a common root cause. In this interconnected risk landscape, "contagion" [8] poses additional challenges. Both non-systemic risks and systemic risks can lead to the accumulation of losses, but the pathways to failure are different [9].

Current concerns and the respective priorities of the insurance/reinsurance industry would be:

- Close the increasing protection gap between insured and non-insured assets/losses amplified for the cases of XTs (Fig. 4).
- Improve understanding and definition of XTs, e.g., including complex and compound (multi-risk) character of perils/ threats and their cascading effects.

[2] https://www.emdat.be, *Guha-Sapir et al. [7]—data from 2023, data updated annually.*

- Improve methods and tools for reliable forecasting of complex emerging (e.g., climate-related) XTs.
- Improve methods and tools for assessing the impacts of XTs and the resulting potential for major disruptions of insurance coverage (e.g., for insured communities or infrastructures).
- Improve methods for assessing resilience against XT-related disasters.
- Obtain better and more direct engagement from the insured parties and other stakeholders in the process of dealing with XTs.
- Align better with standardized approaches to stress-testing.
- Make the PPP more efficient and effective.
- Establish and improve independent rating and certification facilities.
- Develop new incentives for continuous improvements in risk pricing.

The above was also confirmed in the survey performed among 100+ insurance professionals as part of the Insurance workshop at the CERIS Disaster Resilient

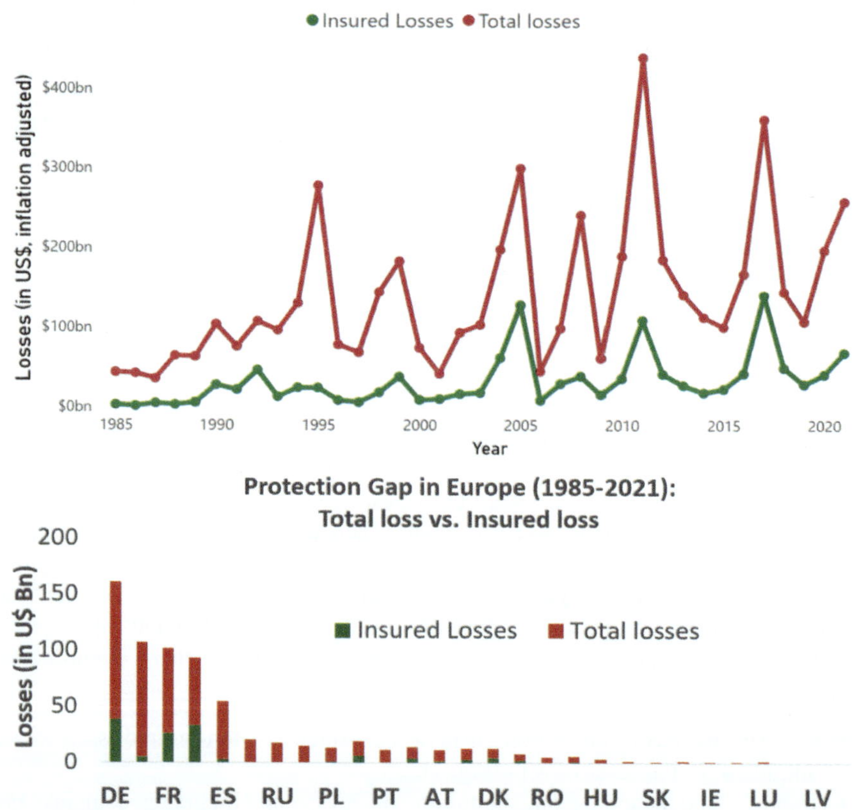

Fig. 4 Protection gap - uninsured losses: Gap between the overall (red) and insured losses (green) – Global (top) and Europe (bottom). EMDAT data processed by the author

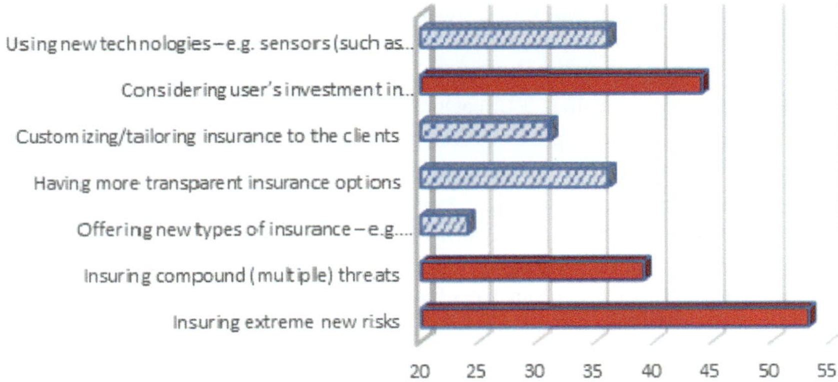

Fig. 5 Needs and top priorities resulting from the survey undertaken amongst ~100 insurance and reinsurance experts at the EU workshop

Societies Cluster Conference, in November 2022,[3] showing the top priorities as in Fig. 5. The topic "stress-testing" is referenced under one point above explicitly, but it appears in most of the other ones, too: e.g., under "resilience against XTs" or "assessing impacts".

Increasing resilience by means of insurance-related actions is a realistic option, as shown by a number of examples collected in the work of Jovanović, Schernberg, and Sansavini [10]. Although the insurance today can still cope with the challenges of XT [6], there is no reason for complacency, as the trend in pace of XT exposures and that of the cost of capital over a longer time period will probably change toward more critical.

2.3 *XTs: The Public Side, Including the PPP*

The insurance stakeholders are certainly aware of their responsibilities in toward society (Fig. 6), and this interaction is certainly a way to be further explored in the future, as confirmed already in 2017, with an update in 2020,[4] of the EU publication "Overview of natural and man-made disaster risks the European Union may face". The main 11 threats considered are flooding, extreme weather, extreme forest fires, earthquakes, pandemic, epizootic diseases, unprecedented/atypical industrial accidents, loss of functionality of critical infrastructures, nuclear accidents, terrorism, and cybercrime. The threats are cross-sectoral, often hybrid, unpredictable, systemic, and constantly emerging or evolving. In addition, they may interrelate with processes like globalization, digitalization, innovation, cross-boundary operations, and many

[3] https://link.springer.com/article/10.1007/s10669-023-09952-9.

[4] *EU: Overview of natural and man-made disaster risks the European Union may face, 2020 edition. Publications Office of the European Union, Luxembourg 2021.*

others, thus influencing each other. But the above taxonomy has not been accepted broadly and uniformly, not even in all EU documents; therefore, the update of the above document, from 2020, does not follow the "11 threats taxonomy" strictly, acknowledging de facto that a threat depends on the scenario. Hence, instead of providing data on losses, it provides data on claims made to the EU Solidarity Fund. These data are useful, but not necessarily a fully representative indicator of the damage costs or losses (Table 3).

The main priority on the public side of the PPP (Fig. 7) and as related to users-needs is clearly on better alignment of policies, especially the EU policies related to the XTs. The comments to the answers have indicated that inconsistencies in that are hampering many efforts and leading to unnecessary difficulties in enhancing the CI resilience. In addition, as the EU-CER [2] follows the all-hazards approach, which, in turn, makes the selection of the event to be used as the reference for the stress-testing more difficult.

Traditionally, governments engage with the private sector in three main ways:

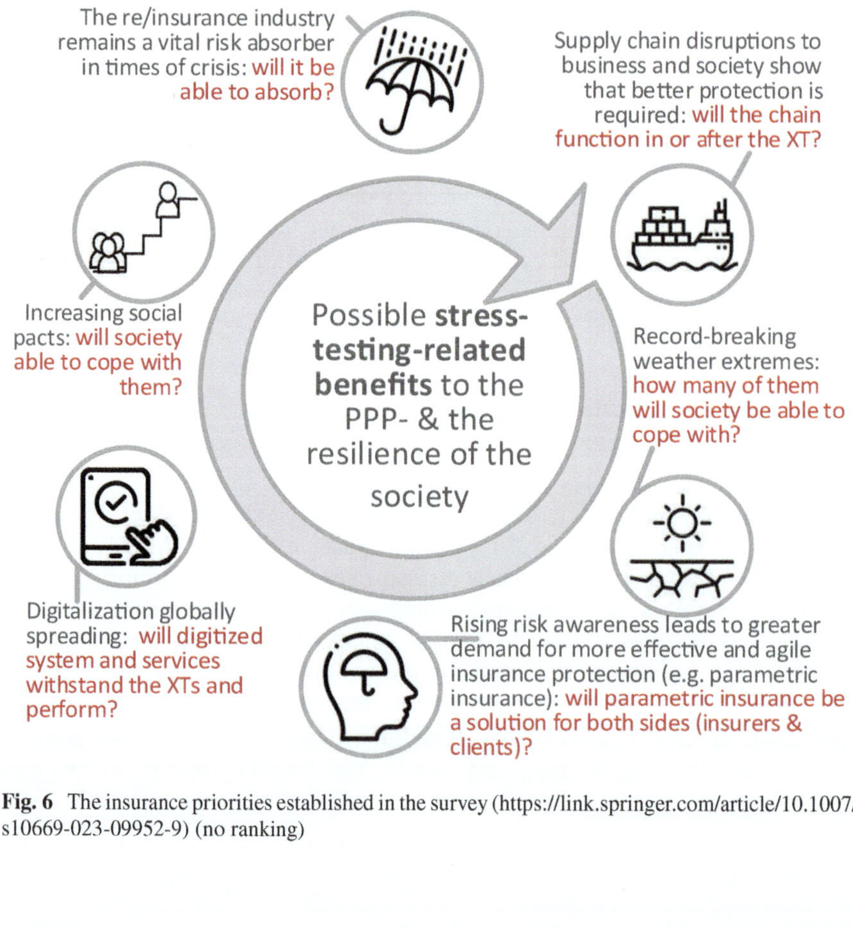

Fig. 6 The insurance priorities established in the survey (https://link.springer.com/article/10.1007/s10669-023-09952-9) (no ranking)

Table 3 Sample XT-damages in the EUF

XTs examples	Physical drivers	Societal drivers and amplifiers	Damages M€ EU (Total direct damage incurred in disasters, as declared in applications to the EU Solidarity Fund, 2002–2019; 2020 edition. EU 2021.):
Concurrent heat & drought	Precipitation, evapotranspiration, antecedent soil moisture, temperature	Water management, land-use change	988
Heatwave & other (e.g., pollution)	Temperature, atmospheric humidity, solar radiation, SOx, NOx, ozone, particulate matter	Urbanization, irrigation, agricultural & industrial activities	436
Extreme weather	Wind speed, humidity, large-scale atmospheric circulation	Urbanization, deforestation	16758
Coastal & river flooding	Coastal flooding, river flow, precipitation, coastal water level, surge, wind speed	Removal of barriers, water management, urbanization	61574

Total direct damage incurred in disasters, as declared in applications to the EU Solidarity Fund, 2002–2019; 2020 edition. EU 2021

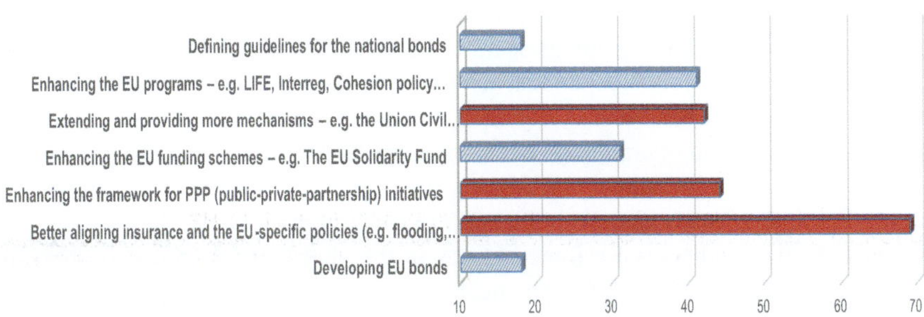

Fig. 7 Needs prioritized by menas of a survey https://link.springer.com/article/10.1007/s10669-023-09952-9 showing the priorities of end-users vis-à-vis EU policies: Alignment, new mechanisms, PPP; PPP highly ranked (scale: number of answers)

Fig. 8 Different ways of how public and private sector can interact within the public-private partnership (PPP), and possibly create dynamic ecosystems in which sustainable PPP value is jointly created (adapted from BCG https://www.bcg.com/publications/2022/how-can-public-private-ecosystems-help-societal-problems)

1. *Through specific regulation*, e.g., in the regulation directed toward companies within a single industry,
2. *Through framework regulation* including performance-based regulation—such as some of the EU directives, e.g., the General Data Protection Regulation imposes strict rules on how all companies can collect and use the personal data of millions of citizens, and.
3. *Through the conventional PPP* e.g., based on formalized agreements between governments and industry.

Recently, however, the traditional PPPs have evolved more and more often into something that BCG calls a public–private ecosystem (PPE), defined as a dynamic network comprising a government or other public body and an evolving and ever-changing group of largely independent economic participants. For instance, a government may invite a large, dynamically changing number of private companies to join a digital platform and operate together in such an "ecosystem", (Fig. 8).

Large-scale infrastructure resilience projects often require partnerships of multiple public and private stakeholders (the PPPs)[5]. For example, the City of New Orleans, Veolia, and SwissRe joined forces in the aftermath of Hurricane Katrina to increase resilience through sophisticated analysis and detailed planning of an efficient resilience strategy. As a second example, the Terrorism Risk Insurance Act (TRIA) is a loss-sharing arrangement between insurers and the government for losses in a future terrorism scenario. It specifies that commercial losses resulting from a terrorist attack will be paid by insurers and uninsured firms until they exceed 60 bn USD. Beyond that threshold, losses are paid by the government [11]. Such PPPs are generally an opportunity for the private sector and insurance industry, in particular, to extend the limits of insurability and thereby open new fields of business. As per

[5] *Total direct damage incurred in disasters, as declared in applications to the EU Solidarity Fund, 2002–2019; 2020 edition. EU 2021.*

the World Bank [12], such large-scale infrastructure resilience PPPs involve three key challenges:

- Contractual allocation of risks between different public and private stakeholders.
- Management of long-term contracts in changing environments.
- Commercial uncertainty in the costs of resilience investments.

Stress-testing can contribute to each of these three challenges through its experience with both long-term and multiple-stakeholder contracts. In addition, stress-testing can provide transparency to PPPs. The risk-based premiums can provide quantified and easy-to-understand information on infrastructure risks and resilience against XTs, also to complex systems (infrastructures-of-infrastructures). In such systems, individual risk is heavily dependent on external variables. For instance, a logistics company's business continuity relies on the availability of transportation and energy systems and so forth[6].

2.4 Linking Climate Change-Related XTs to Stress-Testing

Although XTs are usually the core of larger stress-testing exercises (see Sect. 3.1 and Table 4, indicating different combinations of the types of the CIs and the types of the XTs), the climate-related stress-testing, involving climate change as a global XT, challenging the resilience of society as a whole, raises particular public interest and are in the focus of research of many projects and institutions (e.g., Sparccle,[7] PIK[8] or Cascade Institute). The particular reason for this lies in the fact that the global climate development projections affect all sectors of society, and, at the same time, they are extremely uncertain, as the projections based on past trends and their extrapolations are uncertain and often mistrusted or disputed. For this reason, Steinmann et al. [13] propose a generalized framework for designing open-source natural hazard parametric insurance, the issue we discuss further below.

3 Stress-Testing Critical Infrastructures Against XTs

3.1 General

This section deals with the stress-testing (ST) elements (6–10) of the framework in Fig. 1. A number of important actions, undertaken by governments and multinational institutions, have taken place in the past (some of them still running as continuous

[6] *Boston Consulting Group 2022*. https://www.bcg.com/publications/2022/how-can-public-private-ecosystems-help-societal-problems.

[7] https://iiasa.ac.at/projects/sparccle.

[8] https://publications.pik-potsdam.de/pubman/faces/ViewItemFullPage.jsp?itemId=item_26587_2.

Table 4 Some relevant examples of stress-testing against XTs

Exercise	XTs considered/stress-tested
2001 dark winter, USA	Bioterrorism and pandemic readiness
2005 talisman sabre, Australia/USA	Defense and infrastructure logistics
2006 cyber storm I, USA	National cyber defense
2010 cyber Europe (inception), EU	Pan-European cyber resilience
2010 locked shields (NATO), NATO, Estonia	Live cyber-fire defense
2011 GridEx I, USA/Canada	Power grid cyber/physical attack test
2012 Japan nuclear/earthquake drills	Energy/seismic infrastructure
2018 UK National Grid WINTEX drill	Gas and electricity emergency test
2022 Swiss blackout test	Blackout and cascading systems test
2023 GridEx VI/cyber storm VIII, USA+	Energy and multisector cyber resilience
2023 exercise NorthStar XI, Singapore	Terrorism + urban critical systems

or periodic processes) in order to evaluate the resilience of critical infrastructure against natural disasters, cyberattacks, bioterrorism, systemic failures, and similar XTs. Some notable examples are summarized in Table 4.

The GridEx stress-testing is a recurring electric grid security drill organized by NERC since 2011. The Cyber Storm stress-testing series, led by DHS, has been taking place since 2006. The Cyber Europe stress-testing has been taking place since 2010 through ENISA, focusing on coordinated cyberattacks across health care, finance, and telecom sectors. The Dark Winter stress-testing simulation explores the XTs in the area of biosecurity and bioterrorism. The Japan's post-Fukushima nuclear and earthquake drills (since 2012) stress-test nuclear and urban resilience. Locked Shields is the NATO live-fire cyber exercise, started in 2010 and ongoing, testing real-time defense of SCADA and telecom systems. The UK's National Grid and Switzerland's exercises stress-test prolonged blackout simulations. The Singapore's Exercise NorthStar focuses on civil infrastructure under stress from terrorism, cyber disruptions, or cascading failures. The operation Talisman Sabre in Australia has taken place biennially since 2005. Switzerland's blackout stress-testing (2022) focuses on multi-domain hybrid threats, including infrastructure warfare, energy, and cyber-XTs. A more comprehensive overview of the exercises, including the links to the references, can be found in Linkov et al. [14], and more details about some of the exercises, e.g., in EBA [15], ENSREG [16] or ERCOT [17].

3.2 Stress-Testing Organization and Procedure(s)

The framework in Fig. 1 as a generic stress-testing (ST) framework provides context for most stress-testing exercises, but does not provide details needed for practical applications. For these applications, it is also important to define the organization of the teams, the workflow, and the application details (Fig. 9).

As an example, the STREST-approach foresees "multiple experts" team including the project manager (PM), the technical integrator (TI), the evaluation team (ET), the pool of experts (PoE), and the internal reviewers (IR). All these actors are supposed to interact along the stress-testing in order to assure transparency and accountability of the different actors. The PM, the TI, and the IR should be independent.

In the example of TIBER-EU [18] and DIN 91461 [19], the teams are organized following the "colored", "3 teams" concept, involving the main stakeholders, namely:

(a) *The Blue Team* is the representing the entity which is the subject of the stress-test; the entity's prevention measures, preparedness, and response capabilities are tested;
(b) *The Red Team* carries out the simulated attack by attempting to compromise the critical functions of the entity; and.
(c) *The White Team* is responsible for overseeing the test, managing the stress-testing exercise, analyzing and reporting its results, and making sure the test meets the requirements of relevant authorities.

DIN 91461 foresees also that the management of the CI owners/operators, and the Competent Authorities involved in the implementation of the EU-CER Directive be informed about the stress-test, too. The supporting actors, e.g., those providing, the threat intelligence acc. to ISO 56000 and ISO TS 31050, for identifying and analyzing possible XTs, or the actor providing AI and IT services, can be included into the teams as appropriate.

In terms of the stress-testing workflow, the examples of ST@STREST and DIN 91461 exhibit both general similarity and differences in overall character and the

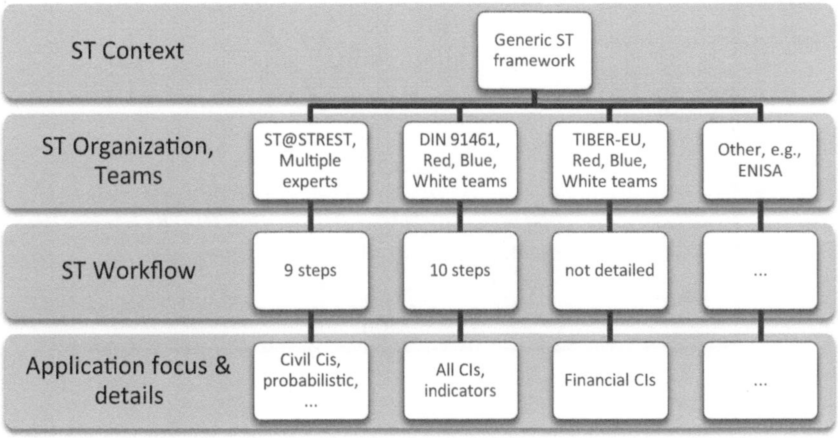

Fig. 9 The main context, organization of the teams, the workflow, and the application details – examples

workflow details: both following the same general logic (from preparation to final outcome), both organized in steps and phases, both assigning the tasks in the workflow to designated teams/persons, but the first (Fig. 11) being more analysis-focused and open to considering complexity and detailed analysis of grading of outcomes, the latter (Fig. 12) more outcome and indicators focused, and what-if analysis and engineering oriented. In addition, DIN 91461 is more directly oriented toward the resilience cycle (Fig. 10, [3]).

The steps of both workflows are listed in parallel below:

DIN 91461 (Fig. 12)	STREST (Fig. 11)
Step 1: Select the infrastructures to be considered in the analysis	Step 1: Data collection
Step 2: Identify the relevant threats, their combinations and interaction/dependencies, and indicators	Step 2: Risk measure and objective
Step 3: Define functionality elements	Step 3: Setup of stress-test

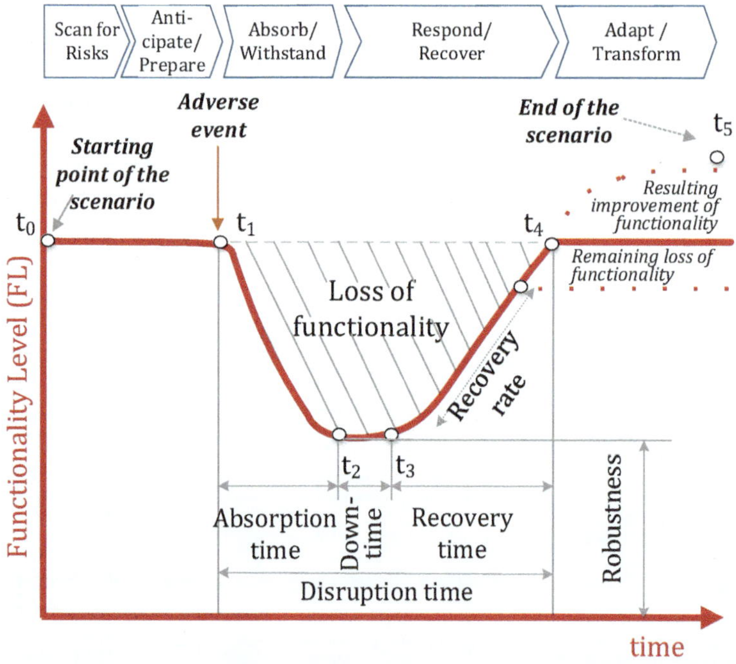

Fig. 10 The macro-indicators of the resilience FL (functionality level resilience curve in DIN 91461)

DIN 91461 (Fig. 12)	STREST (Fig. 11)
Step 4: Provide values for the respective functional elements; calculate elements	Step 4: Component-level assessment
Step 5: Define allowable values for all indicators and elements	Step 5: System-level assessment
Step 6: Assign time limits values t_0 to t_5 for the scenarios (Fig. 10)	Step 6: Risk objectives check
Step7: Provide values for the indicators at given times.	Step 7: Disaggregation/sensitivity analysis
Step 8: Generate and evaluate the functionality curve and impact	Step 8: Guidelines and critical events
Step 9: Consider if interdependencies may affect the results	Step 9: Results presentation
Step 10: Report the results	

3.3 *Outcomes of a Stress-Testing Exercise*

As for the outcome of the stress-testing (element (9)), the yes/no classification can be oversimplified, especially if the testing case is more complex. When considering/analyzing

- Single or multiple threats (exposures to multiple hazards).
- Single or multiple components within an infrastructure.
- Single or multiple infrastructures.
- Single or multiple time horizons.
- Uncertainties and ambiguities (e.g., including frequencies or probabilities).
- Different types of risks (e.g., systemic, emerging, compound…).
- Different types or different levels of load (e.g., the intensity of an XT).
- Different types of contexts (e.g., polycrisis, multi-XTs).

One can end up with extremely complex stress-testing contexts. In these complex contexts, a more result classification (grading) scheme, as the one in Fig. 13, allows grading between full PASS and full FAIL results (i.e., different grades PARTIALLY PASS). Agreeing on such a grading scheme, e.g., in standards or regulation, or at least in standardized practices, is still outstanding, and it will be an essential step in establishing a CI resilience rating system similar to the ones used in assessing the credit worthiness (e.g., S&P, Moody or Fitch rating scales).[9] As the credit worthiness in its core indicates "the likelihood that an issuer will default on a loan", the resilience rating would indicate "the likelihood that an infrastructure will default on a stress-test".

[9] https://www.investopedia.com/terms/c/creditrating.asp#toc-credit-ratings-scale.

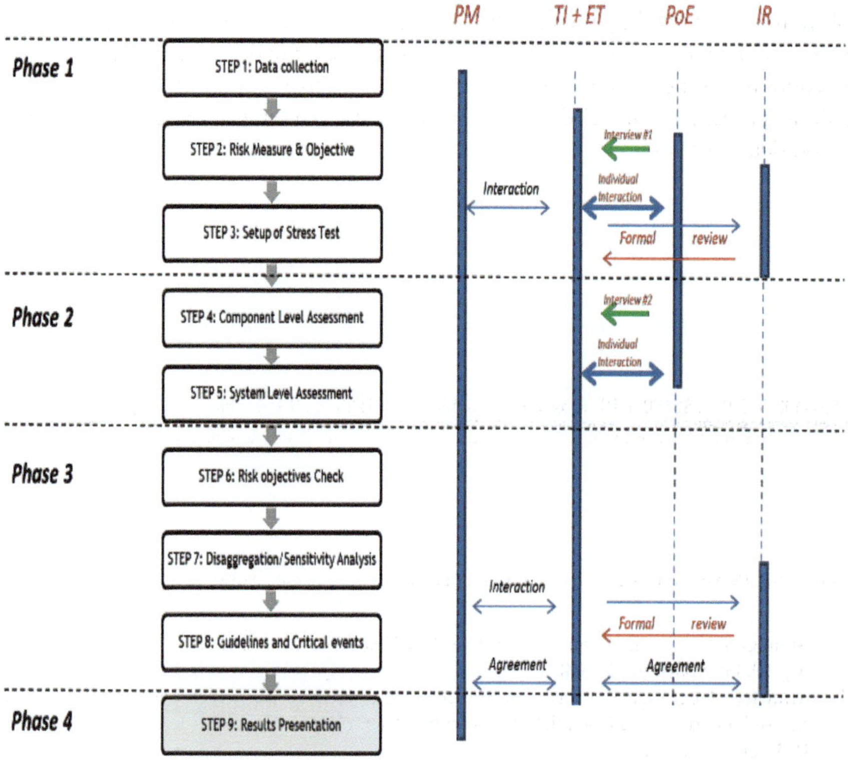

Fig. 11 Workflow of the ST@STREST methodology (Esposito et al. 2017)

On the other hand, DIN 91461 allows to assess and use them in a stress-test, the single macro-indicators, which are defined as:

Robustness: the absorb capacity of the system;

Disruption time: total time taken by the infrastructure to recover to the desired functionality level;

Absorption time: the time to the maximum loss of functionality level;

Downtime: the time between the end of the absorption and the beginning of the recovery;

Recovery time: the time at which the infrastructure recovers from the disruptive event and gains its initial or desired functionality level;

Recovery rate (speed): defined as the rate at which the infrastructure recovers from a disruptive event and gets back to its initial or desired functionality level;

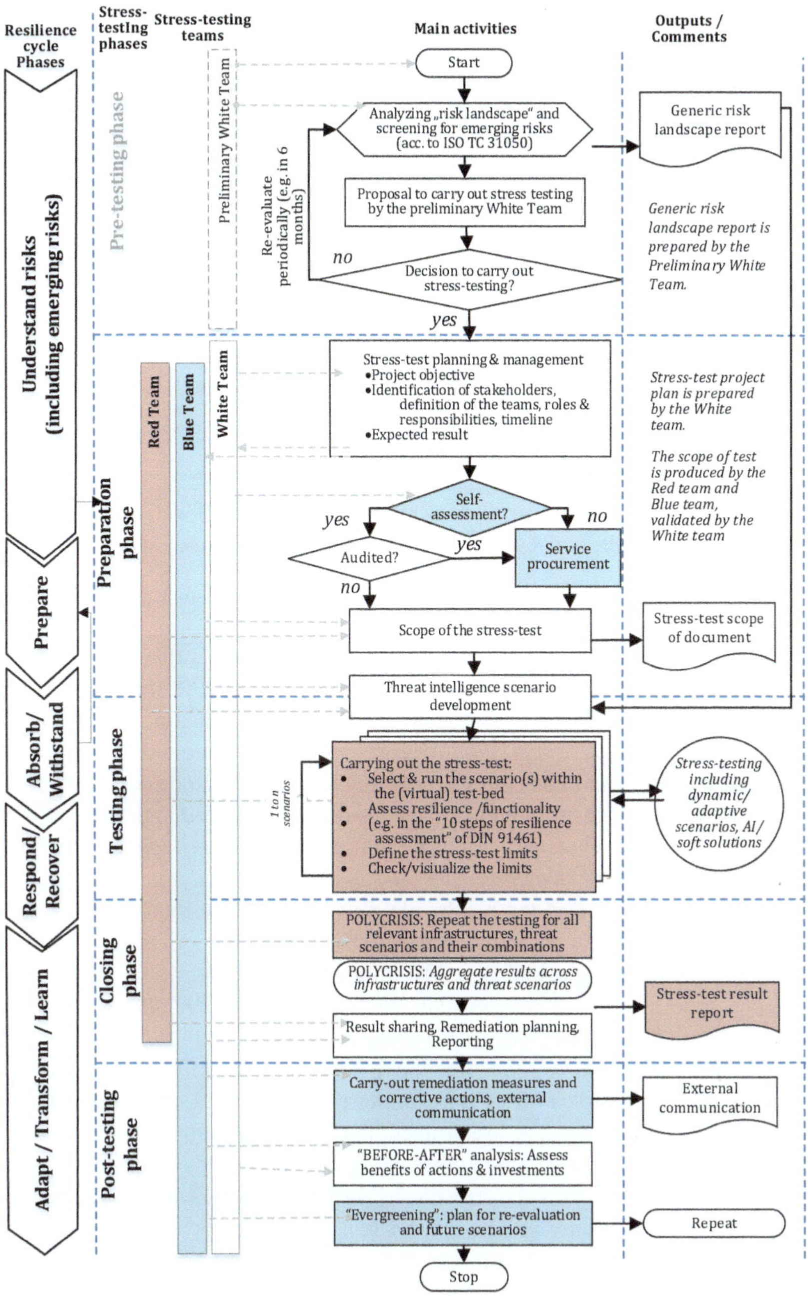

Fig. 12 DIN 91461 procedure for stress-testing

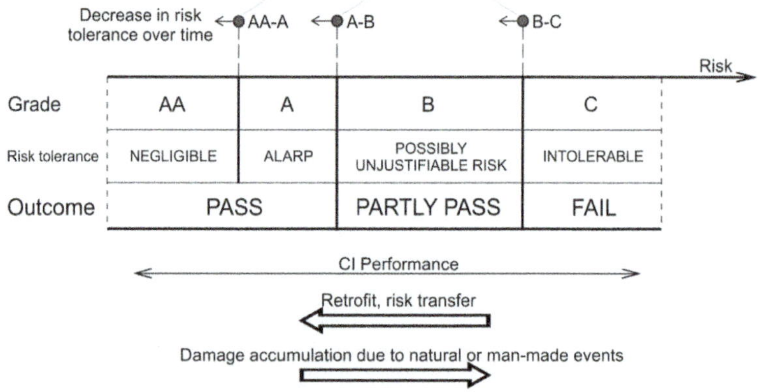

Fig. 13 Classifying results of a stress-test [20]

Improvement/adaptation/transformation: the functionality level at the end of the scenario can be lower, equal to, or higher than the original one;

Loss of functionality: is the integral functionality of the infrastructure maintained or lost in the stress-testing scenario?

These macro-indicators allow for setting "natural" acceptability limits, e.g., when one says that "*an XT cannot lead to a loss of more than 40% loss of the nominal capacity of the electric grid*", it is mirrored in the macro-indicator *Robustness*. Similarly, one can say that "*the population cannot stay without drinking water longer than, e.g., 8 hours*", and this is mirrored in the macro-indicator *downtime*.

The stress-testing results can also be considered in a tiered approach, like in the 3 × 3 one of Linkov et al. [14], yielding the categories "*quick win improvements*", "*system wide resilience strategy*", and "*targeted changes and interventions*".

Finally, visualization of the results plays an increasingly important role. In this context, especially for communication with different types of stakeholders, it is important to ensure right and possibly intuitive understanding of all the results, visually supporting the statement that "*the infrastructure 'A' has NOT met, and infrastructure 'B' HAS met the stress-testing criteria*" (Fig. 14, left—that case, "A" is the IT network, and "B" the pipelines). The example shows the stress-testing results involving multiple critical infrastructures (✖ pipelines, △ safety sub-system, □ reservoirs, ◎ IT network, △ fire protection, ◘ video surveillance, and ○ intrusion prevention) showing that the thick pink line is the "acceptance level line" met by all CIs, except the IT network and barely met by the CI pipelines. The case is described in Jovanović et al. [46], considering a real XT (catastrophic flood in the city of Lucerne in 2005 and the assumed XT, a more extreme flood event). On the other side, it is important to have the quantitative engineering analysis in the background of the final conclusions, allowing for addressing the causes directly (Fig. 14, right—the damage of a warehouse in the case of inundation).

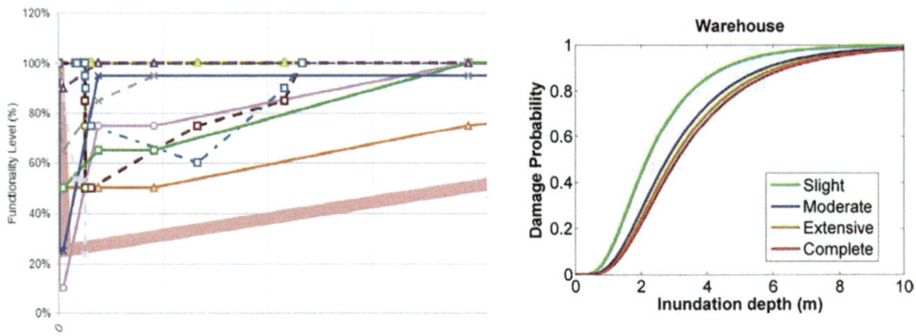

Fig. 14 Two ways on looking at the results of stress-testing: looking at the functionality of the CIs tested (over the time of an XT scenario, left, [1]) or looking at just certain aspects such as damage probability of a warehouse exposed different XT loads (Esposito et al. 2017)

3.4 *Resilience Stress-Testing in ISO Standards*

Out of a number of the ISO standards referenced here (see references, ISO 9001 to ISO 56000), most interesting for the issue of stress-testing are:

- ISO 22316 ([21] edition, security and resilience—Organizational resilience—Principles and attributes) and.
- ISO TS 31050 (Risk Management—Guidelines for managing emerging risk to enhance resilience).

Neither of the two documents addresses stress-testing directly, but the 22,316 document recommends that the ("organizational") resilience[10] should be continuously monitored and enhanced. In particular, the standard asks for the commitment of the organizations to

- Improve the ability to anticipate and address risks and vulnerabilities,
- Increase coordination and integration of management disciplines to improve coherence and performance, and.
- Ensure greater understanding of interested parties and dependencies that support strategic goals and objectives of the organization.

In the revision of the 2017 document, which is currently underway, the stress-testing is expected to be addressed explicitly, primarily in the context identification of gaps and weaknesses in the resilience of the organization. In that context, the revision is expected (inter alia) to elaborate the requirements of the stress-testing as related to:

- The scenarios and the thresholds.

---

[10] *is defined as "... the ability of an organization to absorb and adapt in a changing environment to enable it to deliver its objectives and to survive and prosper."; the organization in ISP terminology is "... a group of people and facilities with an arrangement of responsibilities, authorities and relationships",*

- The knowledge, evidence, and data used (risk intelligence).
- The impacts if the threshold cannot be achieved (e.g., in adverse events like XTs).
- The system approach to stress-testing, and.
- The collaboration with other interested parties.

The DIN SPEC 91461 document addresses most of the above aspects of stress-testing in technical, practical terms and for the CIs, in particular (the 22,316 is intended for all types of organizations, not only CIs). The principles, framework, and procedures for managing emerging risks are provided in ISO TS 31050, thus applicable to the XTs as these are often the most interesting emerging risks (see the lists of [22], as an example).

To conclude, recognized standards are crucial to the application of stress-testing. Only if based on the broadly accepted standards, the answer to the question "are we resilient enough against XTs?" will be accepted, too. In particular, for the global XTs such as those in the reports of WEF [22] or JRC [23]—the latter including XTs like global disruption of critical supply chains or the decline of the EU economy.

4 Use of Stress-Testing in Parametric Insurance

4.1 Overall Insurance Landscape (XTs on the Horizon)

The main question related to the use and usefulness of any new mechanism or insurance products in the insurers'/reinsurers' practice is the one asking how to avoid insuring bad risks and not miss the new insurance opportunity (Fig. 15). For example, insurance against new AI-supported cyber-threats could raise a lot of interest at the insurance market, but, at the same time, it could be a great risk for the insurer. In the same sense, one could also consider the potential products which are going to incentivize the client's efforts to improve resilience could: "buying resilience" can sound better than "buying insurance". It is in line with the fact that risk managers and policymakers increasingly take the resilience perspective in addition to their previous focus on risks [24].

Enhancing resilience against XTs as emerging risks is also suggested in the Geneva Association report (2023) and in the background paper[11] of the same report that the new risks/threats landscape should be regrouped not according to their "labels" (e.g., "AI risks"), but according to their main characteristics (the emerging risks in clusters), which both the report and the background chapter see as:

- *Novelty of risks*, involving new risk sources and new risk consequences.
- *Changing nature of existing risks*, in terms of risk/threat frequency and severity of impacts.
- Knowledge gaps, considering.

[11] *Paper under preparation, Emerging Risks: Clouds on the Risk Landscape, Schernberg, Jovanović.*

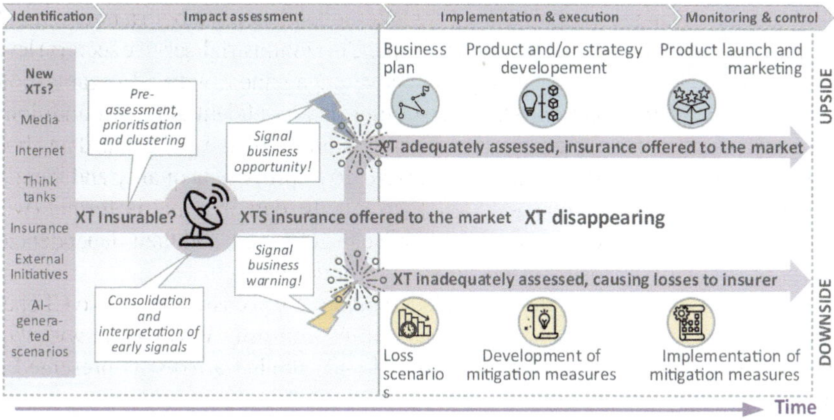

Fig. 15 Insuring/reinsuring XTs: Business opportunity or "risk of insuring a bad risk?

- Complexity and connectedness,
- Uncertainty and ambiguity,
- Information and communication,
- Governance of risks,
- Management of complexity,
- Management of uncertainty, and,
- Management of the pace of change.

- *Governance* slooking also at how one can assess and manage these risks.

4.2 *Insurance Products Enhancing Resilience*

Depending on the approach and the type of insurance, resilience in insurance can be understood as reduced loss (reduced total loss of performance [25], reduced expected recovery time (reduction in business interruption) and reduced loss area on the functionality curve (Fig. 16). It is important that the loss of functionality, integrated over time, remains small in case of XT events. Such a goal corresponds to other concepts in risk management, e.g., the Global Risk Index from the Cambridge Centre for Risk Studies, assessing the resilience of around 300 cities worldwide.[12] An overview of what insurance products and mechanisms can provide in enhancing resilience is shown in Fig. 16. The return on investment in resilience improvement can be assessed acc. to the formula:

[12] *Cambridge Centre for Risk Studies [26], Cambridge Global Risk Index 2018 Methodology.*

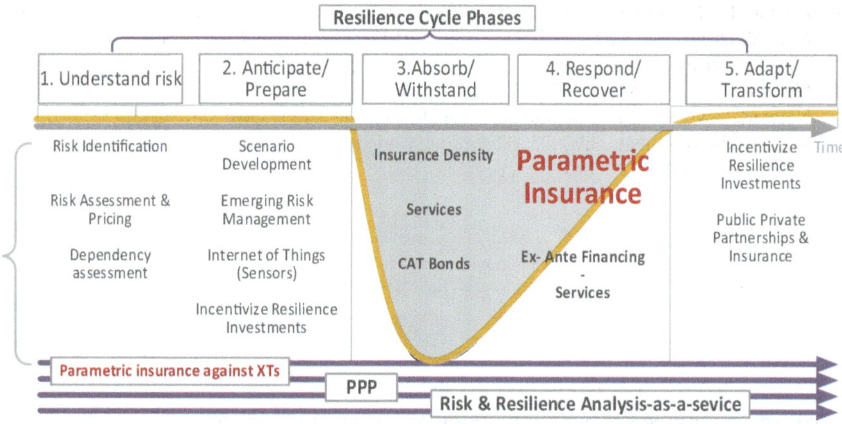

Fig. 16 Enhancing CI resilience by reinsurance/insurance products and mechanisms

4.3 Parametric Insurance against XTs

Parametric insurance against XTs is an important part of the proposed framework. In insurance, it is an increasingly accepted way to include innovation and data analytics into the enlarged scope of commercial insurance solutions, offering coverage for a wider range of threats, exposures, and perils.[13] That is what makes it particularly interesting for the issue of insuring and insurability of XTs. It generally leads to more transparent and faster claims payment and the ability to offer a payout without actual physical damage to an asset. In that context, parametric or index-based solutions are often considered when dealing with "hard-to-insure risks", such as XTs. When compared to traditional property insurance, where paying a premium is made in return for a promise to cover the actual loss incurred by an incident or well-specified peril. The pre-agreed payout is made if the parameter or index threshold is reached or exceeded, regardless of actual physical loss sustained.[14]

The threshold is usually set in such a way that aligns with a client's business continuity plan and risk tolerance. For example, a client may know that with the risk mitigation measures currently in place, their business can sustain the effects of a magnitude 7.0 earthquake. The key differences between traditional and parametric insurance are outlined in Table 5.

Parametric insurance requires more sophisticated structures than traditional insurance to avoid basis risk, which is the risk that the trigger parameter does not perfectly correlate with the underlying insurance risk exposure. Compared to traditional insurance products, parametric insurance is more customizable and flexible.

[13] https://corporatesolutions.swissre.com/insights/knowledge/what_is_parametric_insurance.html..

[14] *NOTE: For example, USD ten million if a magnitude 7.0 earthquake occurs in a defined geographical area, or USD 30 million if a category 5 tropical cyclone occurs in a defined area, or USD 50,000 for every mm of cumulative rainfall above a certain threshold.*

Table 5 Key differences between traditional and parametric insurance

Key differences	Traditional insurance	Parametric insurance
Payments triggered by	Actual loss of or damage to a physical asset	Event occurrence exceeding the parametric threshold
Recovery	Reimbursement of actual loss	Pre-agreed payment structure based on parameter/index
Basis risk	Policy conditions, deductibles, and exclusions	Correlation of chosen index, payout, and the loss sustained
Claims: Assessment and settlement	Complex and based on loss adjuster assessment	Transparent, predictable, based on a parameter/index, quick
Term	Usually annually	Single or multi-year
Structure	Standard products, weak customization	Customized product with high structuring flexibility

As the parametric insurance needs an recognized and independent index (e.g., the categories defined by the meteorological institutes, some sort of "insurance Beaufort scale for XTs"), in Jovanović, Schernberg, Jelić [1], we propose a composite index (CIRIX, aggregation based on OECD, [27]), which can be used for this type of insurance, in a similar way as those for parametric wind/haze/earthquake/etc. insurance.

5 Conclusions—Optimizing the CI Protection Against XTs by Insurance-Related and Resilience Indicator-Based Stress-Testing

The main message of the chapter is.

"Stress-testing resilience of critical infrastructures against XTs is needed and useful, and we propose a solution for meeting this need".

Main elements of the solution and of the framework in Fig. 1 are as follows:

- Emphasizing the need for PPP for creating alignment of stakeholders' opinions, absolutely needed for the success and acceptance of stress-testing resilience (promoting the "private–public ecosystem")—elements (1) to (4) of the framework;
- Using resilience indicators, in particular the ones needed to describe XTs and resilience of critical infrastructures (such as CIRIX, Jovanović [1, 5]) exposed to the XTs in the future (e.g., in the case of a global polycrisis)—elements (5) to (7) of the framework;
- Introducing the parametric insurance against XTs into the reinsurance and insurance practices (based on the accepted indicator-based indices like CIRIX)—elements (7) to (9) of the framework;

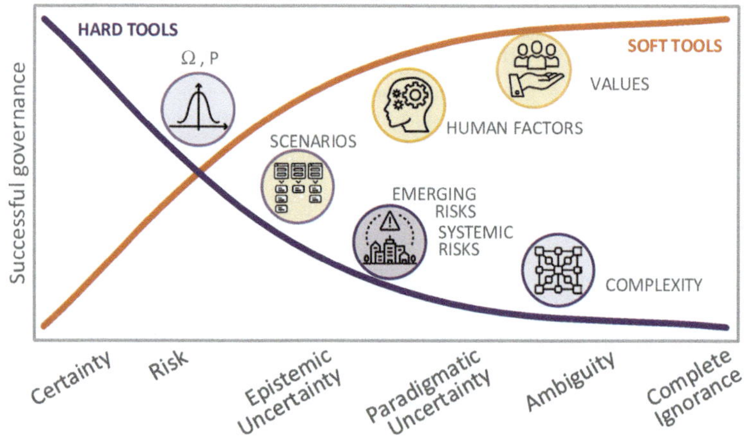

Fig. 17 Increased need for using advanced soft technologies for better dealing with XTs by supporting anticipatory thinking, preparedness for surprises and mastering complexity (adapted from Sachs [10])

- Further development and embedding the stress-testing aspects into the international (primarily ISO), EU (primarily CEN), and national standards—element (10) of the framework.

The above elements are, so far, barely tackled and not explicitly present in the EU regulation and policies (e.g., in CER and NIS2 Directives). We propose to include them in the future, not only in directives, but also in the other regulatory documents and acts. Evidence of a sufficient level of resilience has been declared as one of the highest political goals[15]—stress-testing can provide it. Especially in the new, more complex risk landscape, involving always more XTs. Only by including XTs into all resilience considerations can one ensure that the aspects of anticipation and adaptivity (corresponding, respectively, to phase I and phase V in the SmartResilience approach) are properly included in resilience management.

For all the considerations above, use of advanced (soft) tools and technologies is, in the context of this book, a way to go forward (Fig. 17, Sachs [10]). This is needed to ensure better dealing with XTs, to support anticipatory thinking, to better prepare for surprises, and to master the complexity.

Use of soft smart tools will, at the same time, help enhance accountability in managing complex XTs by ensuring better governance and avoiding unambiguous mandates or distributed responsibilities. The same applies to communication, where these tools can achieve better, bi-directional communication and more objective framing of the XTs (often affected by misrepresentations in fake news and false

[15] *February 12, 2021: Joint Press conference on Recovery and Resilience Facility: Statements by David Sassoli, EP President; Ursula von der Leyen, President of the EC,* https://multimedia.europarl.europa.eu/en/video/joint-press-conference-on-recovery-and-resilience-facility-statements-by-david-sassoli-ep-president-ursula-von-der-leyen-president-of-the-ec-and-antonio-costa-portuguese-prime-minister-on-behalf-of-the-presidency-of-the-council_I202125..

narratives). Finally, the tools can also help develop better policies, e.g., through adaptive regulation, interoperable platforms for knowledge generation, and transfer or collaborative stakeholders' environments/platforms.

Including XTs and stress-testing into current policies and standards requires, however, constant review and alignment of existing ones. An additional challenge lies in the fact that the documents developed within the same organization or institutions often exhibit misalignments and inconsistencies.

Acknowledgments The chapter of the book is leaning on the author's work as the convenor for ISO TS 31050 and DIN SPEC 91461 standard documents, on his work as the project manager and a partner, respectively, in the EU projects Smart Resilience and Infra Stress, as well as on the author's work as a visiting professor and member of the Risk Center at ETH Zurich 2022–24. No AI-tools have been used for the text creation.

References

1. Jovanović, A., Schernberg, H., Jelić, M. (2023). Indicator-based assessment of extreme threats (XTs) and their impacts on the resilience of geographical areas and critical infrastructures. pp 599–624, Springer Nature, Special Issue: Disaster Resilience and Insurance, vol. 43, 4, December 2023, eds. A. Jovanović, H. Schernberg, G. Sansavini.
2. EU-CER (2022). The Critical Entities Resilience Directive (CER) https://www.critical-entities-resilience-directive.com/.
3. Jovanovic, A. S., Chakravarty, S., & Jelic, M. (2021). Resilience and situational awareness in critical infrastructure protection: An indicator-based approach. In Issues on Risk Analysis for Critical Infrastructure Protection. J. Soldatos, I. Praça and A. Jovanovic (eds.). (2021). Cyber-Physical Threat Intelligence for Critical Infrastructures Security. Boston–Delft: Now Publishers.
4. IRGC (2015). Guidelines for Emerging Risk Governance. Lausanne: International Risk Governance Council (IRGC).
5. Jovanović, A., et al. (2020). Assessing resilience of healthcare infrastructure exposed to COVID-19: emerging risks, resilience indicators, interdependencies and international standards. Environment Systems and Decisions, 40, 252-286.
6. Sigma (2025). Natural catastrophes: insured losses on trend to USD 145 billion in 2025, Sigma 1/2025 report. SwissRe Institute.
7. Guha-Sapir, D., Below, R., & Hoyois, P. H. (2009). EM-DAT: The CRED/OFDA International Disaster Database—www.emdat.be. Brussels. Belgium: Université Catholique de Louvain.
8. Wadé, M. & Sachs, R. (2013). Managing complex risks successfully. Emerging Risk Discussion Paper, Munich Re.
9. Florin, M.-F. & Sachs, R. (2019). Critical Infrastructure Resilience. Lessons from Insurance. Lausanne: EPFL International Risk Governance Center https://infoscience.epfl.ch/entities/publication/82ebe4a9-4d38-40c9-8487-d78cbf9b817b/reportdetails.
10. Sachs, R. (2023). The governance of uncertainty: how to respond to non-quantifiable risk? *Environ Syst Decis* 43, 537–543 (2023). https://doi.org/10.1007/s10669-023-09920-3
11. Kunreuther, H., Michel-Kerjan, E., & Tonn, G. (2016). Insurance, Economic Incentives and other Policy Tools for Strengthening Critical Infrastructure Resilience: 20 Proposals for Action. The Wharton School, University of Pennsylvania.
12. World Bank (2017). Resilient Infrastructure Public-Private Partnerships: Contracts and Procurement—The Case of Japan.

13. Steinmann, C., Guillod, B., Fairless, Ch., Bresch, D. (2023). A generalized framework for designing open-source natural hazard parametric insurance. *Environ Syst Decis* **43**, Special Issue: Disaster Resilience and Insurance, vol. 43, 4, December 2023, eds. A. Jovanović, H. Schernberg, G. Sansavini, Springer Nature.
14. Linkov, I., Trump, B., Trump J., Pescaroli G., Hynes, W., Mavrodieva, A. Panda, A. (2023). Resilience stress testing for critical infrastructure. International Journal of Disaster Risk Reduction, https://doi.org/10.1016/j.ijdrr.2022.103323.
15. EBA (2025). The EBA launches its 2025 EU-wide stress test. https://www.eba.europa.eu/publications-and-media/press-releases/eba-launches-its-2025-eu-wide-stress-test
16. ENSREG. (2011). Declaration of ENSREG on EU "Stress tests" specifications. http://www.ensreg.eu/sites/default/files/EU%20Stress%20tests%20specifications_0.pdf
17. ERCOT (2021). The Timeline and Events of the February 2021 Texas Electric Grid Blackouts
18. TIBER (2018). TIBER-EU framework—How to implement the European framework for Threat Intelligence-based Ethical Red Teaming, European Central Bank (Governing Council), Germany. https://www.ecb.europa.eu/pub/pdf/other/ecb.tiber_eu_framework.en.pdf.
19. DIN SPEC 91461 (2023). Framework for stress-testing resilience of industrial plants and sites exposed to cyber-physical attacks, DIN Berlin https://www.beuth.de/de/technische-regel/din-spec-91461/346873248.
20. Esposito S, Stojadinovic B, Mignan A, Dolšek M, Babič A, Selva J, Iqbal S, Cotton F, Iervolino I (2016): Report on the proposed engineering risk assessment methodology for stress tests of non-nuclear CIs. Deliverable 5.1 STREST Project.
21. ISO 22316:2017(en) Security and resilience—Organizational resilience—Principles and attributes, https://www.iso.org/obp/ui/en/#iso:std:iso:22316:ed-1:v1:en
22. WEF World Economic Forum Global Risks Report (2025). The Global Risks Report, 20th edition. Geneva
23. JRC (2024). Risks on the Horizon—Insights from Horizon Scanning. EU Policy Lab, JRC 137493 EUR 31927 EN, European Union, 2024 https://publications.jrc.ec.europa.eu/repository/handle/JRC137493
24. Linkov, I., Trump, B. D., & Fox-Lent, C. (2016). Resilience: Approaches to risk analysis and governance. An edited collection of authored pieces comparing, contrasting, and integrating risk and resilience with an emphasis on ways to measure resilience.
25. Moteff, D. (2012). Critical Infrastructure Resilience: The Evolution of Policy and Programs and Issues for Congress. CRS Report for Congress, https://sgp.fas.org/crs/homesec/R42683.pdf.
26. Cambridge Centre for Risk Studies (2018), Cambridge Global Risk Index 2018 Methodology, https://www.jbs.cam.ac.uk/wp-content/uploads/2020/08/crs-cambridge-global-risk-index-2018-methodology.pdf.
27. OECD (2008). Handbook on Constructing Composite Indicators, Methodology and User Guide, OECD publishing ISBN 978-92-64-04345-9
28. DIN SPEC 91299 (2013). Managing emerging technology-related risks; English version CWA 16649:2013.
29. EU-ECHO (2017). European Commission, Directorate-General for European Civil Protection and Humanitarian Aid Operations (ECHO), Overview of natural and man-made disaster risks the European Union may face, Publications Office, 2017, https://data.europa.eu/doi/10.2795/560045
30. FOCP (2020a). Federal Office for Civil Protection: National risk analysis methodology. Disasters and Emergencies in Switzerland 2020. Version 2.0. FOCP, Bern.
31. FOCP (2020b). Federal Office for Civil Protection: National risk analysis report. Disasters and Emergencies in Switzerland 2020. FOCP, Bern.
32. FOCP (2020c). Federal Office for Civil Protection: What risks is Switzerland exposed to? Disasters and Emergencies in Switzerland 2020, FOCP Bern.
33. Geneva Association (2023): Schanz K.-U., A. Braun, M. Eling, Ch. Jaenicke, A. Jovanović, H. Schernberg, P. Treccani. (2023). The Value of Insurance in a Changing Risk Landscape. Geneva Association, November 2023

34. IPPC (2023). AR6 Synthesis Report: Climate Change 2023. https://www.ipcc.ch/report/sixth-assessment-report-cycle/.
35. IRGC (2010a). Emerging Risks: Sources, Drivers and Governance Issues. Geneva. https://doi.org/10.5075/epfl-irgc-228190.
36. IRGC (2010b). The Emergence of Risks: Contributing Factors. Geneva.
37. IRGC (2011) Improving the Management of Emerging Risks: Risks from new technologies, system interactions and unforeseen or changing circumstances, 2nd edition. Geneva. https://irgc.org/wp-content/uploads/2018/09/irgc_er2conceptnote_2011.pdf.
38. ISO 9001:2015 Quality management systems—Requirements
39. ISO 22300: Security and resilience—Vocabulary (referred in the text also as "ISO 223xx series" for other standards in the same group) 2000
40. Security and resilience—Organizational resilience—Principles and attributes
41. ISO 31000:2018 Risk management Guidelines www.iso.org/iso-31000-risk-management.html and https://www.iso.org/obp/ui/#iso:std:iso:guide:73:ed-1:v1:en
42. ISO/IEC 31010*IEC/ISO 31010: Risk management—Risk assessment techniques
43. ISO TS 31050:2023. Risk Management—Guidelines for managing emerging risk to enhance resilience, https://www.iso.org/standard/54224.html
44. ISO 56000:2020, Innovation management—Fundamentals and vocabulary
45. Jovanovic, A. (2020). Critical Infrastructure Protection: Assessing resilience of European critical infrastructures by means of indicators: Wishful thinking vs. engineering challenge, ECSCI (European Cluster for Securing Critical Infrastructures) e-Workshop, June 24-25. https://120313dd-acf0-424f-8519-f9644d8765ee.filesusr.com/ugd/b2d070_2dd579fec49348068e2009cd1c98b583.pdf
46. Jovanović, A., Schernberg, H. & Sansavini, G. Improving disaster resilience: a common goal for the EU, insurers and researchers. *Environ Syst Decis* **43**, 531–532 (2023). Springer Nature https://doi.org/10.1007/s10669-023-09951-w.
47. (2022). Directive (EU) 2022/2555 of the European Parliament And Of The Council on measures for a high common level of cybersecurity across the Union, https://eur-lex.europa.eu/legal-content/EN/TXT/PDF/?uri=CELEX:32022L2555
48. Smart Resilience (2019). EU Project Smart Resilience Indicators for Smart Critical Infrastructures, Project No. 700621. http://www.smartresilience.eu-vri.eu/
49. STREST (2016). Final Report Summary of the EU project—STREST (Harmonized approach to stress tests for critical infrastructures against natural hazards), project 603389 https://cordis.europa.eu/project/id/603389/reporting?form=MG0AV3.
50. Tsionis G, Pinto A, Giardini D, Mignan A, Cotton F, Danciu L, Iervolino I, Pitilakis K, Stojadinović B, Zwicky P, (2016). Harmonized approach to stress tests for critical infrastructures against natural hazards, EUR 28343 EN, https://doi.org/10.2788/17066.

Analysing the Impact of Compromising IoT Consumer Devices in Power Systems

Néstor Rodríguez-Pérez, Javier Matanza Domingo, Lukas Sigrist, Jose Luis Rueda Torres, and Gregorio López López

Abstract This chapter explores the potential risk in power systems due to the increasing penetration of Internet of Things (IoT) devices at the consumer level. It focuses on the potential for Manipulation of Demand through IoT (MaDIoT) attacks, where attackers exploit these devices to disrupt power system operations. For this, a comparative analysis of the impact of MaDIoT attacks on two distinct power system models is provided: the IEEE 39-Bus (New England) system and the PST-16 system (a simplified European model). By assuming that attackers lack advanced knowledge of the grid, the analysis highlights significant differences in the success probability and impact of attacks between these systems, providing insights into the varying responses of different power systems to MaDIoT attacks, and emphasising the need for system operators to consider these in their cybersecurity risk assessment in the future.

Keywords Cyberattack · MaDIoT · Load altering attacks · Power system dynamics · Power system stability

Abbreviations

DSO	Distribution System Operator
EUCC	European Cybersecurity Certification scheme
IoT	Internet of Things
MaDIoT	Manipulation of Demand through Internet of Things
OFGR	Over-Frequency Generator Rejection
SCADA	Supervisory Control and Data Acquisition
TSO	Transmission System Operator

N. Rodríguez-Pérez (✉) · J. Matanza Domingo · L. Sigrist · G. López López
Institute for Research in Technology, Comillas Pontifical University, Madrid, Spain
e-mail: nestor.rodriguez@iit.comillas.edu

J. L. Rueda Torres
Faculty of EEMCS, Delft University of Technology, Delft, The Netherlands

J. Soldatos and H. Abie (eds.), *Technology-Enabled Resilience: Innovations in Critical Infrastructure Protection*, https://doi.org/10.1007/978-3-032-25917-2_5

UFLS Under-Frequency Load Shedding

1 Introduction

In recent years, the cybersecurity of power systems has become a growing concern among countries. Usually, when we think about cyberattacks to power systems, it comes to our mind cyberattacks to Supervisory Control And Data Acquisition (SCADA) systems [8] that may take over the control centre of system operators or that may compromise the employed communication protocols to open circuit breakers and switchgears, such as in the case of the Industroyer attack in Ukraine [22]. However, there are other elements connected to electricity systems that may be vectors of attacks. The increasing penetration of Internet of Things (IoT) devices at the consumer level (such as home assistants, energy management systems, etc.) introduces new vulnerabilities that can be exploited [40].

Most IoT devices have low security levels [2] and, if an attacker manages to compromise a large amount of high-wattage IoT devices, the power system may see its security margins reduce or even experience load shedding and cascading failures leading to wide-area blackouts [12, 17, 37]. Furthermore, it must be considered that the attack surface of electricity demand is larger than that of SCADA systems, the standard user does not have an advanced level of cybersecurity awareness, and high-wattage devices like electric vehicle charging points are not continuously monitored by system operators (SO) [1, 31].

The idea of compromising demand taking advantage of its growing internet connectivity was first introduced in [30], which identified loads that could be compromised such as data centres, demand side management loads and customer-managed loads (e.g., air conditioning, washing machines). This type of attack received the generic name of Load Altering Attack (LAA) [30] and since then has attracted the attention of the research community with different studies [12, 15, 16, 23–26, 32].

Since the term "load" is very generic, Soltan et al. [37] developed the concept of Manipulation of Demand through Internet of Things (MaDIoT) to describe an attack that disrupts normal operation of the power grid by altering the power demand using compromised IoT devices. Soltan et al. studied the Polish grid model and showed that MaDIoT attacks could cause local outages and large-scale blackouts. However, Huang et al. [19] suggested that the Polish grid model analysed might not be N-1 secure, potentially leading to an overestimation of the impact of the attack.

Huang et al. [19] showed that instigating a widespread blackout in a major North American regional system through widespread MaDIoT attacks is quite difficult. Even if the system is initially susceptible, these attacks are likely to cause only partial blackouts, as some loads and generators would be disconnected, allowing the system to stabilise.

Later research by Shekari et al. [35] developed and studied the concept of MaDIoT 2.0 attacks in the IEEE 39-Bus system. MaDIoT 2.0 is a version of MaDIoT where it is assumed that the attacker had advanced knowledge of the system's topology and the estimated generation/demand for each node. This would enable more sophisticated attacks targeting the most vulnerable nodes. The results obtained by Shekari et al. [35] showed elevated success rates in causing widespread blackouts. However, the criteria for considering a MaDIoT attack successful were unclear, and the likelihood that an attacker possessing the necessary knowledge and system resources is presumably low.

This chapter examines and compares the impact of MaDIoT attacks on two different power systems: the IEEE 39 test system (as in [35]), and the PST-16 system, which is a simplified model of the European power system [34]. These systems differ mainly in size (network and demand) and generation mix, so the PowerFactory simulation results provide insight into the different consequences of MaDIoT attacks in different systems. The results in this chapter have been obtained within the framework of the eFORT project [13], funded by the European Union's Horizon Europe Research and Innovation programme.

2 Methodology

In line with the literature of MaDIoT attacks [19, 35, 37] and given the impossibility of conducting MaDIoT experiments in actual power systems for obvious reasons, a simulation-based approach with DIgSILENT PowerFactory 2022 SP3 (22.0.6.0) was used to evaluate the impact of MaDIoT attacks on two different power systems: the IEEE 39-Bus system, also used in [35], and the PST-16 system. These two system models allow for the evaluation of the impact of MaDIoT attacks on both an American and a European grid, considering differences in electrical topologies, demand distributions, generation structures and dynamic behaviours. The simulations were carried out calling a Python script from PowerFactory, as shown in Fig. 1.

The following sections provide a brief overview of these test systems, detail the protection schemes and outline the assumptions and scenarios considered for the analysis.

2.1 *Test Systems*

Two power systems are used for the analysis: the IEEE 39-bus system and the PST-16 benchmark system. Table 1 summarises the characteristics of these models.

2.1.1 IEEE 39-Bus

The New England power system, consisting of 39 buses, has a total base load of 6097.1 MW of active power and 1408.9 Mvar of reactive power, which are the

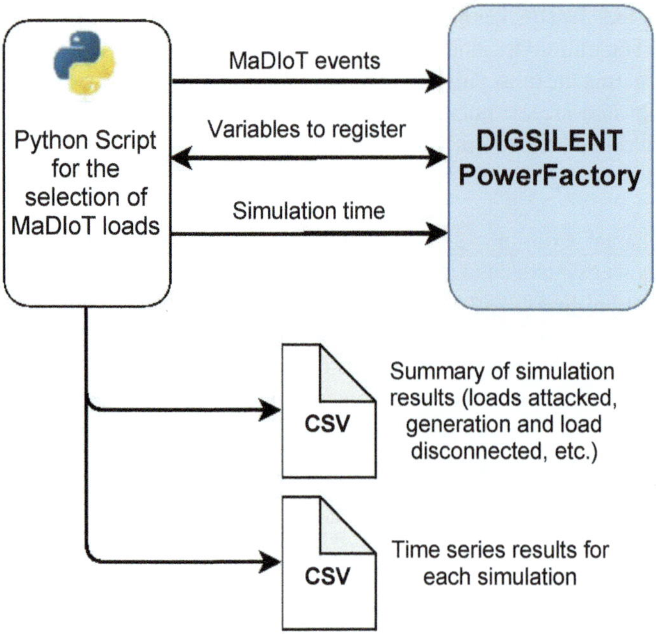

Fig. 1 Diagram showing the interaction between PowerFactory and the python script used to simulate the MaDIoT attacks

Table 1 Summary of the characteristics of the IEEE 39-BUS and the PST-16 models

	IEEE 39-BUS	PST-16
Frequency (Hz)	60	50
Areas	1	3
Number of buses	39	66
Base active load (MW)	6097.1	15,565
Base reactive load (Mvar)	1408.9	2225
Generators	10	16
Generation type	Hydro, thermal	Hydro, thermal and nuclear
Max. active power generation (MW)	14,535	18,220
Line capacity (MVA)	With USA/Canada: 1231	Areas A-C: 1572; Areas B-C: 2476; Areas A-B: 2585
Voltage level (kV)	345	380, 220, 110

default initial conditions of the system before the attacks. The electrical frequency is 60 Hz. The system's maximum active power generation capacity is 14,535 MW, with 8500 MW provided by the generator at bus 39, which represents the interconnection USA/Canada. However, the support from this interconnection is fixed at 1000 MW. Consequently, the actual maximum active power capacity of the IEEE-39 system is 6035 MW, with an available reserve of 938 MW, since 1000 MW of demand are supplied by the USA/Canada interconnection. The interconnection bus (bus 39) connects to the rest of the system through two power lines with a total capacity of 1231 MVA. Initially, the load connected to bus 39 consumes 1104 MW, utilising only 19% of the interconnection lines' capacity.

Regarding the load model, the default dynamic load model of the IEEE-39 system in PowerFactory is used.

2.1.2 PST-16 (Simplified European Model)

The PST-16 Benchmark System [34] comprises three areas (A, B and C) and 66 buses, with a total base load of 15,565 MW of active power and 2225 Mvar of reactive power. These values serve as initial conditions for the system before the simulated MaDIoT attacks. The maximum active power generation capacity of the PST-16 system is 18,220 MW. As it represents a European system, the electrical frequency is set to 50 Hz.

For this system, the load model is constant impedance [34]. The generator models from the base model remain unchanged. Detailed information on the generator model, the grid diagram and the complete PowerFactory model can be found in [34].

Figure 2 presents a simplified diagram of the PST-16 system. Area A represents northern Europe, characterised by a high share of hydro generation, while areas B and C represent central and southern Europe, respectively, with significant shares of thermal and nuclear generation. As shown in Fig. 2, area C requires power transfer from areas A and B through two long links to satisfy its demand. The capacity of the line connecting areas A-C is 1572 MVA, for B-C it is 2476 MVA and for A-B it is 2585 MVA. Under the initial conditions (before the MaDIoT attacks), the loading of these interconnection lines is 110.1%, 17.3% and 37.2%, respectively. These initial conditions for the PST-16 system correspond to peak demand conditions, since it is 85% of the generation capacity.

2.2 *Protections*

Performing MaDIoT attacks on power systems would trigger electrical protections. For the study presented in this chapter, four types of protection were implemented in PowerFactory: overvoltage protections, undervoltage protections, an Under-Frequency Load Shedding (UFLS) scheme and an Over-Frequency Generator Rejection (OFGR) protection scheme.

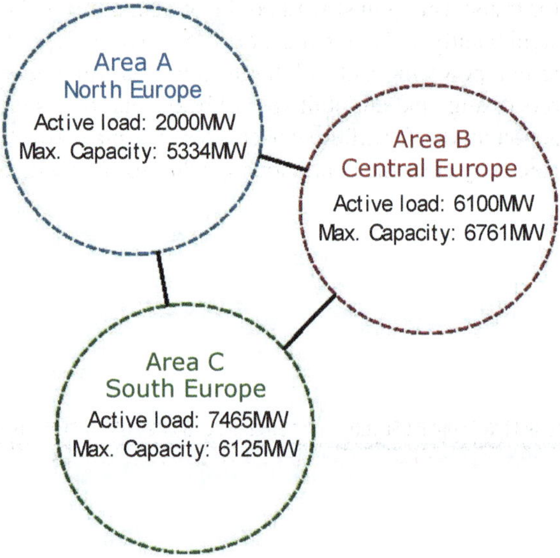

Fig. 2 Simplified diagram of the PST-16 benchmark model [33]

2.2.1 Voltage Protections

These protections disconnect loads when the voltage exceeds (phase overvoltage protection) or falls below (phase undervoltage protection) a predefined threshold.

The overvoltage protections were set to trip when the voltage exceeds 1.1 p.u. for 10 s, while the undervoltage protections trip when the voltage drops below 0.85 p.u. for 10 s.

2.2.2 Frequency Protections

Two type of protections are included in the power system models to be activated when frequency drops or goes above certain thresholds. The UFLS protection scheme gradually disconnects loads from the system as the frequency drops below the levels shown in Table 2.

However, OFGR protections are activated when the frequency of the generation bus reaches 51.7 Hz (PST-16) or 61.7 Hz (IEEE 39), similar to the values used in [19]. When activated, the corresponding generator is disconnected from the system, to safeguard it.

Table 2 UFLS scheme applied for the 50 and 60 Hz models. Based on [18]

Frequency threshold (Hz)	49	48.8	48.6	48.4	48.2	48
	59	58.8	58.6	58.4	58.2	58
Load-shed (%)	5	10	10	10	10	10

2.3 Attack Models and Simulation Scenarios

For the study, it is assumed that each compromised load (i.e. bot) consumes 3 kW of active power, which is a common assumption in previous research [35]. To maintain similar power factors to those in the baseline test systems, the attack is also considered to alter the reactive power of the load. The power factor of the demand (inductive) for the IEEE 39-bus and PST-16 systems is 0.97 and 0.99, respectively. Therefore, the reactive power of a bot is 0.69 kvar for the IEEE 39-bus system and 0.42 kvar for the PST-16 system. Only MaDIoT attacks that increase power consumption in the demand are considered.

The study in this chapter focusses on the impact of MaDIoT attacks on power systems. As in previous works [19, 35, 37], it is assumed that target devices can be successfully compromised in some manner (e.g., they are accessible from the Internet and retain default passwords), and malware can enable remote command and control. This is similar to the famous DDoS attack against the Dyn DNS provider in 2016, which used millions of IoT devices infected with Mirai malware [4, 7].

The attack model is described by Table 3 following the guidelines provided by [41].

Regarding the attacker, previous studies such as [35] assume that the attacker has advanced knowledge of the details of the grid, such as its topology and power flows, allowing the calculation of voltage stability indices to identify the most vulnerable nodes to attack. This assumption is supported by studies indicating that much information can be obtained from open sources [20] or by satellite images (for example, Google Maps) [5, 21, 29]. This process can be very time-consuming and extremely complex, making it difficult to verify its accuracy to perform a power flow analysis. Additionally, the attacker would need to have access to devices at all nodes of the system or target specific nodes and find devices connected to those nodes that can be compromised.

However, an attacker may already possess a botnet to exploit without knowing the exact electrical connections of the bots but having a good idea of their proximity (e.g., by mapping the IPs). Therefore, the study in this chapter assumes that the attacker does not have advanced knowledge of the grid, significantly reducing the

Table 3 Considered MaDIoT attack model based on the modelling guidelines by [41]

	Attack model
Frequency	Iterative
Reproducibility and discoverability	Multiple-times
Functional level	Level 1 or 2
Compromised asset	Field controllers, human–machine interfaces
Potential techniques	Modify control logic, wireless compromise and Denial-of-Service to the power grid
Attack premise	Cyber: communications and protocols and asset control commands

preparatory work required for the attack and, consequently, increasing the likelihood of a MaDIoT attack attempt on the power system.

Table 4 describes the adversary model following the guidelines provided by [41]. As previously mentioned, it is assumed that the adversary's knowledge is limited, and the attacker does not have physical access to the assets (non-possession adversary access). MaDIoT attacks are targeted attacks (the objective is to compromise high-wattage IoT devices), and the attacker is considered to have substantial resources, tools and skills to successfully execute the attack (class II). It is assumed that the attacker has successfully compromised the devices and installed malware that enables command and control, allowing the control of a large number of devices (botnet).

To ensure that bots are in close proximity and to study a worst-case-like scenario, the simulated attacks only affect three nodes. These nodes are selected randomly for each simulation, following a Monte Carlo-like approach. For the PST-16 system, the attacked nodes are within the same area, as closer proximity is expected to have a higher impact on the system. Attacks are initiated at $t = 1$ s in the simulation.

An attack is deemed successful in both systems when, by the conclusion of the simulation, loads are disconnected due to any activated protection mechanisms, or if generators are disconnected (OFGR protections). This same standard was applied by [3], which classified an attack as successful if it triggered at least one over/under frequency protection relay, irrespective of whether this resulted in a full system blackout.

Table 5 outlines the scenarios considered for the analysis of each test system. For each scenario, the botnet size varies from 50,000 to 500,000 in increments of 50,000, and the simulation time is set to 21 seconds to maintain acceptable computational load. For the PST-16 system, nearly 1500 simulations are performed, while the IEEE 39-Bus system accounts for 424 simulations.

Table 4 Considered MaDIoT adversary model based on the modelling guidelines by [41]

	Attack model
Adversary knowledge	Oblivious
Adversary access	Non-possession
Adversary specificity	Targeted attack
Adversary resources	Class II

Table 5 MaDIoT attack scenarios for the IEEE 39-Bus model (New England) and the PST-16 model (Europe)

Scenario	Model	Area	Botnet size	# Nodes attacked
US39	IEEE 39	–	[50 k, 500 k]	3
EU-A	PST-16	A	[50 k, 500 k]	3
EU-B		B		
EU-C		C		

3 MaDIoT Attacks Simulation Results

In this section, the simulation results for the scenarios considered in Table 5 are presented and discussed.

3.1 *How Successful Are the MaDIoT Attacks in Activating Protections?*

Before analysing the impact of MaDIoT on the two power systems analysed, it is interesting to first inspect how successful these attacks are in each system. For this, the success ratio is defined as in Eq. 1.

$$\text{Success ratio} = \frac{\#\text{Successful MaDIoT attacks}}{\#\text{Attacks}} \tag{1}$$

It is good to remember that, following the criteria of previous research, an attack is considered successful if it manages to activate at least one of the electrical protections installed in the system.

To facilitate the comparison between the two power system models, Fig. 3 illustrates the success ratio of the MaDIoT attacks simulated for the scenarios outlined in Table 5. This figure highlights the noticeable differences in success ratios between the US39 scenario and the EU scenarios.

In the US39 scenario, it should be noted that all simulated attacks compromising more than 150 k bots were successful. The difference between 150 k and 200 k bots (theoretically increasing from 450 MW to 600 MW) is significant, with the success ratio increasing from 10% to 100%. The IEEE-39 system, without interconnection with the USA/Canada, theoretically has a margin of 938 MW before reaching its generation limit, which should be sufficient to withstand attacks of up to 300 k bots from a frequency perspective. However, simulations for 300 k bots revealed that

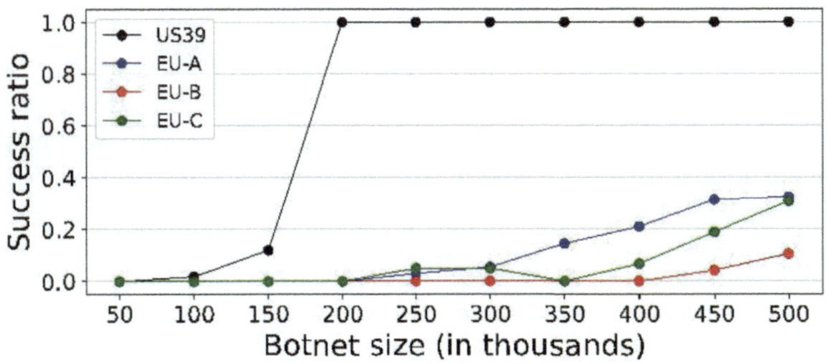

Fig. 3 Success ratio for different scenarios when increasing the size of the botnet

generation does not ramp up quickly enough post-attack to prevent the activation of first-step UFLS protections in some loads, as the attack compromises the loads simultaneously. The high success ratios that are observed are mainly due to the fact that an attack is considered successful if it trips at least one protection. Thus, under the assumed conditions, the proximity of affected buses is irrelevant when more than 150 k bots are compromised; the attack will always trigger UFLS protections in the IEEE-39 system. Consequently, the attacker does not require advanced grid knowledge; performing the attack during peak demand hours, when the system's generation margin is minimal, could result in a high success ratio. Although this may appear to contradict the findings of previous research [35], it is important to note that [35] considered a daily load pattern for the grid, potentially aggregating the success ratios of MaDIoT attacks during valley and peak demand hours.

In contrast, within the PST-16 system, MaDIoT attacks start to succeed in the EUA and EU-C scenarios when botnets surpass 200 k bots, and in the EU-B scenario for botnets larger than 400 k bots. Although EU-A and EU-C eventually show comparable success rates (around 30%) at the largest considered botnet size, the peak success rate for the EU-B scenario is notably lower, at roughly 10%. As illustrated in Fig. 2, areas A and C have the greatest imbalance between generation capacity and demand: area A experiences a generation surplus, while region C depends on power imports from outside the area.

Therefore, Fig. 3 shows that the number of bots required for a successful attack is lower in the IEEE 39-Bus system compared to the PST-16, as it is a smaller system with less generation capacity.

3.2 *Beyond Success: Impact of MaDIoT Attacks on Test Systems*

Although the IEEE 39-Bus and PST-16 grid models exhibit different success ratios for MaDIoT attacks, the success ratio does not equate to the degree of impact, which can be measured as the amount of loads and/or generation disconnected from the system. Table 6 presents the average generation and demand disconnected in successful MaDIoT attacks involving 500 k bots in the US39 and EU-C scenarios (with EU-C being the highest impact scenario for the PST-16 model). While the average demand affected is similar in both scenarios, generation is not disconnected in the US39 scenario. For comparison, two high-impact cases (one per model) have been selected for a more detailed analysis. The results of these cases are illustrated in Figs. 4 and 5.

Figure 4 shows the frequency (Hz), voltages (p.u) and the relative rotor angle of generators (relative to the reference generator) over time when 500 k bots are compromised, in total, in loads 30, 31 and 34 of the PST-16 model (a high-impact EU-C scenario). The attack time ($t = 1$ s) is marked by an asterisk (*) on the x-axis. For visual clarity, the frequency and voltage data for only six buses are plotted,

Table 6 Average impact on the system when successfully attacking 500 k bots in the US39 and EU-C scenarios

Scenario	Botnet size (K)	Average generation disconnected (MW)	Average load loss (MW)
US39	500	0	983.64
EU-C	500	1515.28	938.84

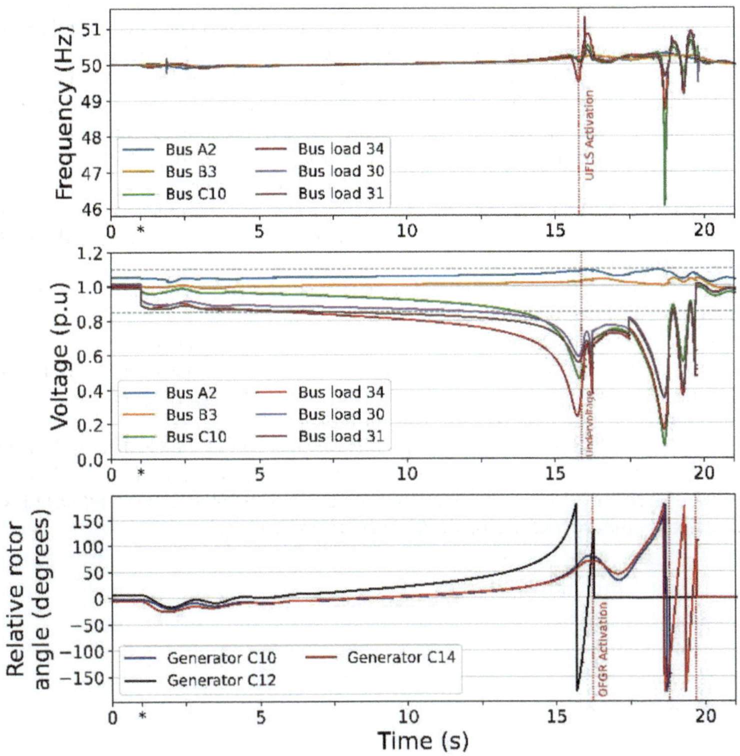

Fig. 4 Frequency, voltages and relative rotor angle of generators when attacking 500 k bots in loads 30, 31 and 34 in the PST-16 system (high-impact case)

including those to which the compromised loads are connected. For the relative rotor angle, only three generators from area C are depicted.

Figure 4 shows the significant destabilisation of the system caused by the attack. Initially, the attack has a minimal impact, noticeable only for a few seconds; a slight oscillation between the areas is observed, but the system manages to confine frequency variations and appears stable. However, by $t = 15$ s, area C begins to diverge from the other two areas. The frequency at bus C10, with generation connected,

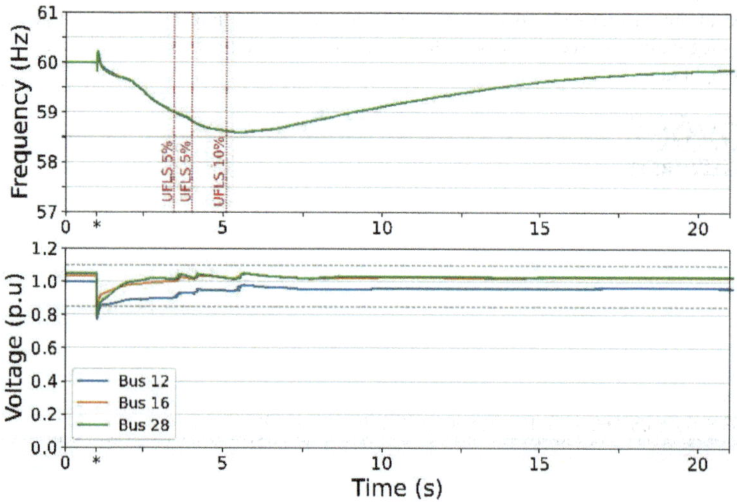

Fig. 5 Frequency and voltages when attacking 500 k bots in loads 12, 16 and 28 in the IEEE 39-Bus system (high-impact case)

suddenly drops to 46 Hz around $t = 18.5$ s. These frequency variations approximately 12 seconds after the attack are attributed to the loss of rotor angle stability in the system.

The middle plot in Fig. 4 clearly illustrates the immediate and significant impact of the attack on the voltages in area C. It is important to note that, before the attack, the system was already operating under peak demand conditions, with area C reliant on power imports from the other two areas. The voltages at the attacked buses drop significantly, nearing the threshold set for the tripping of undervoltage protections. However, due to the increased demand caused by the attack, the system loses rotor angle stability and subsequently experiences voltage collapse. The bottom plot of Fig. 4 shows that the rotor angles of the generators in area C begin to diverge from the reference generator after some initial oscillations. Thus, the system first encounters a rotor angle stability issue, which then leads to voltage collapse.

As voltages drop below 0.85 p.u for more than 10 s (Fig. 4), undervoltage protections begin to trip, disconnecting loads from the system. The activation of these protections, along with Under-Frequency Load Shedding (UFLS) and Over-Frequency Generator Rejection (OFGR) protections, are the primary causes of the oscillations observed in the 15–20 s interval. Following the activation of these protections, the system appears to recover by t = 20 s, albeit with rather low voltage levels (e.g., at Bus C10). By this time, nearly 2.9 GW of generation has been disconnected from the system due to the OFGR scheme. The impact could vary if additional protections were implemented (e.g., distance protection with or without out-of-step protection). Despite the increased demand caused by the attack, the system eventually ends up with approximately 3 GW less demand than before the attack (around a 20% decrease) due to load disconnections (UFLS and undervoltage protections).

This indicates that not only the extraordinary demand caused by the attack had to be disconnected, but additional loads also had to be shed to recover the system.

Similar to Fig. 4, Fig. 5 plots the frequency and voltages when 500 k bots are compromised in loads 12, 16 and 28 in the IEEE 39-Bus system, which constitutes a high-impact US39 scenario.

In this case, the immediate impact of the attack on the system's frequency and voltages is significant. The frequency drops by 1 Hz in approximately two seconds. Below 59 Hz, the UFLS scheme begins to actuate, as detailed in Table 2. This mitigates the frequency drop; only when it reaches approximately 58.6 Hz does the system start to increase the frequency. However, the recovery is slow. The system manages to keep all voltages within limits, so the UFLS protections are the only ones activated. These protections shed about 1.1 GW of loads across the system. Despite this, by the end of the simulation, the total system demand increases by 76 MW compared to the demand before the attack (approximately a 1.2% increase). This means that the amount of demand disconnected is roughly equivalent to the demand increase caused by the MaDIoT attack. However, it must be noted that the shedding would also affect legitimate loads, as UFLS protections do not differentiate between legitimate and compromised loads. Compared to the EU-C case analysed in Fig. 4, the relative impact is smaller, as the system manages to maintain its stability.

Therefore, although success ratios suggest that any attack compromising any three buses in the IEEE 39-Bus system may be successful, system dynamics results suggest its impact could be relatively low, proportional to the magnitude of the attack. In contrast, destabilising the PST-16 system is more challenging due to its larger size and greater generation resources to counter the effects of the attack; however, as discussed, a successful attack can significantly impact system stability, leading to partial disconnection of loads and generation which would also have economic effects.

The simulation results also highlight the different types of impact that MaDIoT attacks can have on grids that differ significantly. In the PST-16 system, the attack mainly affects rotor angle instability in area C and voltages, whereas in the IEEE 39-Bus system, the main impact is on frequency, driven by the inertia of generation in the model.

It is important to note that these results correspond to a worst-case scenario where demand is high and the attack targets only three closely located electrical nodes. The success ratio and impact of the attack are expected to be lower when the attack is distributed across a greater number of nodes (while maintaining the same botnet size), when demand is low (as more line capacity is available and the relative attack size per botnet is smaller), or when distant nodes are affected (e.g., one node per area in the PST-16 model). Both the size and location of the attack influence the system's survival, determining whether the attack can destabilise or not the power system. For instance, while distributing the attacks across different locations may reduce frequency-stability issues, it could exacerbate voltage-stability problems.

4 Potential Countermeasures

While completely eradicating the risk of a MaDIoT-type attack is unfeasible, implementing certain measures might substantially reduce its likelihood or, at least, mitigate its negative effects.

Enhancing the security of IoT devices fundamentally starts with their design. Consumer-level IoT products often lack substantial security features, a problem exacerbated by the general users' insufficient understanding of effective cybersecurity practices necessary for device protection [2].

In 2024, the European Cybersecurity Certification scheme (EUCC) [14] in Europe introduced baseline criteria for certifying devices from a cybersecurity perspective. This encompasses adherence to global standards like the Common Criteria for Information Technology Security Evaluation (ISO 15408), appointing an independent accreditation entity, ensuring that certification holders can identify and report security vulnerabilities, having well-defined procedures for managing and disclosing vulnerabilities, and establishing a system of penalties for noncompliance with security requirements.

To reduce the likelihood of being targeted in a MaDIoT attack, software and firmware updates on IoT devices must be verifiable through cryptographic means and allow for roll-back updates if needed. Implementing user authentication and role-based access control is essential for IoT devices, alongside data encryption whenever feasible.

Since Distribution System Operator (DSO)s manage the systems in which high-wattage IoT devices are predominantly connected, they should aim to develop and deploy intrusion detection and prevention systems that utilise cutting-edge techniques and technologies [28]. These may include deep learning [39], data-driven algorithms [25], edge computing [36] or preventive algorithms designed to prevent line failures [38]. Previous research has also recommended mitigation strategies such as cyber-resilient economic dispatch [9], the use of Stackelberg games for system reconfiguration [27], or implementing adaptive protections capable of identifying and counteracting attacks [6].

Moreover, it is crucial for DSOs, TSOs and other stakeholders, such as DER operators, to collaborate to enhance systemic resilience and reduce the impact of attacks. This collaboration relies on comprehensive real-time data sharing and a coordinated response to foster collective situational awareness and mitigate the risk of cascading failures. In this context, European regulations are progressively adapting to promote initiatives for information exchange and communication of cyber incidents, notably through the Network and Information Security 2 (NIS2) Directive [10]. Specifically addressing the electricity sector, the Network Code on Cybersecurity [11] (adopted in 2024) outlines baseline cybersecurity requirements, cross-border risk management strategies, cybersecurity controls, a system for sharing cybersecurity information, and sets out roles and responsibilities for involved parties, among other provisions.

5 Conclusions

The proliferation of high-wattage IoT devices at the consumer level introduces a novel attack vector, as these devices are distributed across numerous nodes within power systems. Understanding the potential impact of MaDIoT attacks on various systems under different conditions is crucial as the prevalence of these devices increases.

This chapter has examined and compared the effects of MaDIoT attacks on two models: the PST-16, which is a simplified European model, and the IEEE 39-Bus model, representing New England's power system. The study found that attack success rates vary widely depending on the targeted model. For the IEEE 39-Bus system, success rates reached up to 100%, while for the PST-16 system, the maximum success rate was around 30%. Nevertheless, further analysis of scenarios with high impact revealed that a higher success rate does not necessarily equate to a greater impact. The PST-16 system, which is larger and has more resources, can better accommodate the increased load from an attack. Still, a severe attack on this system could culminate in a blackout affecting 20% of the initial demand, having an impact exceeding the attack itself. On the other hand, in the IEEE 39-Bus system, the high-impact scenario assessed showed an impact matching the magnitude of the attack.

One significant limitation of this type of studies is the omission of the dynamics and protection mechanisms of large electricity distribution systems linked to transmission networks. To overcome this, it would be essential to create a combined model of transmission and distribution systems, which would demand considerable computational power. Future studies should explore MaDIoT attacks while taking into account both the transmission and distribution systems. Furthermore, integrating automatic or manually prompted operator actions, like re-dispatching to avert line overloads and the resulting cascading failures, offers another promising research line.

Acknowledgements This work was developed within the framework of the eFORT Project. This project has received funding from the European Union's Horizon Europe Research and Innovation programme under Grant Agreement No 101075665. Views and opinions expressed are those of the author(s) only and do not necessarily reflect those of the European Union or CINEA.

References

1. Samrat Acharya, Yury Dvorkin, and Ramesh Karri. Public plug-in electric vehicles+ grid data: Is a new cyberattack vector viable? *IEEE Transactions on Smart Grid*, 11(6):5099–5113, 2020.
2. Omar Alrawi, Chaz Lever, Manos Antonakakis, and Fabian Monrose. SoK: Security evaluation of home-based IoT deployments. In *2019 IEEE Symposium on Security and Privacy (SP)*, pages 1362–1380, 2019.
3. Sajjad Amini, Fabio Pasqualetti, and Hamed Mohsenian-Rad. Dynamic load altering attacks against power system stability: Attack models and protection schemes. *IEEE Transactions on Smart Grid*, 9(4):2862–2872, 2018.

4. Manos Antonakakis, Tim April, Michael Bailey, Matt Bernhard, Elie Bursztein, Jaime Cochran, Zakir Durumeric, J. Alex Halderman, Luca Invernizzi, Michalis Kallitsis, Deepak Kumar, Chaz Lever, Zane Ma, Joshua Mason, Damian Menscher, Chad Seaman, Nick Sullivan, Kurt Thomas, and Yi Zhou. Understanding the mirai botnet. In *26th USENIX Security Symposium (USENIX Security 17)*, pages 1093–1110, Vancouver, BC, August 2017. USENIX Association.
5. C. Arderne, C. Zorn, C. Nicolas, and E. E. Koks. Predictive mapping of the global power system using open data. *Scientific Data*, 7(1):19, 2020.
6. Anjana Balabhaskara, Sunandan Adhikary, Ipsita Koley, Soumyajit Dey, and Ashish R Hota. Adaptive protection of power grids against stealthy load alterations. *2024 ACM/IEEE 15th International Conference on Cyber-Physical Systems (ICCPS)*, 00:285–286, 2024.
7. Adam Borys, Abu Kamruzzaman, Hasnain Nizam Thakur, Joseph C. Brickley, Md L. Ali, and Kutub Thakur. An evaluation of iot ddos cryptojacking malware and mirai botnet. In *2022 IEEE World AI IoT Congress (AIIoT)*, pages 725–729, 2022.
8. Clementina Bruno, Luca Guidi, Azahara Lorite-Espejo, and Daniela Pestonesi. Assessing a potential cyberattack on the Italian electric system. *IEEE Security & Privacy*, 13(5):42–51, 2015.
9. Zhongda Chu, Subhash Lakshminarayana, Balarko Chaudhuri, and Fei Teng. Mitigating load-altering attacks against power grids using cyber-resilient economic dispatch. *IEEE Transactions on Smart Grid*, 14(4):3164–3175, 2023.
10. European Commission. Eu directive 2022/2555 (nis 2 directive), December 2022.
11. European Commission. Eu electricity supply—sector-specific rules on cybersecurity (network code), October 2023.
12. Adrian Dabrowski, Johanna Ullrich, and Edgar R. Weippl. Grid shock: Coordinated load-changing attacks on power grids: The non-smart power grid is vulnerable to cyber attacks as well. In *Proceedings of the 33rd Annual Computer Security Applications Conference*, pages 303–314. ACM, 2017.
13. eFORT Project. D2.2 Cascading effects analysis and related actions to increase resilience, August 2025.
14. European Union Agency for Cybersecurity (ENISA). Cybersecurity—security requirements for ICT product certification, October 2023.
15. Maldon Patrice Goodridge, Subhash Lakshminarayana, and Christopher Few. Analysis of load-altering attacks against power grids: A rare-event sampling approach. In *2022 17th International Conference on Probabilistic Methods Applied to Power Systems (PMAPS)*, pages 1–6. IEEE, 2022.
16. Maldon Patrice Goodridge, Subhash Lakshminarayana, and Alessandro Zocca. Uncovering Load-Altering Attacks Against N-1 Secure Power Grids: A Rare-Event Sampling Approach. *IEEE Transactions on Power Systems*, pages 1–13, 2024.
17. Maldon Patrice Goodridge, Alessandro Zocca, and Subhash Lakshminarayana. Analysis of cascading failures due to dynamic load-altering attacks. In *14th IEEE International Conference on Communications, Control, and Computing Technologies for Smart Grids (IEEE SmartGridComm 2023)*, 2023.
18. Hassan Haes Alhelou, Mohamad Esmail Hamedani Golshan, Takawira Cuthbert Njenda, and Nikos D. Hatziargyriou. An Overview of UFLS in Conventional, Modern, and Future Smart Power Systems: Challenges and Opportunities. *Electric Power Systems Research*, 179:106054, 2020.
19. Bing Huang, Alvaro A Cardenas, and Ross Baldick. Not everything is dark and gloomy: Power grid protections against iot demand attacks. In *28th USENIX security symposium (USENIX Security 19)*, pages 1115–1132, 2019.
20. Anastasis Keliris, Charalambos Konstantinou, Marios Sazos, and Michail Maniatakos. Open source intelligence for energy sector cyberattacks. In Dimitris Gritzalis, Marianthi Theocharidou, and George Stergiopoulos, editors, *Critical Infrastructure Security and Resilience: Theories, Methods, Tools and Technologies*, pages 261–281. Springer International Publishing, 2019.

21. Heetae Kim, David Olave-Rojas, Eduardo Alvarez Miranda, and Seung-Woo Son. In-depth data on the network structure and hourly activity of the central chilean power grid. *Scientific Data*, 5(1):180209, 2018.
22. Pavel Kozak, Ivo Klaban, and Tomáš Slajs. Industroyer cyber-attacks on Ukraine's critical infrastructure. In *2023 International Conference on Military Technologies (ICMT)*, pages 1–6, 2023.
23. Subhash Lakshminarayana, Sondipon Adhikari, and Carsten Maple. Analysis of IoT-Based Load Altering Attacks Against Power Grids Using the Theory of Second-Order Dynamical Systems. *IEEE Transactions on Smart Grid*, 12(5):4415–4425, September 2021.
24. Subhash Lakshminarayana, Yexiang Chen, Carsten Maple, Andrew Larkins, Daryl Flack, Christopher Few, Kenny Awuson-David, and Anurag K. Srivastava. Threats to power grid operations from the demand-side response ecosystem: A comprehensive overview. *IEEE Industrial Electronics Magazine*, PP(99):2–12, 2025.
25. Subhash Lakshminarayana, Juan Ospina, and Charalambos Konstantinou. Load-Altering Attacks Against Power Grids Under COVID-19 Low-Inertia Conditions. *IEEE Open Access Journal of Power and Energy*, 9:226–240, 2022.
26. Mahrukh Mahrukh and Mini Shaji Thomas. Load Altering Attacks- a Review of Impact and Mitigation Strategies. In *2023 International Conference on Recent Advances in Electrical, Electronics & Digital Healthcare Technologies (REEDCON)*, pages 397–402, May 2023.
27. Sajjad Maleki, Subhash Lakshminarayana, Charalambos Konstantinou, and E Veronica Belmaga. Distribution system reconfiguration to mitigate load altering attacks via stackelberg games. *arXiv*, 2024.
28. Sajjad Maleki, Shijie Pan, Subhash Lakshminarayana, and Charalambos Konstantinou. Survey of load-altering attacks against power grids: Attack impact, detection and mitigation. *arXiv*, 2024.
29. Wided Medjroubi, Ulf Philipp Müller, Malte Scharf, Carsten Matke, and David Kleinhans. Open data in power grid modelling: New approaches towards transparent grid models. *Energy Reports*, 3:14–21, 2017.
30. Amir-Hamed Mohsenian-Rad and Alberto Leon-Garcia. Distributed internet-based load altering attacks against smart power grids. *IEEE Transactions on Smart Grid*, 2(4):667–674, 2011.
31. Afroz Mokarim, Giovanni Battista Gaggero, and Mario Marchese. Evaluation of the impact of cyber-attacks against electric vehicle charging stations in a low voltage distribution grid. In *14th IEEE International Conference on Communications, Control, and Computing Technologies for Smart Grids (IEEE SmartGridComm 2023)*, 2023.
32. Juan Ospina, Xiaorui Liu, Charalambos Konstantinou, and Yury Dvorkin. On the Feasibility of Load-Changing Attacks in Power Systems During the COVID-19 Pandemic. *IEEE Access*, 9:2545–2563, 2021.
33. Néstor Rodríguez-Pérez, Javier Matanza Domingo, Lukas Sigrist, Jose Luis Rueda Torres, and Gregorio López López. Confronting the threat: Analysis of the impact of madiot attacks in two power system models. *Energies*, 16(23):7732, 2023.
34. José Luis Rueda, Jaime C. Cepeda, István Erlich, Abdul W. Korai, and Francisco M. Gonzalez-Longatt. Probabilistic approach for risk evaluation of oscillatory stability in power systems. In Francisco M. Gonzalez-Longatt and José Luis Rueda, editors, *PowerFactory Applications for Power System Analysis, Power Systems*, pages 249–266. Springer International Publishing, 2014.
35. Tohid Shekari, Alvaro A Cardenas, and Raheem Beyah. Madiot 2.0: Modern high-wattage iot botnet attacks and defenses. In *31st USENIX Security Symposium (USENIX Security 22)*, pages 3539–3556, 2022.
36. Bibek Shrestha and Hui Lin. Data-centric edge computing to defend power grids against iot-based attacks. *Computer*, 53(5):35–43, 2020.
37. Saleh Soltan, Prateek Mittal, and H Vincent Poor. Blackiot: Iot botnet of high wattage devices can disrupt the power grid. In *27th USENIX Security Symposium (USENIX Security 18)*, pages 15–32, 2018.

38. Saleh Soltan, Prateek Mittal, and H. Vincent Poor. Protecting the grid against MAD attacks. *IEEE Transactions on Network Science and Engineering*, 7(3):1310–1326, 2020.
39. Lizhi Wang, Lynn Pepin, Yan Li, Fei Miao, Amir Herzberg, and Peng Zhang. Securing power distribution grid against power botnet attacks. In *2019 IEEE Power & Energy Society General Meeting (PESGM)*, pages 1–5, 2019.
40. Christos Xenofontos, Ioannis Zografopoulos, Charalambos Konstantinou, Alireza Jolfaei, Muhammad Khurram Khan, and Kim-Kwang Raymond Choo. Consumer, commercial, and industrial IoT (in)security: Attack taxonomy and case studies. *IEEE Internet of Things Journal*, 9(1):199–221, 2022.
41. Ioannis Zografopoulos, Juan Ospina, Xiaorui Liu, and Charalambos Konstantinou. Cyberphysical energy systems security: Threat modeling, risk assessment, resources, metrics, and case studies. *IEEE Access*, 9:29775–29818, 2021.

ATLANTIS LSP#1 Safety and Security Risks: Assessment Approach Testing, and Results Obtained

Marko Gerbec, Denis Čaleta, Jolanda Modic, Abla Mimi Edjossan-Sossou, Emiliano Altobelli, Dario Ogrizović, and Gabriele Giunta

Abstract Critical Infrastructures (CIs) are exposed to a number of safety and security hazards and threats that can compromise their operations and the necessary services and goods provided to society. The EU has funded in the past years a number of Critical Infrastructure Projects (CIP/INFRA) aiming to improve the security of CIs. The ATLANTIS project builds on past projects, extending the risk and resilience assessment and management scope from the individual CI to the consideration of their (inter)dependencies, also cross-CI domains and cross-border within the EU member states. The chapter covers the ATLANTIS approach to the systemic risk assessment involving its system requirements (architecture), sensor data management, CI-to-CI dependency modelling, risk modelling methods used, and organisation and results of case studies (Large-Scale Pilots—LSPs) where the methods and tools developed had been tested at the CI operator partners via simulations of the specific threat

M. Gerbec (✉)
Jožef Stefan Institute, Ljubljana, Slovenia
e-mail: marko.gerbec@ijs.si

D. Čaleta · J. Modic
Institute for Corporative Security Studies, Ljubljana, Slovenia
e-mail: denis.caleta@ics-institut.si

J. Modic
e-mail: jolanda.modic@ics-institut.si

A. M. Edjossan-Sossou
Sixense Engineering, Nanterre, France
e-mail: abla-mimi.edjossan-sossou@resallience.com

E. Altobelli
FS Technology SpA, Rome, Italy
e-mail: e.altobelli@fstechnology.it

D. Ogrizović
Faculty of Maritime Studies, University of Rijeka, Rijeka, Croatia
e-mail: dario@pfri.uniri.hr

G. Giunta
Engineering Ingegneria Informatica S.p.a., Palermo, Italy
e-mail: Gabriele.Giunta@eng.it

J. Soldatos and H. Abie (eds.), *Technology-Enabled Resilience: Innovations in Critical Infrastructure Protection*, https://doi.org/10.1007/978-3-032-25917-2_6

scenarios. The results obtained using structured questionnaires suggest that the CI operators highly value the enabled effective communication across CI organisations, the usefulness of the shared situational awareness, as well as obtaining access to the innovative solutions being thus available both cross-CI sectors and cross-national borders.

Keywords EU · Critical infrastructure · Sensors · Resilience · Risk assessment · Situational awareness · Case studies · Scenarios · Simulations

1 Introduction

Critical Infrastructures (CIs) are increasingly interconnected and are becoming targets of new categories of hybrid threats and attacks, which come with significant challenges. These include hybrid threats and risks empowered by technological innovations such as the (mis)use/attack of Artificial Intelligence (AI) tools/methods (e.g., risk of adversarial attacks against AI models, vulnerability of digital assets that contain AI, risks from unprecedented effects when AI fails) and the rapid development of other sophisticated technologies. In a cross-sectoral and cross-border domain, the need for effective collaboration and cooperation among all the involved stakeholders (CI operators, public authorities, standardisation bodies, policy makers, citizens) becomes imperative. Disruptions in one domain, e.g., energy, telecommunication, or transportation, could rapidly propagate into others, generating systemic consequences. It could have become worse considering the lack of common standards, as it can lead to fragmentation, inefficiencies, and vulnerabilities. Moreover, given that a significant portion of European CIs is owned and operated by private entities, finding a balance between the overarching need for safety and security of CIs is one of the most challenging and complex tasks the relevant stakeholders are facing.

During the last ten (10) years, the European Commission has funded several CIP/INFRA research projects: DEFENDER [1], InfraStress [2], STOP-IT [3], SATIE [4], SecureGAS [5], PRAETORIAN [6], FINSEC [7], 7SHIELD [8], SAFECARE [9], and RESISTO [10]. These projects were aimed at understanding sector-specific risks and security correlations between infrastructures within a given sector or domain, and even more narrowly, within the context of a single country. Limited research has been conducted in such projects on large-scale, transnational, and cross-domain coordinated attacks. More importantly, large-scale vulnerability assessment and risk analysis of CI at the systemic level, facing the risk of the entire system's collapse, has never been addressed before. While threats (both physical and cyber) have been handled with different levels of effectiveness, systemic risks derived from major hazards and complex threats have not been sufficiently considered and understood. They can be cyber-physical, coordinated (among different actors, even in different countries), mixed (using different tactics), disruptive (leading to the collapse of entire systems, sectors, or regions), unexpected, subversive, and difficult to identify early.

They can also attribute it to novel criminal business models, such as crime-as-a-service. Growing major natural hazards (e.g., floods, wildfires), which are often unexpected and unpredictable, can create disruption to CIs, thus affecting wider functions of our society. It was only in recent projects, beginning with PRECINCT [11], that a serious analysis of the broader cross-sectoral perspective of cascading effects and interdependencies between different infrastructures across various domains was undertaken. The scope of the assessment was significantly expanded to include a regional perspective on potential risks affecting the resilience and continuity of key infrastructure operations. In addition to exploring possible cascading effects, the foundations for the need to analyse systemic risks also began to emerge. This could require long-term analysis and foresight. It is important to note that the systemic risks do not follow normal risk distribution strongly, but tend to propagate with extremely high rates, carrying a high likelihood to trigger anomaly, instability, and potentially catastrophic, cascading effects to nested or interconnected systems, organisations, regional economy, countries, or even at the EU level. A fundamental challenge to governing systemic risk in such a networked scenario is to understand the system as a complex network of individual and institutional actors with different and often conflicting interests, values, and views.

In order to answer the increasing demand of safety, security and resilience improvement of interconnected and interdependent CIs, the following set of actionable insights should be proposed at a strategical level: (a) proper strategies to protect public and private interests of CI owners without undermining corporate rights, competitiveness, or innovation must be drafted; (b) collaborative mechanisms must be improved for enhancing resilience, particularly in light of emerging threats like large-scale systemic risks (such as pandemics or climate-induced events); and (c) the adoption of standardised protocols, procedures, and benchmarks, organisations which can ensure very high levels of protection, enabling interoperability across systems, assets and among CIs from different sectors.

The European Commission drew important insights from the implementation of Critical Infrastructure (INFRA) and Cyber Security (CS) projects. These insights were later reflected in the adoption of two key directives on the resilience of critical entities (CER Directive) [12] and on measures for a high common level of cybersecurity across the Union (NIS2 Directive) [13] which must be understood as a unified whole, promoting a more holistic and integrated approach to EU wide resilience, with each representing an inseparable part of the broader framework. This also represented a clear response by relevant institutions to the emergence of systemic and hybrid risks that could no longer be addressed through the tools of a single-sector approach. Part of the effort to find such responses was also the funding of the ATLANTIS [14] project, which, through three highly complex pilot deployments and carefully selected partners, began to assess whether the mechanisms put in place could be effective in identifying and managing systemic risks. In one of the three extensive and collaborative Large-Scale Pilot projects (LSPs), four interdependent and neighbouring countries, France, Italy, Slovenia, and Croatia, are participating in one of the most complex and ambitious pilots ever undertaken in Critical Infrastructure Protection (CIP) initiatives. This pilot represents a unique opportunity to explore and enhance cross-border cooperation, resilience strategies, and risk mitigation measures through the perspective of different interdependent CI sectors, to diversify between various types of organisations, such as governmental or CI operators, and between different use case threat scenarios implications.

The project places a strong emphasis on innovative approaches to systemic risk assessment for CI, recognising the evolving nature of safety and security threats in an increasingly interconnected environment. By integrating cutting-edge methodologies, advanced technologies, and real-world testing scenarios, the pilot seeks to redefine risk evaluation models, ensuring more robust and adaptive protection strategies.

Through comprehensive assessment, testing, and validation of results, the project aims to uncover new insights into the vulnerabilities and interdependencies of critical sectors. The findings will serve as a foundation for enhanced risk management frameworks, supporting both national and international efforts to strengthen infrastructure resilience against emerging threats.

In the rest of the chapter, the ATLANTIS approach will be presented considering its system requirements, data management, CI dependency modelling, and a set of risk assessment methods used. Next, the testing and validation of the approach and tools prepared in a case study will be presented, including the testing and validation results obtained, followed by conclusions.

2 Approach

2.1 The System Requirements

From a logical perspective, the technologies to process all collected data and provide CI-related services for threat detection, cross-sectoral situation awareness, decision making, and incident response are featured by a three-layer architecture, as shown in Fig. 1.

- The Detection and Adaptation Layer (Layer 1) is the CI-specific Intrusion Detection and Prevention System (IDS/IPS). It is responsible for CI-specific information gathering and CI observation, including physical (i.e., sensors, video surveillance, drones), cyber (i.e., complex data from PLC, SCADA, IDC and network connectivity systems such as routers and firewalls) and human sensors (i.e., HiVIC).
- Awareness and Mitigation Layer (Layer 2) is responsible for local incidents processing, systemic risks patterns extraction, situation awareness, threats prediction/early-detection and mitigation proposition at variable timeframes, while.
- Federated Cross-CI Layer (Layer 3) is a collaborative Cross-CI Knowledge Sharing, Risk Assessment, State Awareness and Incidents Mitigation (CCI-SAAM) platform between collaborative cross-border and cross-domain CIs. On the right side, the ATLANTIS platform offers interconnection with external

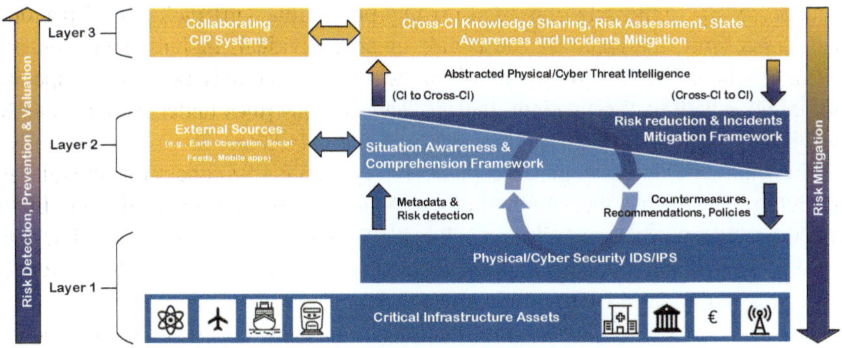

Fig. 1 ATLANTIS high level logical architecture view

systems (e.g., other IDS and CI Security Systems, weather forecast, Earth Observation systems) along with external Information Sharing and Analysis Center (ISAC) systems.

The logical view of the ATLANTIS architecture is generic, focusing on fundamental, domain-independent concepts such as risk detection, situational awareness, and risk mitigation. By abstracting these core functions, it creates a logical framework that can be applied across a wide range of CI sectors, regardless of the specific technologies or systems in use. This abstraction allows for broad applicability without being tied to toolchains or platforms. At the same time, it holds a modular design that separates different concerns—like cyber-physical security, situational awareness and risk mitigation—while providing clear interactions between them. This separation of concerns ensures that each sub-framework and its components can be tailored or swapped based on the specific needs of the CI sector, making it adaptable without disrupting the overall architecture. Each layer represents a logical function that can be customised according to the unique operational or regulatory requirements of different CIs. Further, since the view is logical, it is not tied to any specific technological solutions or vendor products.

Detection and Adaptation Layer

It plays a crucial role within the ATLANTIS architecture, serving as the first checking point for data collection from a wide range of sensors and tools that are deployed at CI to monitor cyber and physical environments. Specifically, the detection layer identifies potential threats or anomalies, while the Adaptation Layer ensures that the data collected from different sensors and tools are normalised into the Intrusion Detection Message Exchange Format (IDMEF) [15] format to be published for further analysis. The detection layer is responsible for capturing data from the monitoring systems implemented in different critical infrastructures. These systems employ various types of sensors that monitor both the cyber environment (e.g., network activity) and the physical environment (e.g., environmental conditions). The primary goal of this layer is to detect anomalies or threats that could compromise the CI integrity. There are several key types of sensors involved:

- **Cyber sensors**: include tools like firewalls, network monitoring systems, and intrusion detection systems that identify unusual network activity, potential intrusions, or unauthorised access. For instance, multiple CIs use cyber sensors to monitor and protect their network infrastructure, which is critical for maintaining the security of their operations. Some cyber sensors used in ATLANTIS are the following:

    (a) Health cybersecurity monitoring tools.
    (b) Banking cybersecurity monitoring tools.
    (c) LUR (Port of Rijeka) Intrusion Prevention System.
    (d) LUK (Port of Koper) Intrusion Prevention System.

- **Physical sensors**: These sensors gather data from the physical world. Examples include air quality sensors and radiological sensors that monitor environmental conditions like CO_2 levels or radiation. These sensors play a vital role in detecting changes in the environment that could signal a physical threat to CIs. Physical sensors used in ATLANTIS are radiological and air quality sensors.
- **Human sensors**: These sensors are represented by the Human-in-Vicinity (HiVIC) tool and app, which gather real-time information from people on the ground during incidents. This information is then sent back to the framework, contributing to the situational awareness needed to assess the severity of incidents and make informed decisions.
- **Geospatial platforms**: These include Global Navigation Satellite System (GNSS) and Earth Observation (EO) sources, which provide satellite data, modelled climate data, GNSS alerts, and geographical information that can be used to monitor large-scale environmental changes or physical events that could affect critical infrastructure.

While the detection layer captures data from these diverse sensors, the Adaptation Layer transforms this data into a standardised format. Given that data is collected in different formats from various sources, the Adaptation Layer ensures it can be processed and shared uniformly across the ATLANTIS framework. The Adaptation Layer components convert raw data into IDMEF messages, enabling real-time analysis and situational awareness.

This standardisation applies to cyber, physical, and human data, as well as GNSS alerts:

- Cyber data from network security systems is transformed into IDMEF messages to ensure real-time detection of cyber threats.
- Physical data from sensors such as those monitoring air quality or environmental conditions are also adapted into IDMEF messages. This allows the platform to assess physical incidents like environmental hazards alongside potential cyber threats, ensuring a comprehensive risk assessment.

- Similarly, human-produced data and GNSS alerts are adapted into IDMEF messages in order to be analysed in the following Layer 2.

Awareness and Mitigation Layer

The ability to identify the perceived threat is a feature of the ATLANTIS Awareness and Comprehension Framework (ACF). It's responsible for constructing a situational overview by applying correlation rules as well as supervised and unsupervised machine learning algorithms to the input data. In particular, systemic and continuous risk analysis, along with classification, predictive, and clustering models, are crucial for the enhancement of the overall situation picture. The ATLANTIS ACF focuses on the ongoing assessment and monitoring of risks within CI networks by integrating a Complex Event Processor (CEP) to process alerts coming from the Adaptation Layer. This component processes real-time and historical data to detect patterns and anomalies indicative of potential threats and enables the system to respond swiftly to emerging risks by correlating events across diverse data streams.

The use of interdependency graph analysis enhances the ACF capability to understand and visualise the relationships between different components of the infrastructure. This analysis helps to identify vulnerabilities and interdependencies that could lead to cascading failures and systemic disruptions. By combining these advanced analytical techniques with real-time data from various sources, including external systems and intelligence feeds, ACF enables proactive risk mitigation. It ensures CI operators are well-informed and prepared to respond effectively to emerging threats and evolving risk scenarios. This approach empowers organisations to maintain a resilient and secure infrastructure in the face of dynamic and persistent challenges.

The enriched situational picture is then utilised by the Risk Reduction and Incident Mitigation (RRIM) framework to gather all the relevant information for implementing the containment, mitigation, and emergency plan application phase to minimise the damage associated with the incident. In this phase, policies and strategies derived from domain study and agreed upon with end-users concerning the specific risk will be applied to support risk reduction and the implementation of countermeasures. The Adaptation Layer at the end of the risk management process aims to reduce the negative impacts and vulnerabilities associated with the specific risk or threat identified. The adaptation phase can involve modifying existing processes and structures in the real world to make them more resilient to risk management and to prevent damage.

Federated Cross-CI Layer

In this layer of the ATLANTIS logical architecture (Fig. 1), the Cross-domain, Cross-CI, Cross-border knowledge Sharing, risk Assessment, threat analysis and countermeasures Mitigation (CCI-SAAM) Coordinated Framework is located. The ATLANTIS CCI-SAAM Framework acts as a policy-based information broker of cyber/physical/operational threat/risk intelligence data among interested parties at cross-border, cross-CI level, operated by a European Institution or Competent Authority [13].

It aims at exchanging physical/cyber threat intelligence (PCTI) with all ATLANTIS-enabled CIs as well as external, collaborating CIP systems, as well as identifying and timely forecasting coordinated and/or cascading systemic risks affecting the CIs in Europe. In this way, CI operators and stakeholders can effectively collaborate, assessing risks and threats, sharing information and defining mitigation strategies. In addition, the CCI-SAAM Framework integrates capabilities for systemic risk analysis, threat intelligence, explainable AI, and cooperative response to strengthen resilience against cascading failures. Lastly, the CCI-SAAM Framework facilitates seamless communication and coordinated action across CI domains and national borders.

In the CCI-SAAM Framework context, CI Owners and CI Operators serve as Data Owners, which are entities that share information. They need only define the information to be shared and agree on processing results shared with relevant Data Receivers, which are entities that obtain information. Data Receivers may include, e.g., CERTs, CSIRTs, and other relevant national or EU-level security organisations, such as ENISA. Therefore, the ATLANTIS CCI-SAAM Framework acts as the Data Facilitator/Processor. It could be depicted as a Pan-European platform designed for sharing PCTI data and alerts among CIP stakeholders, CERTs, and CSIRTs, which could be defined as a group of entities, also called Trusted Circle (TC), that generate threat intelligence information and agree to share it with the CCI-SAAM Framework to benefit from the results of its data processing. A TC, therefore, consists of entities committed to providing input data for CCI-SAAM Framework, which then uses this data to fuel its data processing.

In this light, the CCI-SAAM Framework employs various data pipelines in order to:

- Suggest mitigation actions and policy recommendations.
- Identify potential targets of attacks, considering:
  - (a) Geographical locations/areas at risk.
  - (b) Timeframes for potential attacks.
  - (c) Administrative entities likely to be impacted.
- Detect coordinated or cascading attacks and alert relevant Data Receivers.

The CCI-SAAM Framework consists of **backbone** and **peripheral** components:

- Backbone within the context of the CCI-SAAM Framework is interpreted as a set of components that provide functionality related to the data management and the relevant APIs (utilising Trusted Automated Exchange of Intelligence Information (TAXII) [16]/Malware Information Sharing Platform (MISP) [17] server deployments), data model conversion (e.g., to/from IDMEF), and data persistence. Some of the backbone components of CCI-SAAM are the following: PCTI & Cross-CI Systemic Risk analysis, Systemic Risk Mitigation Recommendations Extraction, Risk Awareness & Mitigation Recommendations Sharing Engine.
- Peripheral within the context of the CCI-SAAM Framework is interpreted as a set of components that offer functionality at a higher level, applying business

logic to process, closer to the end-user. Such components include eXplainable Artificial Intelligence (XAI) functionality, the components that apply Machine Learning (ML) technologies on data from the CIs, the components that perform risk analysis, and the components that visualise the functionality provided by the CCI-SAAM components.

The aforementioned components of the CCI-SAAM Framework interact through standardised interfaces and protocols to enable seamless sharing of risk intelligence, situational awareness data, and mitigation recommendations. This allows proactive anticipation of threats, coordinated response plans, and accelerated information exchange across organisational boundaries. In addition, automated workflows and human-in-the-loop processes facilitate smooth handovers between analysis, assessment, planning, and implementation stages. Overall, the component interactions within CCI-SAAM Framework create an integrated environment for cross-CI risk management.

Moreover, for modelling the communication between the CCI-SAAM Framework and the local ATLANTIS-enabled CIs, the OASIS-standardised Structured Threat Information eXpression (STIX) [18] data language has been selected, coupled with the Trusted Automated eXchange of Intelligence Information (TAXII) data model. Using the STIX object notation, it is possible to model the relationships between assets, vulnerabilities, threats and countermeasures. As a result, the information regarding identified risks at the local level effectively allows CCI-SAAM Framework to combine information from multiple CIs, elaborate on the generated insights and come up with actionable alerts and risk mitigation/alleviation strategies so that the residual risk of the identified risk may be minimised. This data (could be related to reports on both physical and cyber threat intelligence—PCTI—data) may contain information on CI-related identified risks, threats, vulnerabilities, attacks, attack campaigns, as well as identified courses of action (mitigation strategies) that were taken at the local CI level.

2.2 *Data Management*

The general strategy for data management in ATLANTIS relies on the principles pointed out in the "Open Research Data and Data Management Plans" of the European Research Council (ERC). Specifically, in the ATLANTIS project context, the Data Management uses a FAIR approach [19].

Concerning the processing of personal data, the rules relating to the protection of natural persons with regard to the processing of data and the free movement of such data are laid down in the General Data Protection Regulation (GDPR) [20], which also contains obligations regulating the security of personal data. The legislator of the new legal framework for the protection of critical infrastructures at EU level was well aware that security incidents at a critical infrastructure are most likely to adversely affect its processing of personal data. Therefore, CER Directive [12] and especially the NIS2 Directive [13] include provisions on the relationship between their security obligations for critical infrastructures and those under GDPR.

Finally, under the legal framework of the CER Directive and the NIS2 Directive, critical infrastructures need to be regularly monitored by competent authorities, but also, they should be observed by society for their ethical and legal behaviour. The ATLANTIS project has integrated ethics into its project plan and has established an external Ethics Board for the ethics review of all tasks and deliverables contributing to the development of the ATLANTIS system. The necessary oversight of the development process requires each tool and functionality, as well as the entire ATLANTIS framework, to be reviewed (auditability) and every decision concerning the technical design to be explained and justified (explicability) on a technical level. Each tool's and feature's functionality and that of the entire ATLANTIS system need to be verifiable and reproducible in all situations (reproducibility).

2.3 Dependencies Modelling

CIs are operated by organisations that can be understood as systems with some specific business goal/function to fulfil. CIs/systems usually do not operate in complete isolation from each other but are highly interconnected, both across borders and sectors. The same applies to the internal structure of each system, which typically consists of some hierarchical elements or assets in general. Analysis can go to any level of detail (in both directions, top-down and bottom-up), and we can generally consider CIs as a system of systems.

The ATLANTIS approach to modelling dependencies among and within CIs follows the hierarchical asset structure[1] adopted from ISO 14224:2016 [21]. In practice, CI can be described either as a set of assets at various levels or in the broader context of CI sectors [12]. However, a balance should be obtained between the necessary analytical complexity and the risk of excessive modelling efforts. The relationships or dependencies among assets should be clearly defined and described. The one-to-one unidirectional dependencies among pairs of assets usually apply collectively; thus, direct and indirect bidirectional dependencies lead to the term **interdependency** (interdependencies), which is a widely used term in the field. Further, the ATLANTIS approach adopts the basic interdependencies concepts from Rinaldi et al. [22], particularly those related to CI operational dimensions and sub-dimensions, as outlined in Table 1.

More detailed definitions and discussions on the types of interdependencies, as well as other relevant dimensions, are available in [22, 23], though it is important to note that these interdependency categories may potentially overlap in some cases.

The adopted data model, designed to facilitate the modelling of interdependencies between CIs and their assets, considers the assets' properties, along with one-to-one dependency pair data, as illustrated in Fig. 2.

[1] One should note that there are many definitions of the term asset.

Table 1 Examples of CI operational dimensions and sub-dimensions from Rinaldi et al. [22]

Dimensions					
Infrastructure characteristics	State of operation	Types of interdepend-dencies	Environment	Coupling and response behaviour	Type of failure
Organisational Operational Temporal Spatial	Normal Repair/ restoration Stressed/ disrupted	Physical Cyber Logical Geographic Logical Functional Policy Shared Economic	Economic Legal Technical Social/ political	Adaptive Inflexible Loose/ tight Linear/ complex	Common cause Cascading Escalating

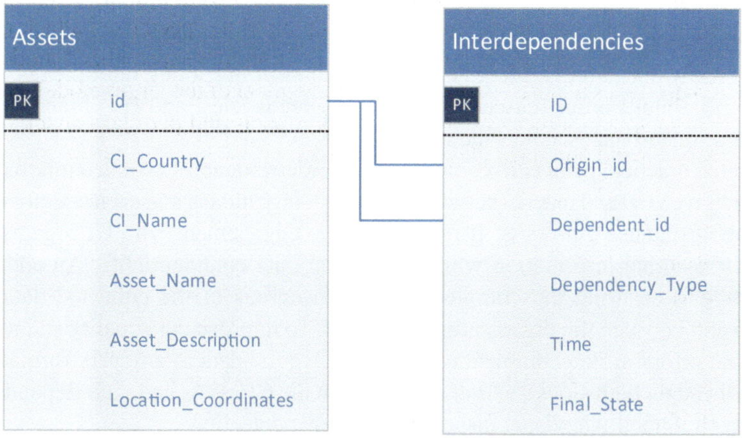

Fig. 2 UML notation of the ATLANTIS assets and interdependency data model

The asset data includes information on the country where it resides, the CI that owns it, its full name and brief description/purpose, and exact location coordinates. This structured approach enables creating a comprehensive and detailed inventory of assets within a given CI and across multiple countries. On the interdependency side, for each origin-to-dependent asset pair, the corresponding interdependency category (see Table 1) can be assigned in each direction. Additionally, the model specifies the timeframe in which the dependent asset is expected to reach its final state (e.g., incapacitation, failure, etc.) if the origin asset fails to provide its required supply or service.

This approach allows us to describe and model interdependencies both at the **macro** level (between the CIs) and the **micro** level (between specific CI assets), depending on the needs and available resources. While the data are mainly qualitative/

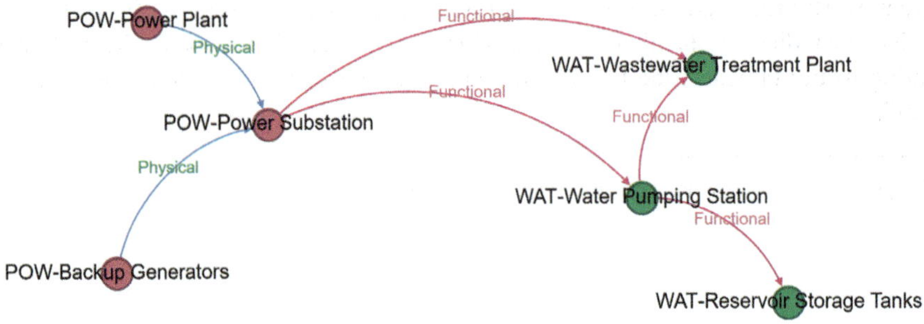

Fig. 3 An illustrative interdependency graph for an example of two CIs (POW and WAT), their assets, and corresponding interdependencies

descriptive, certain elements, such as the time and final state of a specific dependency pair, can provide information on the tight/loose linkages and potential for cascading effects (following Perrow's concepts [24]). Next, a series of origin-to-dependent pairs can effectively represent complex interdependencies among the assets, including: (i) one-to-many dependencies (e.g., a single origin asset impacting multiple dependent assets), (ii) multi-order (cascading) effects dependencies, and (iii) branching and spread of disturbances across the modelled system.

From a practical perspective, modelling interdependencies at both a micro (assets within a given CI) and macro (between multiple CIs or industrial/service sectors) level is a labour-intensive process. It requires active engagement with CI organisations to elicit accurate information while preserving data confidentiality. An additional challenge is ensuring the completeness and accuracy of the collected data, both within and between the organisations involved. To improve data quality and identify potential errors, it is recommended to (i) retrieve the data in a matrix format (with assets listed on both sides), and/or (ii) visualise the CI/assets and their dependencies as a graph to facilitate visual inspection and error detection.

An illustrative interdependency graph demonstrating six assets from two CI operators (an energy provider POW, and a water utility company WAT) along with their six one-to-one dependencies, using coloured and labelled arrows, is presented in Fig. 3.

2.4 Risk Assessment Methods

The risk assessment methodology combines established methods for risk assessment, specific hazard/threat sensing from [25–29], and the interdependency modelling approach presented in the previous sub-chapter. The methodology follows a six-step assessment process, as illustrated in Fig. 4).

The details of each step in the risk assessment methodology are presented below.

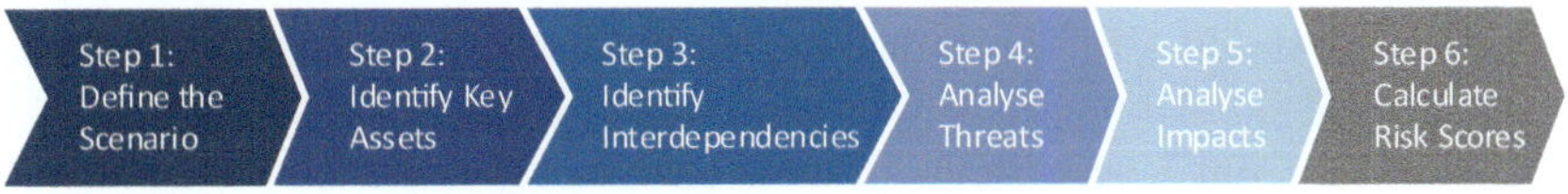

Fig. 4 Illustration of the six methodological steps

Step 1: Define the scenario. The procedure begins with the development of a credible risk scenario, built upon the prior hazard and threat identification and classification activities. This is carried out in close cooperation with the CI operators within the system/service under analysis. The scenario must feature at least one threat or hazard that directly impacts at least one CI operator, with the potential to affect other CIs via some type of interdependency. To capture the risk propagation, the scenario is broken down into sequential steps, mapping how the crisis evolves over time. For each CI involved in each stage, we analyse the following:

- What happened, and what caused this step?
- What actions are being taken by CIs in response?
- What cascading effects can be observed, and what could happen next?

This structured approach ensures that risks are not assessed in isolation, but rather within the context of cross-CI interdependencies, leading to more comprehensive risk management.

Step 2: Identify Key Assets. For each CI operator, its key assets are identified that are either directly impacted by the identified hazard/threat or are indirectly impacted due to cross-CI interdependencies. For each identified asset, we provide a brief description and analysis of the following aspects:

- What is the role and the purpose of this asset?
- What inputs does the asset require to operate?
- What does the asset provide to support other assets?

This step ensures a systematic consideration of the interdependencies, allowing for a more accurate evaluation of cascading and compounding risks.

Step 3: Identify Interdependencies. The interdependency analysis is performed at two levels. First, at macro level, considering the CI-to-CI dependencies that represent system-wide interactions. Second, at micro level, considering the asset-to-asset dependencies that reveal detailed interactions within each involved CI. The analysis is performed following the method presented in Sect. 2.3 Dependencies Modelling. In addition, possible alternatives (spares) for the failed/incapacitated assets and utility systems are considered. Once this stage is completed, the starting failures, their expected propagation, and their potential branching across the CI/asset network can be mapped and visualised via interdependency graphs (an example is illustrated in Fig. 3).

Step 4: Analyse Threats. This is the key step consisting of sub-steps linking sensor data to asset states, defining possible causes for those states, assigning probabilities to each state, identifying relevant threat categories, and assigning their likelihood and impact.

For each asset, at least one sensor must be defined that monitors its status. Common sensor types include a Status Monitoring Sensor, which is a binary sensor that reports whether an asset is "Working" (normal conditions) or "Not working" (issue detected). Another sensor type might be reporting continuous values, e.g., a sensor that reports water tank levels as a percentage of full capacity.

Next, the asset states are related to their causes. For each sensor value indicating an issue, all possible causes and originating assets to that cause are analysed (this involves information from the interdependency analysis mentioned before). Under the originating assets, we consider internal interdependencies (from Step 3) as well as external causes such as natural disasters or technical failures. The resulting states of the dependent (target) asset are, for e.g., "Normal" reflecting normal conditions, and then "Physical damage" and "No Power" representing diverse issues.

To assign probabilities to each possible state/cause, all available specific historical data, statistical trends, or other intelligence data should be used to establish realistic probabilities. This means that we consider conditional probabilities per each asset state related to some issue.

To simplify the risk analysis, probabilities are assigned under the following two assumptions. First, the sum of probabilities for all possible states/causes equals 100%. Second, in the absence of an issue, the probability of normal operation is assumed to be 100%.

Finally, each identified asset is linked to one or more relevant threat categories (see Table 2). This ensures that the risk assessment accounts not only for the asset's state but also for the nature of the originating threat/hazard.

The sum of probabilities for a given asset state across all associated threats must equal unity (1 or 100%). The normal state is not linked to any threat category. An illustrative example, based on the CI asset analysis presented in Fig. 3, is provided in Table 3.

Step 5: Analyse Impacts. In this step, we analyse the potential impacts of the CI failures across multiple impact categories. The structured approach, adapted from [39], ensures that all relevant categories are semi-quantified, aggregated, and weighted to appropriately reflect the scenario's real-world implications. For each failure scenario (i.e., for each row in Table 3), we score impacts using a scoring scale from 0 to 4 across the categories and scoring criteria presented in Table 4.

The impact categories are subject to prioritisation, as their relative importance varies across different CI sectors. The following weights and criteria are suggested to be applied at the CI level:

- **1** (low priority): The category has minimal influence on risk mitigation decisions.
- **2** (moderate priority): The category is important but balanced with other high-priority factors.

- **3** (high priority): This category is a critical factor in risk mitigation; failure would have severe implications for the business process.

Table 2 Overview of the adopted threat/hazard categories

Threat/hazard category	Brief description and references
Technology-human-organisational (THO)	Unintentional industrial site failures due to human error, technological faults, or hazardous substance releases. May include nuclear and radiological events [27, 30]
NaTech and climate-related	Natural hazards (e.g., floods) that trigger failures in CI due to weakness in THO measures. Also includes extreme weather phenomena linked to climate change [31, 32]
Physical attack	Intentional human-caused disruption, such as unauthorised access or direct attacks on CI sites (e.g., terrorist attack, sabotage) [33]
Cyber-attack	Malicious cyber intrusions or conditions that lead to asset loss or operational failures, including hacking, malware, data breaches, and system disruptions [34]
Technology trends related	Emerging disruptive technologies that could create vulnerabilities or security concerns within CI systems [35]
Foreign direct investments (FDI)	Security risks associated with foreign ownership or investment in CI, including potential denial of access, espionage, and technology leakage [36, 37]
Critical supplies (non-EU)	Risks related to supply chain dependencies on non-EU countries, potentially causing disruptions in essential materials, technology, or expertise [38]

Table 3 Example of assets, states, events, threat categories, and probabilities based on the interdependencies illustrated in Fig. 3

Asset (CI)	Sensor value	Value interpretation	Origin asset (CI)	Asset state	State prob.	Event	Threat category	Threat prob.
Power plant (POW)	Working	Normal state	/	Normal	/	Normal	/	/
	Not working	Power failure	External	Electricity unavailable	100%	Threat	THO	70%
						Threat	NaTech	30%
Power substation (POW)	Working	Normal state	/	Normal	/	Normal	/	/
	Not working	Power disruption	Power plant (POW)	Electricity unavailable	90%	Threat	THO	100%
	Not working	Power disruption	Backup generators (POW)	Backup power failure	10%	Threat	Physical attack	100%

Note / not applicable

Table 4 Adopted impact categories, scores, and criteria

Impact category	Score and criteria
1. People exposed: the number of individuals affected	0: None exposed 1: 1–50 people exposed 2: 51–250 people exposed 3: 251–1000 people exposed 4: 1001+ people exposed
2. Economic impact (repair or replacement costs): the financial burden of restoring services	0: No significant economic effect 1: Restoring cost is less than 250.000 € 2: Restoring cost is between 250.000 and 1.000.000 € 3: Restoring cost is between 1.000.000 and 10.000.000 € 4: Restoring cost is greater than 10.000.000 €
3. Economic impact (contribution to the economy): wider economic consequences	0: No significant economic effect 1: Impact on the individual critical asset's profitability is >10% 2: Impact on the organisation's profitability is >10% 3: Impact on the regional economy 4: Impact on the national economy
4. Business or service interruption: duration and severity of operational downtime	0: Critical assets could operate with minimal operational changes or repair 1: Critical assets could partially operate 2: Critical asset is shut down or unable to operate for <6 months 3: Critical asset is shut down or unable to operate for >6 months 4: Critical asset is not expected to be restored
5. Interdependencies: effects on interconnected infrastructure	0: No effect on the critical asset's normal operations 1: Critical asset is a stand-alone facility and is not interdependent on other assets; adverse effects would not extend beyond this single asset 2: Critical asset is part of a larger system; however, adverse effects would not extend beyond this single asset 3: Critical asset is part of a larger system, and at least one other asset depends on its outputs 4: Critical asset is part of a larger system, and many other assets depend on its outputs

(continued)

Table 4 (continued)

Impact category	Score and criteria
6. Criticality: the importance of the asset in maintaining essential services	0: No adverse effect 1: Minor adverse effects would occur, limited to a local environment 2: Significant adverse effects would occur, limited to a local environment 3: Significant adverse effects would occur in a wider environment 4: Significant adverse effects would occur nationally or worldwide
7. Environmental impact: potential damage to water, air, soil, and biodiversity	0: None 1: Limited damage 2: Short-term damage to a limited extension of the surrounding environment 3: Long-term damage to a limited extension of the surrounding environment or short-term damage to a significant extension of the surrounding environment 4: Permanent or long-term damage to a significant extension of the surrounding environment

The total impact TI at the specific CI and threat/hazard is calculated as follows:

$$\mathrm{TI} = \sum_{\substack{\text{Sum Across all}\\ \text{Impact Categories}}} (\text{Category Weight} \times \text{Impact Score}) \tag{1}$$

To make results comparable across different assets and scenarios, *Total Impact* is normalised (TI_N) to [0,1] as:

$$\mathrm{TI}_N = \frac{\mathrm{TI}}{\sum_{\substack{\text{For all}\\ \text{Impact}\\ \text{Categories}}} \text{Category Weight} \times 4} \tag{2}$$

Step 6: Calculate Risk Scores

After the total normalised impact is determined, the final *Risk Score* is calculated considering also conditional probabilities of the specific asset's state and threat using a risk value between [0–100]:

$$\begin{aligned}\text{Risk Score} = {} & \text{State Probability} \times \text{Threat Probability}\\ & \times \mathrm{TI}_N \times 100\end{aligned} \tag{3}$$

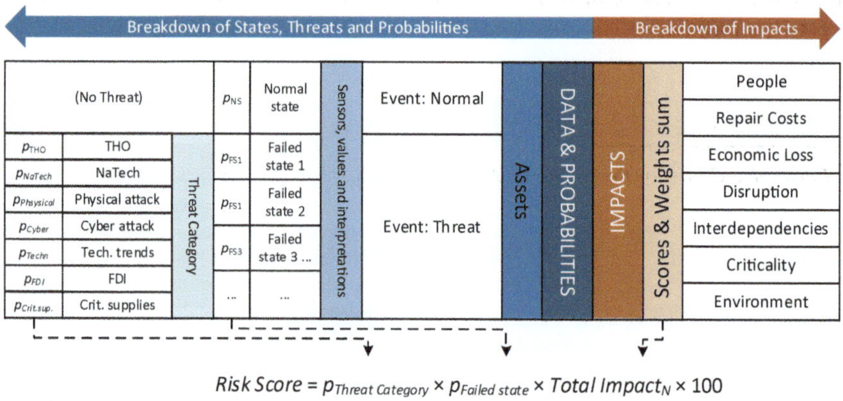

Fig. 5 Graphical summary of the risk assessment method

The calculated *Risk Score* represents the likelihood and severity of the particular threat scenario, expressed as a percentage. *Risk Score* can be a subject of risk interpretation for risk management-based decisions. Examples of possible risk levels and pertaining criteria on the *Risk Score* are:

- < 25%: Low Risk
- 25–50%: Medium Risk
- 50–75%: High Risk
- >75%: Critical Risk.

To summarise, the overall breakdown of the CI's assets, their respective states, associated threats, and corresponding probabilities, along with the impact assessment across relevant categories, can be effectively and visually presented in Fig. 5.

2.5 *Testing and Validation*

As part of the ATLANTIS testing and validation process, comprehensive **simulation scripts** were developed for each scenario to evaluate the performance and effectiveness of the proposed risk assessment methodology and supporting technologies. These scripts are essential for replicating real-world incidents, allowing stakeholders to assess their potential impacts on interconnected CIs and evaluate response strategies.

Each simulation script follows a structured format designed to ensure clarity, consistency, and practical implementation. The scenario description outlines the context and background of the simulated incident, detailing the initiating event, the CI operators involved, and the anticipated sequence of events. To provide a clear timeline of developments, the scenario is divided into **sequential steps**, with each capturing a distinct stage of the unfolding event.

For each step, a concise **name** is assigned to describe the type of event or transition. This is followed by a detailed **description** outlining what occurred, which CI operators were affected, and how the situation evolved. The script further identifies the **triggering conditions** that caused the step to occur, ensuring the logical flow throughout the scenario.

In addition to outlining the incident itself, each step specifies the **response actions** taken by the affected CI operators to mitigate the situation. The **cascading effects** are also examined to demonstrate how disruptions can propagate across interconnected services and infrastructures. Potential challenges encountered during the response phase, such as resource limitations, communication gaps, or technical difficulties, are also highlighted. Finally, each step concludes with a description of **anticipated next steps**, outlining how the scenario is expected to develop.

To facilitate practical implementation and align with the relevant ATLANTIS technologies, the simulation scripts are transformed into **structured tables** as shown in Table 5. These tables define the expected step-by-step system behaviour in the form of sensor data values for each involved asset within each involved CI. This step ensures that the simulation identifies critical moments and decision points where CI operators must assess information, implement countermeasures, and manage emerging systemic risks.

Once the simulation scripts are prepared and integrated into relevant technologies (e.g., Digital Twin), they should be initially tested in controlled laboratory environments. Following this, **demonstration and validation workshops** should be conducted with the end-users, including key personnel from CI operators. During these sessions, the simulations are executed, and participants are encouraged to provide feedback on system performance, usability, and potential improvements. These insights are gathered through structured discussions as well as a dedicated survey designed to evaluate the potential of ATLANTIS to enhance CI operators' awareness, preparedness, and response capabilities.

Table 5 Simulation script template

CI scenario steps—sensors observation table								
Scenario steps			CIs, their key assets, and associated sensors					
			CI	CI_1			CI_2	…
			Asset	$KA_{1.1}$	$KA_{1.2}$		$KA_{2.1}$	…
Step #	Step name	Time	Sensor	$S_{1.1.1}$	$S_{1.2.1}$	$S_{1.2.2}$	$S_{2.1.1}$	…
0	$Condition_0$	$T_0 + t_0$		$s_{1.1.1}(T_0 + t_0)$	$s_{1.2.1}(T_0 + t_0)$	$s_{1.2.2}(T_0 + t_0)$	$s_{2.1.1}(T_0 + t_0)$	…
1	$Condition_1$	$T_0 + t_1$		$s_{1.1.1}(T_0 + t_1)$	$s_{1.2.1}(T_0 + t_1)$	$s_{1.2.2}(T_0 + t_1)$	$s_{2.1.1}(T_0 + t_1)$	…
2	$Condition_2$	$T_0 + t_2$		$s_{1.1.1}(T_0 + t_2)$	$s_{1.2.1}(T_0 + t_2)$	$s_{1.2.2}(T_0 + t_2)$	$s_{2.1.1}(T_0 + t_2)$	…
…	…	…		…	…	…	…	…

3 Case Study: Large-Scale Pilot #1

3.1 Overall Pilot Description

The Large-Scale Pilot #1 (LSP#1) focuses on enhancing the resilience of CIs that support the seamless, secure, and safe delivery of essential multi-modal transportation services (sea, rail, road), energy supply, and telecommunications, thus spanning multiple CI domains [12]. The pilot extends both within and across the national borders of four EU countries, namely Slovenia, Croatia, Italy, and France. It brings together CI operators and relevant authorities along the Mediterranean trade route, specifically the *Mediterranean Corridor* [40]. This corridor (illustrated in Fig. 6 with relevant LSP#1 CI partners) is one of the key priority axes of the Trans-European Transport Network (TEN-T), linking the Mediterranean Basin with Central Europe and Ukraine.

All LSP#1 CI partner organisations are integral members of the TEN-T network:

- Seaports *Port of Rijeka* (**LUR**), Croatia, and *Port of Koper* (**LUK**), Slovenia.
- Rail operators *Slovenske železnice* (Slovenian railways, **SZ**), Slovenia, and FS Technology S.p.A. (Part of the *Ferrovie dello Stato Italiane* Group) italian railways, **FST**), Italy.
- Highway operators **DARS** (*Družba za avtoceste v Republiki Sloveniji*—national highways operator), Slovenia, and the cross-border Fréjus road tunnel stakeholders **SINA-SITAF** (*Societa Italiana Traforo Autostradale del Frejus spa* as tunnel operators on the Italian side of the tunnel) and **SDIS73** (fire and rescue service providers on the French side of the tunnel).

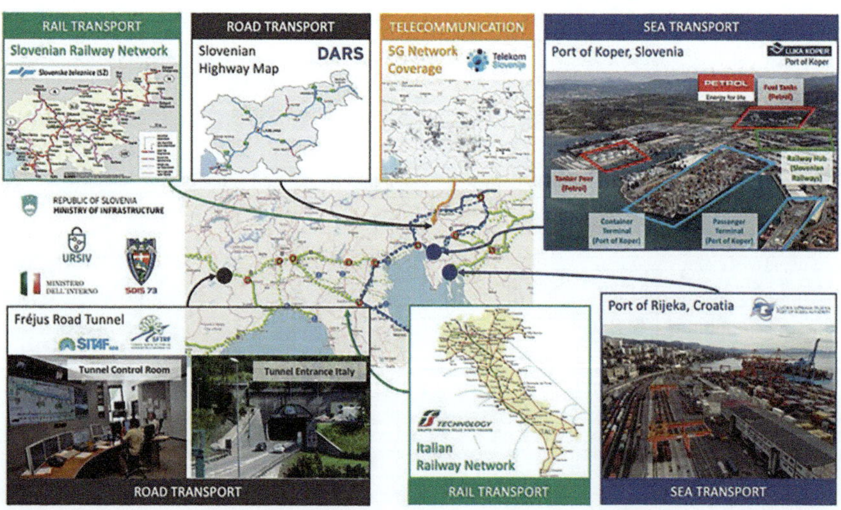

Fig. 6 Stakeholders in LSP#1 along the Mediterranean Corridor (green line)

- Oil products distributor *Petrol* (**PET**), Slovenia, operating along the corridor.
- Communications service provider *Telekom Slovenije* (**TS**), serving stakeholders in the Slovenian region of the Corridor.
- The CI operators in LSP#1 are supported by the following CI authorities:
- Ministry for Infrastructure (**MZI**), Slovenia.
- Government Information Security Office (**URSIV**), Slovenia.
- Ministry of Interior, Railway State Police (**MDI**), Italy.

3.2 Potential Threat Scenarios

Each of the CI organisations mentioned above faces its own set of unique challenges, shaped by its internal processes, capabilities, and overall preparedness. These challenges are further influenced by threats and hazards arising from various factors, such as their physical environment (e.g., floods or landslides due to geographic location), cultural specifics (e.g., large protests), political context (e.g., physical or cyber terrorism), technological accidents (e.g., human error, equipment failure, organisational deficiencies), and supply chain disruptions. However, it is important to recognise that the challenges faced by each organisation are not isolated; they often share common vulnerabilities, hazards, threats, and other interdependencies.

Within the ATLANTIS project, the initial in-depth screening analysis was conducted for existing cyber-physical-human (CPH) threats across each CI organisation. This comprehensive list of the potential hazards and threats was used to identify and develop a set of pilot scenarios through a series of hands-on workshops and meetings. The selection of these scenarios was based on the following criteria:

- The most pressing safety and security threats are currently faced by the organisations.
- Key collaboration and communication challenges among the CIs, particularly those at the cross-sector and/or cross-border level, that are critical for identifying, analysing, and managing such threats.
- The availability of the specific solutions provided by ATLANTIS to address these security threats and enhance cooperation in managing them.

The outcome of the analysis was the identification and development of **four detailed pilot scenarios** in which ATLANTIS solutions were applied, tested, and ultimately evaluated:

1. *HR-SI-Cyber scenario*: This scenario simulates a cyber-attack on the IT and communication systems of the Port of Rijeka Authority. The attack's objective is to: (i) reduce the safety and efficiency of navigation at Rijeka's maritime gateway; (ii) increase port congestion through schedule disruptions, longer transit times, blank sailings, and alternative ports of entry; and (iii) incapacitate communication among the port stakeholders, thus disrupting normal operations.
2. *SI-Fire scenario*: This scenario revolves around a potential physical attack involving a bomb planted on a train arriving at the Port of Koper's shunting

station. Alternatively, it involves a railway car collision, derailment, spill, and subsequent fire. In both cases, a large fire ignites on the train carrying flammable materials, spreading across the shunting station tracks, and potentially affecting and incapacitating surrounding CIs.
3. *IT-SI-Cyber-Physical scenario*: This scenario focuses on the rail transport of hazardous waste in Italy and Slovenia. The convoy departs from a Slovenian station near the border and heads towards Italy. Upon crossing the border, the tracking system for the freight convoy is compromised, causing it to disappear from the railway staff's radar. In response, the railway personnel implement safety protocols to slow down the train and prevent collisions. Simultaneously, a cyber-attack on the Italian rail operator's tracking system leads them to alert their Slovenian counterpart and the rail police, resulting in widespread transport delays and, consequently, public disturbance.
4. *FR-IT-Natural-Hazards scenario*: Following a major earthquake in the Fréjus tunnel region during the snowmelt period, a landslide blocks the tunnel entrance, disrupting traffic on the connected highways in France and Italy. Aftershocks trigger rockfalls that strike a dangerous goods truck, causing a spill and igniting a fire within the tunnel. This disrupts transport on both primary and secondary roads, creating congestion and hindering emergency access. The disruption delays emergency response, complicates coordination efforts, and causes shortages of fuel as well as water for firefighting. Utility services for residents, businesses, and industries located on both sides of the region are also affected.

The core of these pilot scenarios centres around two key elements, namely: (i) enhancing **common situational awareness** for logistically interdependent CIs; and (ii) ensuring the efficient and effective **engagement and coordination** of emergency response teams, along with seamless **cross-CI communication**. These elements are enabled through ATLANTIS solutions, which facilitate the monitoring and forecasting of CIs' operational status informed by modelled interdependencies.

3.3 Dependencies, Risk Assessment, and Simulation Scripts

Following the cross-CI risk assessment methodology presented in Sect. 2.4, the key **assets** for each CI have been identified in the context of each defined scenario, and then further identified all their (cross-organisational/-sectorial) **interdependencies**. Using this basic information and considering the specifics and contexts of each scenario, we further analysed **threats**, **impacts** of their realisation, and overall **risk scores**. To prepare for the final simulation of the scenarios with ATLANTIS technologies, we prepared a **simulation script** for each.

Below, the results of this exercise for the *SI-Fire* scenario are presented. This scenario is also considered as a representative case to test the practical applicability and usability of ATLANTIS technologies, as further discussed in the next section.

In total, **26 key assets** have been identified for all CI operators involved in the scenario, and the network of interdependencies among their assets is presented in Fig. 7. In the diagram, the CIs are logically separated by colours, namely green is used for LUK, orange for DARS, purple for PET, green for TS, and blue for SZ. Their assets are represented as graph nodes and the interdependencies among them are illustrated as graph edges. Label on each edge describes the type of interdependency identified.

Fig. 7 Interdependencies among the assets of the CIs involved in the SI-Fire scenario. Green colour is used for LUK, orange for DARS, purple for PET, green for TS, and blue for SZ

Table 6 Part of the analysis of threats and impacts for some of the PET assets

Asset	Sensor	Possible values	Value interpretation	Origin asset	Asset state	Impact [value 0–1]	Risk score [value 0–100]	Risk level
Tanker terminal (PET)	Status monitoring sensor	0, 1	1: Working	/	Normal state	0	0	No risk
			0: Not working	Storage (PET)	Storage damaged, fuel is lost	0.55	2.77	Low
			0: Not working	Ship mooring (LUK)	Ships mooring not available, new ships are not accepted	0.55	1.66	Low
			…	…	…	…	…	…

Table 6, we provide an excerpt of the analysis of the different internal and external threats relevant for one of the PET assets that show the interconnected nature of CIs in the Port of Koper logistics zone. The table also provides examples of the sensors we have defined to monitor the state of these assets and the results of the impact analysis/risk scores.

A part of the simulation script for the SI-Fire scenario is presented in Table 7. It shows what different sensors report for the associated asset in each step of the scenario.

4 Testing and Validation Results

The SI-Fire scenario served as a testbed to evaluate how two of the ATLANTIS technologies, specifically the Digital Twin platform and the HiVIC (Human-in-Vicinity) mobile application, can support complex, cascading crisis involving multiple CI operators. The main objective of the testing and validation exercise was to assess the potential of these tools to enhance coordinated risk assessment, situational awareness, and cross-organisational communication.

To conduct the validation, a dedicated workshop was held in Koper, bringing together representatives from CI operators, relevant public authorities, and technical partners. The event was structured around the step-by-step simulation of the scenario, based on a previously prepared simulation script (partially shown in Table 7). Throughout the demonstration, each step of the scenario, from the initial fire ignition at the shunting station, to the full recovery of operations, was visualised using the Digital Twin platform. Participants in the role of field personnel also interacted with

Table 7 Part of the simulation script for the SI-Fire scenario

		CI operator	SZ						PET								
		Asset	CVP	Main Tracks		Switch 501	SNEV	SVN	TT	ZP	AP	CRP1	CRP2	REZ		OC	PC
		Sensor	Status	Status	Tracks occupancy (%)	Status	Status	Status	Status	Status	Status	Status	Status	Status	Fuel storage capacity (%)	Status	Status
Step #	Step Name	Time															
0	Normal	Before T0	1	1	66	1	1	1	1	1	1	1	1	1	80	1	1
1	Explosion	T0	0	0	66	1	1	1	1	1	1	1	1	1	80	1	1
2	Fire detection	T0 + some seconds	0	0	66	1	0	1	1	1	1	1	1	1	80	1	1
3	Firefighting	T0 + 10–15 min	0	0	66	1	0	1	0	0	1	1	1	1	80	1	1
4	Fire spread	T0 + up to 1 hour	0	0	66	1	0	0	0	0	1	1	1	1	79	1	1
5	Fire escalation	T0 + up to few hours	0	0	66	0	0	0	0	0	0	0	0	1	79	1	0
…	…	…															
15	Shunting station repaired	T0 + up to 3 days	1	1	34	1	1	1	1	0	1	1	0	1	74	1	1

(continued)

Table 7 (continued)

		CI operator	SZ						PET								
		Asset	CVP	Main Tracks		Switch 501	SNEV	SVN	TT	ZP	AP	CRP1	CRP2	REZ		OC	PC
		Sensor	Status	Status	Tracks occupancy (%)	Status	Status	Status	Status	Status	Status	Status	Status	Status	Fuel storage capacity (%)	Status	Status
Step #	Step Name	Time															
16	PET tracks repaired		1	1	54	1	1	1	1	1	1	1	1	1	74	1	1
17	Normal	T0 + up to few weeks	1	1	42	1	1	1	1	1	1	1	1	1	74	1	1

the HiVIC application to simulate first-hand reports of visible smoke and fire in the area.

The Digital Twin provided a comprehensive visualisation of all relevant CI assets, their interdependencies, and real-time changes in state. This enabled the stakeholders to observe how the incident unfolded and how alerts were triggered as the cascading effects progressed. Meanwhile, the HiVIC app allowed users to report field-level incidents, which were then reflected on the Digital Twin, demonstrating its capacity to support early warning and response coordination across sectors.

Feedback was gathered through structured questionnaires completed by participants following the simulation. The responses revealed a clear recognition of the added value provided by the ATLANTIS tools, particularly in enabling effective communication across CI organisations, which was consistently rated as one of the strongest aspects of the solution. Respondents gave the highest scores to questions relating to cross-organisational communication and the usefulness of shared situational awareness, with average ratings of 4.40 and 4.30 (on a scale from 1 to 5), respectively.

The average scores across the usability assessment were approximately 74%, indicating that while tools were generally well received, there remains room for optimisation. The improvements are expected mainly in the context of enhancing data quantity and quality, which remains an open issue around CI and security in general.

Several important lessons emerged from the exercise. The simulation reaffirmed the importance of **shared data environments and structured communication tools** in managing cross-sector crisis events. Scenario-based demonstrations proved particularly effective in uncovering gaps in preparedness and identifying opportunities for improved communication and coordination. Perhaps most importantly, the participatory format of the workshop allowed for actionable feedback directly from end-users. The feedback will inform the next iteration of ATLANTIS technology development and deployment.

This pilot test marked a significant milestone in the project, demonstrating both the promise of the ATLANTIS approach and the value of co-developing crisis management tools with the direct involvement of their intended users.

The Italian and Slovenian scenario—the IT-SLO-Cyber-attack, outlines a cyber-physical attack on the transportation of hazardous waste between Italy and Slovenia, where a freight convoy disappears from the tracking system due to a cyber compromise. This triggers emergency protocols among railway personnel, leading to slowed rail traffic and public disturbances. Advanced technologies like SNIFFER detect hazardous material leaks, prompting coordinated responses from emergency teams.

In March 2024, a workshop was held to present the scenario to all stakeholders involved, including both end-users and technology providers. Key participants, such as the Italian police and representatives from the Italian and Slovenian railway sectors, had the opportunity to share their insights on the risks and challenges associated with cross-border collaboration in the management of hazardous goods. The discussion highlighted the urgent need for a unified and standardised regulatory framework, as well as digital platforms that could provide real-time updates on infrastructure status and report potential threats or hazards. Such systems would enable simultaneous alerts to both parties involved, thereby enhancing emergency management and coordination. On the technology provider side, participants showcased the technologies they have developed and how these could potentially integrate into the scenario. For instance, CEA's SNIFFER technology for detecting hazardous substances in the event of environmental spills, and ENG's SAFER platform, which enhances situational awareness capabilities. The workshop served as a crucial event to concretely demonstrate the potential of the ATLANTIS tools and platforms to the engaged end-users, fostering collaboration and innovation in the field of hazardous materials management.

The workshop showcased key aspects related to the topic, sparking meaningful discussions and providing participants with valuable insights to guide both the future activities of ATLANTIS and the daily operations of the stakeholders involved in the project. The collaborative environment fostered a sense of shared purpose, enabling attendees to explore innovative solutions and best practices for managing hazardous materials. Participants expressed positive feedback regarding the event, highlighting its effectiveness in facilitating dialogue and enhancing understanding among diverse stakeholders. This engagement not only strengthened relationships but also laid the groundwork for ongoing collaboration and improvement in the field of cross-border hazardous goods management.

5 Conclusions

In this chapter, a holistic and integrated approach to EU-wide resilience against systemic risks derived from major hazards and complex threats have been considered and analysed. Moreover, the main findings coming from the ATLANTIS [14] project, such as the integrated technological framework, the data management plan, the interdependency modelling and the risk assessment methodology have been described and validated within one of the three extensive and collaborative LSPs, which include four (4) neighbouring countries—France, Italy, Slovenia, and Croatia, according to a set of identified threat scenarios.

All the feedback coming from the involved end-users which actively participated in a collaborative work, has been then collected through questionnaire, analysed and duly presented in terms of system usability, cross-CI collaboration and lessons learned. While on one hand the ATLANTIS solutions and methodologies have been distinguished as a valid contribution for effective crisis management and resilience improvement, especially dealing with a cross-border and cross-sectorial dimension, on the other hand, the importance of shared data environments and structured communication tools has been observed, which should be further improved.

Acknowledgements This project has received funding from the European Union's Horizon Europe research and innovation programme under grant agreement No. 101073909. Authors would like to thank to the participating CIs partners enabling the implementation and testing of the ATLANTIS methods and tools. This article reflects only the authors' views and the Research Executive Agency and the European Commission are not responsible for any use that may be made of the information it contains.

Competing Interests The authors declare that they have no known competing financial interests or personal relationships that could have appeared to influence the work reported in this chapter.

Abbreviations

ACF	Awareness and comprehension framework
AI	Artificial intelligence
ATLANTIS	Improved resilience of Critical Infratsructures AgainsT LArge-scale transNational and sysTemic rISks
CCI-SAAM	Cross-CI (CCI) sharing assessment analysis mitigation
CER	Directive (EU) 2022/2557 (CER directive)
CERT	Computer emergency response team
CI	Critical infrastructure
CIP	Critical infrastructure protection
CSIRT	Computer security incident response team
DARS	Družba za avtoceste v Republiki Sloveniji (Slovenian highways operator)
ENISA	European Union agency for cybersecurity
EO	Earth observation
FAIR	Findability, accessibility, interoperability, and reuse
FST	Ferrovie dello Stato Italiane (Italian Railways)
GDPR	General data protection regulation
GNSS	Global navigation satellite system
HiVIC	Humans in vicinity
IDMEF	Intrusion detection message exchange format
IDS/IPS	Intrusion detection and prevention system
ISAC	Information sharing and analysis center
LSP	Large-scale pilot
LUK	Luka Koper (Port of Koper)
LUR	Luka Rijeka (Port of Rijeka)
MDI	Ministry of Interior, Railway State Police, Italy
MISP	Malware Information Sharing Platform
ML	Machine learning
MZI	Ministrstvo za Infrastrukturo (Ministry for Infrastructure)
NIS-2	Directive (EU) 2022/2555 (NIS-2 directive)
OASIS	https://www.oasis-open.org/
PCTI	Physical/cyber threat intelligence
PET	Petrol (Slovenian oil products distributor)
RRIM	Risk reduction and incident mitigation

SAFER	Situation awareness for enhancing resilience
SDIS73	Fire and rescue service providers on the French side of the tunnel
SINA-SITAF	Societa Italiana Traforo Autostradale del Frejus spa (tunnel operator on the Italian side of the tunnel)
SNIFFER	Air quality sensor technology provided by partner CEA
STIX	Structured threat information expression
SZ	Slovenske Železnice (Slovenian Railways)
TAXII	Trusted automated exchange of intelligence information
TEN-T	Trans-European transport network
TS	Telekom Slovenije (Slovenian telecommunication services provider)
URSIV	Government Information Security Office, Slovenia
XAI	Explainable artificial intelligence

References

1. DEFENDER project, https://cordis.europa.eu/project/id/740898
2. InfraStress project, https://cordis.europa.eu/project/id/833088
3. STOP-IT project, https://cordis.europa.eu/project/id/740610
4. SATIE project, https://cordis.europa.eu/project/id/832969
5. SecureGAS project, https://cordis.europa.eu/project/id/833017
6. PRAETORIAN project, https://cordis.europa.eu/project/id/101021274
7. FINSEC project, https://cordis.europa.eu/project/id/786727
8. 7SHIELD project, https://cordis.europa.eu/project/id/883284
9. SAFECARE project, https://cordis.europa.eu/project/id/787002
10. RESISTO project, https://cordis.europa.eu/project/id/786409
11. PRECINCT project, https://cordis.europa.eu/project/id/101021668
12. Directive (EU) 2022/2557 of the European Parliament and of the Council of 14 December 2022 on the resilience of critical entities and repealing Council Directive 2008/114/EC (CER Directive). Available at: https://eur-lex.europa.eu/eli/dir/2022/2557/oj/eng
13. Directive (EU) 2022/2555 of the European Parliament and of the Council of 14 December 2022 on measures for a high common level of cybersecurity across the Union, amending Regulation (EU) No 910/2014 and Directive (EU) 2018/1972, and repealing Directive (EU) 2016/1148 (NIS 2 Directive). Available at: https://eur-lex.europa.eu/eli/dir/2022/2555/oj/eng
14. ATLANTIS project, https://cordis.europa.eu/project/id/101073909
15. Lehmann G., 2024. "The Incident Detection Message Exchange Format version 2 (IDMEFv2)," Network Working Group, [Online]. Available: https://www.ietf.org/archive/id/draft-lehmann-idmefv2-03.html
16. Trusted Automated Exchange of Intelligence Information (TAXII) protocol. https://oasis-open.github.io/cti-documentation/taxii/intro.html
17. Malware Information Sharing Platform. https://www.misp-project.org/
18. Structured Threat Information eXpression (STIX). https://www.oasis-open.org/standard/stix-version-2-1/
19. FAIR Principles. https://www.go-fair.org/fair-principles/
20. Regulation (EU) 2016/679 of the European Parliament and of the Council of 27 April 2016 on the protection of natural persons with regard to the processing of personal data and on the free movement of such data, and repealing Directive 95/46/EC (General Data Protection Regulation). https://eur-lex.europa.eu/eli/reg/2016/679/oj/eng

21. Standard ISO 14224:2016. Petroleum, petrochemical and natural gas industries — Collection and exchange of reliability and maintenance data for equipment. Available at: https://www.iso.org/standard/64076.html (Accessed 11.3.2025)
22. Rinaldi M, Peerenboom JP, Kelly T, 2001. Identifying, understanding and analysing critical infrastructure interdependencies. IEEE Control System Magazine, 11–25. https://doi.org/10.1109/37.969131
23. Ouyang, M, 2014. Review on modelling and simulation of interdependent critical infrastructure systems. Reliability Engineering and System Safety 121, 43–60. https://doi.org/10.1016/j.ress.2013.06.040
24. Charles Perrow, 1984. Normal Accidents: Living with High-Risk Technologies. New York: Basic Books, pp. 89–100. ISBN 0-465-05142-1
25. Standard ISO 31000:2018 Risk management — Guidelines. https://www.iso.org/standard/65694.html
26. Guidelines for chemical process quantitative risk analysis, Second edition. Center for Chemical Process Safety, AIChE, 2000. ISBN 0-8169-0720-X
27. Faisal Khan, Samith Rathnayaka, Salim Ahmed, 2015. Methods and models in process safety and risk management: Past, present and future. Process Safety and Environmental Protection. 98, 116–147. https://doi.org/10.1016/j.psep.2015.07.005
28. CCPS, 2002. Guidelines for analyzing and managing the security vulnerabilities of fixed chemical sites. American Institute of Chemical Engineers: New York. ISBN 978-0-470-92500-3
29. Gerbec, Marko, Giunta, Gabriele, 2021. InfraStress approach on risk modelling of cascading events with live data for decision support. In: Soldatos, John (Ed.), Praça, Isabel (Ed.), Jovanović, Aleksandar (Ed.). Cyber-physical threat intelligence for critical infrastructures security: securing critical infrastructures in air transport, water, gas, healthcare, finance and industry, (NowOpen). Hanover: Now Publishers. 2-21, https://doi.org/10.1561/9781680838237
30. Uijt de Haag P.A.M., Ale B.J.M., 2005. Guideline for quantitative risk assessment 'Purple book', CPR 18E. http://content.publicatiereeksgevaarlijkestoffen.nl/documents/PGS3/PGS3-1999-v0.1-quantitative-risk-assessment.pdf
31. Suarez-Paba Maria Camila, Perreur Mathis, Munoz Felipe, Cruz Ana Mariad, 2019. Systematic literature review and qualitative meta-analysis of Natech research in the past four decades. Safety Science, 116, 58–77
32. Krausmann Elisabeth, Cruz Ana Maria, Salzano Ernesto, (Eds), 2017. Natech Risk Assessment and Management Reducing the Risk of Natural-Hazard Impact on Hazardous Installations. Elsevier. ISBN: 978-0-12-803807-9
33. EC, JRC, Institute for the Protection and Security of the Citizen, 2008. Understanding Malicious Attacks against Infrastructures - Overview on the Assessment and Management of Threats and Attacks to Industrial Control Systems. Bogdan Vamanu, Marcelo Masera. EUR 23681 EN. https://publications.jrc.ec.europa.eu/repository/bitstream/JRC49474/understanding%20malicious%20attacks%20against%20infrastructures.pdf
34. ENISA report on threats in the area of smart grids and good practice guide. 17.12.2013. https://www.enisa.europa.eu/sites/default/files/publications/Smart%20Grid%20Threat%20Lanscape.pdf
35. Camino Kavanagh, 2019. New Tech, New Threats, and New Governance Challenges: An Opportunity to Craft Smarter Responses? Carnegie Endowment for International Peace. https://carnegieendowment.org/files/WP_Camino_Kavanagh___New_Tech_New_Threats1.pdf
36. Chapter 4: FDI and National Security: Separating Legitimate Threats from Implausible Apprehensions. In: Foreign direct investment in the United States: Benefits, Suspicions, and Risks with Special Attention to FDI from China, 3RD EDITION, Edward M. Graham (PIIE) and Paul R. Krugman, January 1995. ISBN: 978-0-88132-204-0
37. REGULATION (EU) 2019/452 OF THE EUROPEAN PARLIAMENT AND OF THE COUNCIL of 19 March 2019 establishing a framework for the screening of foreign direct investments into the Union; https://eur-lex.europa.eu/legal-content/EN/TXT/PDF/?uri=CELEX:32019R0452&from=EN

38. European Commission, Study on the EU's list of Critical Raw Materials—Final Report (2020). ISBN 978-92-76-21049-8, https://doi.org/10.2873/11619
39. Bennett, Brian T., 2007. Understanding, Assessing, and Responding to Terrorism - Protecting Critical Infrastructure and Personnel; Wiley: Hoboken, USA. ISBN: 978-1-119-23781-5
40. Regulation (EU) 2019/941 of the European Parliament and of the Council of 5 June 2019 on risk-preparedness in the electricity sector and repealing Directive 2005/89/EC. Available at: https://eur-lex.europa.eu/eli/reg/2019/941/oj/eng
41. https://transport.ec.europa.eu/transport-themes/infrastructure-and-investment/trans-european-transport-network-ten-t/mediterranean-corridor_en

Enhancing Resilience Based on Artificial Intelligence

Privacy Risk Assessment of AI Models in the Healthcare Domain

Maya Anderson, Abigail Goldsteen, Guy Amit, Shlomit Shachor-Ifergan, Natalia Razinkov, Noam Koren, and Ariel Farkash

Abstract Artificial Intelligence (AI) is increasingly being used to drive business, and specifically in the healthcare domain, it has proven to be a critical factor in improving efficacy and outcomes. However, as the technology advances, and especially with the increased use of Large Language Models (LLMs), existing privacy challenges have been exacerbated, and new privacy issues have arisen. LLMs' tendency to memorize and repeat complete data samples seen during training or tuning has increased their attack surface and given rise to new data leakage vulnerabilities. This, in turn, has slowed the adoption of AI, due to significant privacy concerns, in addition to the need to adhere to strict data protection regulations, standards, and guidelines. To help counter these issues, we present an end-to-end framework for privacy risk assessment in AI models, and demonstrate the use of this framework to assess AI models of different types in concrete healthcare use cases. We thus offer a practical approach to bridge the gap of trust in AI-driven solutions by building privacy risk assessment into the ML pipeline as an integral part of the flow.

Keywords AI privacy · Healthcare · Privacy risk assessment

Abbreviations

AAA	Algorithmic Accountability Act
AI	Artificial Intelligence
AIDA	Artificial Intelligence and Data Act
ARIMA	Autoregressive Integrated Moving Average
ART	The Adversarial Robustness Toolbox

M. Anderson (✉) · A. Goldsteen · G. Amit · S. Shachor-Ifergan · N. Razinkov · N. Koren · A. Farkash
IBM Research, IBM Israel—Science and Technology Ltd (IBM), Haifa, Israel
e-mail: creatress_m@outlook.com

A. Goldsteen
e-mail: abigailt@il.ibm.com

J. Soldatos and H. Abie (eds.), *Technology-Enabled Resilience: Innovations in Critical Infrastructure Protection*, https://doi.org/10.1007/978-3-032-25917-2_7

ATLAS Adversarial Threat Landscape for Artificial-Intelligence Systems
CCPA The California Consumer Protection Act
CGM Continuous Glucose Monitoring
CPPA Consumer Protection Privacy Act
EHDS European Health Data Space Regulation
EHR Electronic Health Record
ENISA The European Union Agency for Cybersecurity
FPR False Positive Rate
FTC The US Federal Trade Commission
GDPR General Data Protection Regulation
HIPAA The US Health Insurance Portability and Accountability Act
LGPD The Brazilian General Data Protection Law
LLM Large Language Model
LSTM Long Short-Term Memory
MDCG The Medical Device Coordination Group
MDR The Medical Devices Regulation
MI Model Inversion
MIA Membership inference attack
ML Machine Learning
NFT Neural Fourier Transform
NIST The American National Institute of Standards and Technology
NLP Natural language processing
OECD Organisation for Economic Co-operation and Development
PET Privacy-enhancing technology
PHI Protected Health Information
PII Personally Identifiable Information
PIPL The Personal Information Protection Law
RAG Retrieval Augmented Generation
RMF Risk Management Framework
ROC Receiver Operating Characteristic
CoSAI The Coalition for Secure AI
TCN Temporal Convolutional Network
TPR True Positive Rate
TS Time-series

1 Introduction

AI privacy is relevant whenever an AI system involves personal data. The concept of personal data is not universally defined and varies between geographies, cultures, and regulatory environments. According to the EU General Data Protection Regulation

(GDPR),[1] personal data is any information relating to an identified or identifiable natural person (called the data subject). This refers to a person who can be identified, directly or indirectly, by reference to an identifier such as a name, an identification number, location data, an online identifier, or one or more factors specific to their physical, physiological, genetic, mental, economic, cultural, or social identity.

The California Consumer Protection Act (CCPA)[2] provides an even wider definition of personal information as information that identifies, relates to, or could reasonably be linked with a person. For example, it could include not only direct identifiers, but also records of products purchased, internet browsing history, geolocation data, and inferences from other personal information.

In general, personal data may include information about demographics (age, gender, place of residence, etc.), location, health, finances, biometrics, employment, and various activities, behaviors, and opinions.

It is crucial to ensure that any AI project complies with all relevant data protection and privacy regulatory requirements. This may include obtaining consent, minimizing data collection, enabling data correction or withdrawal, and more. These requirements apply to all data used in an AI system, and especially the data used to train machine learning (ML) models.

Violating data protection regulations can incur serious fines. GDPR has set fines of up to €20 million, or 4% of the company's worldwide annual revenue from the preceding financial year, whichever is higher. The highest GDPR fine given so far was an amazing 1.2 billion Euros[3] incurred against Meta Platforms Ireland Limited for insufficient legal basis for data processing.

Specifically in the context of machine learning, in 2021, a precedential ruling from the US Federal Trade Commission (FTC) forced an AI company to delete its machine learning models after unlawfully collecting user data.[4]

Recent studies have shown that a malicious third party with access to a trained ML model, even without access to the training data itself, can still reveal sensitive, personal information about the people whose data was used to train the model. For example, it may be possible to reveal if a person's data was part of the model's training set, or even infer sensitive attributes about them, such as their salary [1, 2]. Real-world Large Language Models have been shown to leak individual users' personal information[5] [3].

Factors that may affect the level of privacy risk posed by a machine learning model include the size of the model, the size of the training set, the dimensionality of the data, etc. In general, the more a model overfits, or in other words, memorizes its training data, the higher the probability of the model leaking unintended information.

---

[1] https://ec.europa.eu/info/law/law-topic/data-protection/data-protection-eu_en

[2] https://leginfo.legislature.ca.gov/faces/billTextClient.xhtml?bill_id=201720180AB375

[3] https://www.enforcementtracker.com/

[4] https://jolt.law.harvard.edu/digest/everalbum-inc-in-first-facial-recognition-misuse-settlement-ftc-requires-destruction-of-algorithms-trained-on-deceptively-obtained-photos

[5] https://www.businessupturn.com/world/chatgpts-answer-gives-away-a-journalists-number-to-join-signal/

To mitigate these risks, organizations developing AI systems should implement privacy risk assessment early in the development lifecycle. This approach supports informed decisions regarding the deployment of a model in production, its publication, or sharing it with third parties. An early privacy risk assessment also facilitates the identification of potential mitigations to create a more privacy-preserving model.

1.1 AI Privacy Risks in the Healthcare Domain

There is a clear conflict between the ever-increasing interest in analyzing personal data to enhance and improve processes that affect human life and the need to preserve the privacy of data subjects. This is especially the case in the medical domain, where sensitive health data from real patients is often used to train machine learning (ML) models that aid physicians in efficient diagnostics and personalized treatment. Artificial intelligence (AI) models are also used in medical devices and applications to predict malfunctions and enhance continuous patient care. The deployment of such systems introduces privacy risks, and ensuring patient privacy and protecting their sensitive health information presents a significant challenge.

Assessing the privacy risk of such AI models is crucial for making informed decisions about whether to use a model in production, share it with third parties, or deploy it in patients' homes. Privacy risk assessment is often achieved by running privacy attacks against one or more models and measuring how successful they are in leaking personal or sensitive information. Striking a balance between the need for patient privacy and effective health and disease management through technology is pivotal in the design and implementation of healthcare solutions.

We outline here a few relevant use cases from the healthcare domain that employ AI models. In the next sections, we will give specific examples of such AI models and the privacy risk assessments that may be performed on them.

One example of a system that uses AI models trained on sensitive health data is a medical device for the continuous monitoring of patients with neurodegenerative diseases, such as Parkinson's disease. It is designed to trace, record, and store a variety of motor symptoms commonly associated with these diseases through the continuous use of wearable body sensors. The system is composed of: (1) a set of wearable monitoring devices, (2) a docking station for collecting, processing and uploading the data of the monitoring devices to the cloud, (3) a mobile application that enables patients and caregivers to record complementary information, such as medication, nutrition and non-motor status information, and (4) a web application, called the Physician's Tool, that presents all patient-related information to the healthcare professional. Since the trained AI models may be part of the system's mobile application or other devices deployed at patients' homes, it is especially important to ensure that they do not leak any sensitive information.

Another example is a diabetes support application that uses an AI model to provide personalized recommendations on daily insulin adjustments based on treatment-related information. The system is a Software as a Medical Device (SaMD) that can be installed on the patient's mobile phone and is capable of connecting to external devices and remote servers. It processes sensitive data related to the patient's condition and therapy, such as blood-glucose levels, insulin intakes, etc. The current version of the model is trained on a single patient's data, but an improved model is in development. It will use a model pre-trained on data from multiple patients in a centralized manner and will then be fine-tuned on a specific patient's data once deployed on that patient's device. Since this pre-trained model will be deployed on every patient's local device, it is important to ensure that it does not leak sensitive health data of other patients used to train the model.

The third example is a diagnostic device and AI assistant designed to empower healthcare practitioners to provide improved, personalized diagnoses of skin cancer. Patient data is extracted from various sources such as lesion images, clinical history, and genotypic information, and analyzed by AI algorithms to assign each patient a holistic melanoma risk score. These scores are presented to physicians within a cognitive assistant that also displays an explanation justifying its decisions. These ML models for assessing patients' melanoma risk based on clinical, genetic, and family history information are trained on highly sensitive patient data, making it crucial to assess their risk of leaking this information.

Regulations Related to AI Privacy in Healthcare

In this section, we highlight the main regulations relevant to AI Privacy Risk assessment, including regulations for privacy, AI, and health.

1.2 *Privacy Regulations*

The EU, with its General Data Protection Regulation (GDPR), was the first to enact a comprehensive, enforceable regulation around information privacy. It includes chapters on data subject rights, data controllers' and processors' obligations, and transfers of personal data to third countries. To date, it is one of the strictest privacy regulations in the world, awarding fines of up to 4% of a company's annual revenue in case of a breach. The UK, after breaking away from the European Union in 2020, decided to keep the GDPR in force, coining it the UK GDPR.[6]

The GDPR has become a model for many similar laws around the world, either already enacted or under development, including in Brazil, Japan, Singapore, South Africa, and more.

In the United States, there is no comprehensive national privacy law, apart from the Privacy Act of 1974, that governs the collection, maintenance, use, and dissemination of information about individuals by federal agencies. However, there are a number of sector-specific privacy and data security laws that apply to financial institutions,

---

[6] https://ico.org.uk/for-organisations/data-protection-and-the-eu/data-protection-and-the-eu-in-detail/the-uk-gdpr/

telecommunications companies, credit reporting agencies, and healthcare providers, as well as a multitude of state and local laws.

The landmark California Consumer Privacy Act (CCPA) secures comprehensive privacy rights for California consumers, including the right to know what personal information is collected about them, how it is used and shared, the right to delete personal information, and the right to limit the use and disclosure of sensitive personal information. Many states have followed by enacting different privacy and data protection acts, including Colorado, Connecticut, Delaware, Florida, Virginia, and more.[7]

The American Data Privacy and Protection Act,[8] proposed in 2022, establishes requirements for how companies handle personal data. The proposed bill requires most companies to limit the collection, processing, and transfer of personal data to what is reasonably necessary to provide a requested product or service and prohibits the transfer of personal data without explicit consent.

In China, the Personal Information Protection Law (PIPL),[9] which came into effect in 2021, seeks to protect personal data and regulate its processing.

The Brazilian General Data Protection Law (LGPD)[10] governs the processing of personal data, with the purpose of protecting the fundamental rights of freedom and privacy, and is in line with the GDPR on most articles.

In Canada, the proposed Consumer Protection Privacy Act (CPPA),[11] would replace the existing Personal Information Protection and Electronic Documents Act and establish a new Personal Information and Data Protection Tribunal.

1.3 AI Regulations

In the EU, the Artificial Intelligence Act[12] (EU AI Act), in force as of 2024, harmonizes the different AI rules and regulations across Europe, with penalties for violations greatly exceeding those of GDPR. This landmark regulation organizes AI systems by risk level. Some uses of artificial intelligence are prohibited, such as manipulating people's decisions, classifying people based on social behavior, and predicting a person's risk of committing a crime.

---

[7] https://www.dlapiperdataprotection.com/?t=law&c=US

[8] https://www.congress.gov/bill/117th-congress/house-bill/8152#:~:text=American%20Data%20Privacy%20and%20Protection%20Act,-This%20bill%20establishes&text=The%20bill%20establishes%20consumer%20data,opt%20out%20of%20such%20advertising.

[9] http://en.npc.gov.cn.cdurl.cn/2021-12/29/c_694559.htm

[10] https://iapp.org/resources/article/brazilian-data-protection-law-lgpd-english-translation/

[11] https://cppa.ca.gov/regulations/consumer_privacy_act.html

[12] https://artificialintelligenceact.eu/wp-content/uploads/2022/05/AIA-COM-Proposal-21-April-21.pdf

High-risk applications are those that pose a significant risk of harm to the health, safety, or fundamental rights of individuals. This includes safety components, biometric identification, education, employment, law enforcement, and justice. These are allowed, but subject to strict obligations and requirements, including risk assessment, logging, documentation, human oversight, and security. These applications must have a risk management system in place throughout the entire lifecycle of the AI system, which should identify any known and reasonably foreseeable risks and implement appropriate measures to reduce them. In Article 10 it specifically addresses the training of AI models and mandates that the training, validation, and testing datasets used must meet quality, safety, and fairness criteria, requiring that any special categories of personal data be subject to technical limitations on reuse and state-of-the-art security and privacy measures.

The European Council also opened for signature in September 2024 the Framework Convention on Artificial Intelligence,[13] the first international legally binding treaty aimed at ensuring that activities throughout the lifecycle of AI systems are consistent with human rights, democracy, and the rule of law.

Other countries have recently also begun to promote AI legislation. The proposed Canadian Artificial Intelligence and Data Act (AIDA),[14] introduced as part of the Digital Charter Implementation Act in 2022, would set the foundation for the responsible design, development, and deployment of AI systems that impact the lives of Canadians.

In the United States, the FTC has published multiple reports and guidelines on the use of artificial intelligence in agencies and businesses and the risks concerned. In 2025, they warned against risks of consumer harm from AI and recommended ensuring privacy and security by default in generative AI tools.[15] A few bills, such as the Artificial Intelligence Initiative Act,[16] the Algorithmic Accountability Act (AAA)[17] and the National AI Commission Act[18] have also been proposed to Congress.

In 2022, the White House Office of Science and Technology Policy published a Blueprint for an AI Bill of Rights,[19] with the goal of developing policies and practices for automated systems that protect civil rights and promote democratic values such as privacy In 2023, the Joe Biden administration issued an executive order on "Safe, Secure, and Trustworthy Development and Use of AI" however, it was later revoked by the incoming president Donald Trump.

---

[13] https://www.coe.int/en/web/artificial-intelligence/the-framework-convention-on-artificial-intelligence

[14] https://ised-isde.canada.ca/site/innovation-better-canada/en/artificial-intelligence-and-data-act-aida-companion-document

[15] https://www.ftc.gov/policy/advocacy-research/tech-at-ftc/2025/01/ai-risk-consumer-harm

[16] https://www.congress.gov/bill/116th-congress/senate-bill/1558/text

[17] https://www.congress.gov/bill/117th-congress/house-bill/6580/text

[18] https://www.congress.gov/bill/118th-congress/house-bill/4223/text

[19] https://bidenwhitehouse.archives.gov/ostp/ai-bill-of-rights/

Drawing inspiration from the EU AI Act, The Colorado AI Act of 2024[20] adopts a risk-based approach to regulate the deployment of high-risk AI systems, emphasizing transparency, risk mitigation, and proper documentation.

Many other countries are currently publishing strategies and frameworks around AI, including the UK, Singapore, and Australia. However, these have not yet matured into concrete legislation.

1.4 Healthcare Regulations

The US Health Insurance Portability and Accountability Act (HIPAA) establishes federal standards protecting sensitive health information. HIPAA contains a Privacy Rule[21] ("Standards for Privacy of Individually Identifiable Health Information") that addresses the use and disclosure of individuals' protected health information (PHI). The Privacy Rule protects all "individually identifiable health information" including demographic data, to the individual's physical or mental health or condition, provision of health care, and payment for the provision of health care. This includes many common identifiers such as name, address, and birth date.

While the GDPR is mostly domain-neutral, it recognizes health data as a special category of data requiring enhanced privacy protection. This includes all personal data concerning or linked to the health of an individual, including genetic and biometric data. The GDPR stipulates that health data can only be used for specific purposes or with explicit patient consent, and outlines rules under which it can be processed.

The Medical Devices Regulation (MDR)[22] lays down rules concerning the sale and servicing of medical devices for human use. Regarding privacy and confidentiality of data associated with the use of Medical Devices (MDs), Article 62.4 (h) mandates that the subject's rights to privacy and the protection of the personal data in accordance with Directive 95/46/EC are safeguarded.

In 2019, the Medical Device Coordination Group (MDCG) issued the Guidance on Cybersecurity for medical devices.[23] Some of the key points addressed in this guidance are secure design, security risk management, and implementation of security measures to mitigate risks. It also mandates the application of the GDPR principles in the design and development of medical devices.

ENISA published in 2023 a report on Cybersecurity and privacy in AI—Medical imaging diagnosis.[24] This report covers the use of AI in the health sector and highlights specific privacy and security threats that may arise. The report finds that efforts

[20] https://leg.colorado.gov/bills/sb24-205

[21] https://www.hhs.gov/hipaa/for-professionals/privacy/laws-regulations/index.html

[22] https://eur-lex.europa.eu/eli/reg/2017/745/oj/eng

[23] https://ec.europa.eu/docsroom/documents/41863

[24] https://www.enisa.europa.eu/publications/cybersecurity-and-privacy-in-ai-medical-imaging-diagnosis

to optimize security and privacy can often come at the expense of system performance, and consequently insists on the search for new security measures. The guide places strong emphasis on privacy issues alongside cybersecurity issues, mentioning privacy as one of the most important challenges facing society today.

In January 2025, the EU Council adopted the European Health Data Space Regulation (EHDS)[25] that aims at facilitating cross-border exchange of EU health data and improving individuals' control over how their health data is used.

AI Privacy Risk Assessment

As mentioned earlier, access to a trained ML model may be sufficient to reveal sensitive personal information about individuals whose data was used to train the model [4]. Examples of such attacks include:

Membership inference—a type of attack in which a given trained model and a data sample can deduce whether that sample was part of the model's training set [1]. This can be considered a privacy violation if the mere participation in training a model reveals sensitive information, such as in the case of disease progression prediction.

Attribute inference—an attack in which certain sensitive features may be inferred about individuals who participated in training a model. Given partial information about a person and an AI model that was trained on that person's data, it is possible to infer traditional, unknown information about that person [5].

Model inversion—an attack that attempts to reconstruct the representative value of a certain class of training data by inverting the weights of an AI model [6]. For example, it may be possible to reconstruct a person's facial image from a facial recognition model.

Data leakage—a relatively new type of attack targeting Natural Language Processing (NLP) models, where partial or complete text sequences from the model's training (or fine-tuning) data are reconstructed [3]. This can also lead to a secondary attack involving the leakage of Personally Identifiable Information (PII) or other confidential information if the reconstructed texts contain such data [7].

Prompt leakage—an attack that aims to reconstruct the system prompt, also known as the base prompt or prompt template from a Large Language Model (LLM) [8, 9]. This may reveal confidential information about a model's inner workings, sensitive information in the form of few-shot examples, and can also be used as a first step towards more sophisticated prompt injection attacks.

The existence of such attacks has led many to conclude that, in some cases, ML models themselves should be considered personal data, and therefore, subject to GDPR and similar regulations. It is therefore crucial to assess the privacy risks of AI models that may contain personal data before they are deployed, allowing time to apply appropriate mitigation strategies. Such assessments are also important to support the comparison and selection of different ML models based not only

[25] https://health.ec.europa.eu/ehealth-digital-health-and-care/european-health-data-space_en

on accuracy but also on privacy risk, thus enabling informed decisions about which model is most suitable for a given use case. Privacy risk assessment may be a one-time process, or it could be applied multiple times or even continuously.

Many organizations have in place qualitative risk assessment processes that may include documentation of the model development and datasets used, questionnaires filled by different stakeholders, and even dedicated teams that manually analyse models and report on their findings. However, these model validator teams often constitute a bottleneck in releasing new models to production, since these manual inspections take a lot of time and expertise. More recently, quantitative methods have started to become available that apply automated algorithms to measure different aspects of AI risk.

Most quantitative privacy assessments involve running one or more inference attacks against the model and measuring their success rate. The success rate of an attack may be evaluated using different metrics, such as accuracy, precision, recall, ROC (Receiver Operating Characteristic) curve, True Positive Rate (TPR) at low False Positive Rate (FPR). Such a risk assessment can also result in computing risk scores based on attack results that can be used to compare models or serve as a gatekeeping mechanism for release to production.

1.5 AI Risk Catalogs

The MITRE ATLAS (Adversarial Threat Landscape for Artificial-Intelligence Systems)[26] is a globally accessible, living knowledge base of adversary tactics and techniques against Al-enabled systems, based on real-world attack observations and realistic demonstrations from Al red teams and security groups. It lists 82 techniques as part of 14 tactics to breach the security and privacy of AI systems. One such tactic is Exfiltration, which includes Exfiltration via ML Inference API, including Inferring Training Data Membership, Inverting ML Models, and Extracting ML Models, as well as LLM Meta Prompt Extraction, where an adversary may induce an LLM to reveal its initial instructions, or "meta prompt."; and LLM Data Leakage, which causes the LLM to leak sensitive information. The framework also outlines several mitigations for each such threat.

The OECD (Organisation for Economic Co-operation and Development) is a forum and knowledge hub for data, analysis, and best practices in public policy. It provides the OECD. AI Catalog of Tools & Metrics for Trustworthy AI.[27] These tools and metrics are designed to help AI actors develop and use Trustworthy AI systems and applications that respect human rights and are fair, transparent, explainable, robust, secure and safe.

[26] https://atlas.mitre.org/matrices/ATLAS

[27] https://oecd.ai/en/catalogue/metrics

The OWASP Top 10 for LLM Applications[28] is a community-driven effort to highlight and address security issues specific to AI applications. Among other risks, it lists Sensitive Information Disclosure as risk number 2 (LLM02:2025): LLM applications have the potential to reveal sensitive information, proprietary algorithms, or other confidential details through their output. This can result in unauthorized access to sensitive data, intellectual property, privacy violations, and other security breaches. Examples of listed vulnerabilities include:

- Incomplete or improper filtering of sensitive information in LLM responses.
- Overfitting or memorization of sensitive data in LLM training.
- Unintended disclosure of confidential information due to LLM misinterpretation, lack of data scrubbing methods or errors.

LLM07:2025 System Prompt Leakage refers to the risk that the system prompts or instructions used to steer the behavior of the model may also contain sensitive information that was not intended to be disclosed.

MIT AI Risk Repository[29] is a comprehensive living database of over 1000 AI risks categorized by their cause and risk domain. It lists AI risks related to privacy and security, such as compromise of privacy by obtaining, leaking, or correctly inferring sensitive information.

Google's Secure AI Framework contains both a risk map[30] that displays different risks associated with AI as part of the AI development lifecycle and a description of each risk.[31]

Finally, the IBM AI Risk Atlas[32] details some of the risks of working with generative AI, foundation models, and machine learning models. These are categorized into four types: risks associated with training data, risks associated with model inference, risks associated with model outputs and non-technical risks. In the context of privacy, it lists risks such as confidential data in prompt, exposing personal information, and revealing confidential information.

1.6 AI Risk Assessment Guidelines and Frameworks

The American National Institute of Standards and Technology (NIST) published in 2023 an AI Risk Management Framework (AI RMF)[33] aimed at managing risks to individuals, organizations, and society associated with AI. The framework was developed through an open and collaborative process, involving stakeholders from

[28] https://owasp.org/www-project-top-10-for-large-language-model-applications/assets/PDF/OWASP-Top-10-for-LLMs-2023-v1_1.pdf

[29] https://airisk.mit.edu/

[30] https://saif.google/secure-ai-framework/saif-map

[31] https://saif.google/secure-ai-framework/risks

[32] https://www.ibm.com/docs/en/watsonx-as-a-service?topic=ai-risk-atlas

[33] https://www.nist.gov/itl/ai-risk-management-framework

both the private and public sectors. It aims to improve the ability to incorporate trustworthiness considerations into the design, development, use, and evaluation of AI products, services, and systems. The framework includes aspects of explainability, reliability, privacy, robustness, safety, security, and bias. It states that "Privacy values such as anonymity, confidentiality, and control generally should guide choices for AI system design, development, and deployment. Like safety and security, specific technical features of an AI system may promote or reduce privacy. AI systems can also present new risks to privacy by allowing inference to identify individuals or previously private information about individuals".

An associated RMF Playbook[34] has also been published, providing concrete actions to help achieve the goals laid out in the RMF. It offers guidance in four main categories: Govern, Map, Measure and Manage. As part of the Map section, the document recommends transparency and documentation that includes documenting which assessments were conducted on trustworthiness characteristics, specifically on security and privacy. Measure 2.10 is dedicated entirely to privacy risks and recommends the use of Privacy-enhancing technologies (PETs), such as Differential Privacy (DP).

In April 2019, the European Commission published the Ethics Guidelines for Trustworthy AI.[35] Its second chapter provides guidance on how Trustworthy AI can be realized, by listing seven requirements that AI systems should meet: (1) human agency and oversight, (2) technical robustness and safety, (3) privacy and data governance, (4) transparency, (5) diversity, non-discrimination and fairness, (6) environmental and societal well-being, and (7) accountability.

Recently, the OWASP foundation published a GenAI red-teaming guide[36] that provides recommendations on how to identify vulnerabilities and mitigate risks in AI systems, with a special focus on security and privacy. One type of risk discussed in this document is: Data Risks–Protecting against leakage of sensitive or training data. The guide recommends thoroughly and proactively testing these systems to ensure the early identification and mitigation of data leakage, before this kind of vulnerability can be exploited and cause actual harm.

It also presents some practical techniques to address privacy and data leakage, such as probing the system for potential exposure of sensitive information, assessing the model's risk of data extraction, and ensuring proper access control is in place in RAG systems that contain confidential documents.

Lately, several alliances or coalitions of multiple public and private sector partners have arisen with the goal of fostering trustworthy, responsible, and safe AI. The AI Alliance[37] is a group of technology developers, researchers, industry leaders, and advocates who collaborate to advance safe, responsible AI rooted in open innovation. With Trust and Safety as one of their focus areas, they aim to build tools, refine

[34] https://airc.nist.gov/AI_RMF_Knowledge_Base/Playbook

[35] https://digital-strategy.ec.europa.eu/en/library/ethics-guidelines-trustworthy-ai

[36] https://genai.owasp.org/resource/genai-red-teaming-guide/

[37] https://thealliance.ai/

methods, and create benchmarks for detecting and mitigating those risks. The Coalition for Secure AI (CoSAI)[38] is more focused on security aspects in AI. It is composed of an open ecosystem of AI and security experts from industry-leading organizations dedicated to sharing best practices for secure AI deployment and collaborating on AI security research.

1.7 AI Risk Assessment Tools

Based on the guidelines and recommendations to assess privacy risks associated with AI, there is a pressing need for easy-to-use and informative AI risk assessment tools.

1.7.1 Open-Source Risk Assessment Tools

There are various open-source packages that can help assess AI Privacy risk:

- The Adversarial Robustness Toolbox[39] (ART) [10] provides several privacy attack implementations (membership inference, attribute inference, model inversion, database reconstruction), as well as membership leakage metrics.
- ML-privacy-meter[40] [11, 12] is a Python library that enables quantifying the privacy risks of ML models for classification, regression, computer vision, and NLP based on state-of-the-art membership inference attacks.
- COMP-AI[41] [13] is a compliance-centered evaluation framework for Generative AI that measures LLMs' compliance with the EU AI Act.
- MIMIR[42] [14] is a Python package for measuring memorization in LLMs.
- TensorFlow Privacy[43]—is an open-source library that provides TensorFlow optimizers for training ML models with DP, as well as empirical tests for measuring potential memorization using membership inference attacks.
- AIJack [15] is an easy-to-use open-source simulation tool for testing the security of AI systems against hijackers. AIJack can test and simulate defenses against various attacks such as model poisoning, model inversion, backdoor, and Free-Rider.
- Microsoft's responsible-ai-toolbox-privacy[44] is a library for statistically estimating the privacy of ML pipelines using membership inference attacks.

---

[38] https://www.coalitionforsecureai.org/

[39] https://github.com/Trusted-AI/adversarial-robustness-toolbox

[40] https://github.com/privacytrustlab/ml_privacy_meter

[41] compl-ai/compl-ai: An open-source compliance-centered evaluation framework for Generative AI models

[42] https://github.com/iamgroot42/mimir

[43] https://github.com/tensorflow/privacy/tree/master/tensorflow_privacy/privacy/privacy_tests

[44] https://github.com/microsoft/responsible-ai-toolbox-privacy

1.7.2 End-to-End AI Risk Assessment Tool

Existing tools and frameworks for assessing privacy risk of ML models tend to be low-level, requiring a high degree of expertise to employ, or are tightly coupled with a specific ML framework. Each type of attack may have one or more implementations for different access modes (black-box vs. white-box), model types (classical ML/deep learning/LLMs), and types of model outputs (classification/regression/generative). It is therefore quite complicated to discern, given a specific model and dataset, which attacks to run and with which parameters. It can also be difficult to correctly interpret the results of the attacks and metrics. There is a strong need for more automated solutions that can enable non-experts to perform, and more importantly, understand the results of, such privacy risk assessments.

The AI Privacy Risk Assessment [16] tool is an end-to-end framework for privacy risk assessment of AI models. It aims to alleviate the issues of existing tools and frameworks by automatically selecting which attacks and metrics to run based on Q&A-based interactions with the user, then running the attacks, summarizing and visualizing the results in an easy-to-consume manner.

The first goal of this end-to-end risk assessment tool is to automate many of the decisions around which attacks and metrics to run, and all of the technical preparation required in order to run them. For example, some attacks require specific data preparation or have multiple runtime options and parameters that need to be set in a way that best matches the given model and data.

Secondly, since most non-technical users cannot understand the meaning of each individual attack or score, let alone compare the results of different models. It is therefore crucial to summarize these individual results into an overall privacy risk score.

The tool is generic and enables evaluating models from different ML frameworks (scikitlearn,[45] pytorch,[46] keras[47]), different data modalities (tabular, images, text) and different model types (classification, regression, detection). This is done by decoupling the actual models and datasets from the rest of the framework using generic wrappers. It performs multiple attack runs with different random data splits, collecting performance metrics like accuracy, precision, recall, and F1 score. The results are stored in a JSON structure and aggregated into an overall privacy health score for the model (Fig. 1).

The tool also employs a novel attack framework introduced in [17], significantly enhancing membership inference attacks against classification models. This framework leverages an ensemble method, generating multiple specialized attack models for different data subsets. By aggregating results from these specialized models, the framework provides a more comprehensive assessment of the target model's privacy risks. This approach outperforms single attack models and per-class attack models across classical and language classification tasks.

[45] https://scikit-learn.org/stable/

[46] https://pytorch.org/

[47] https://keras.io/

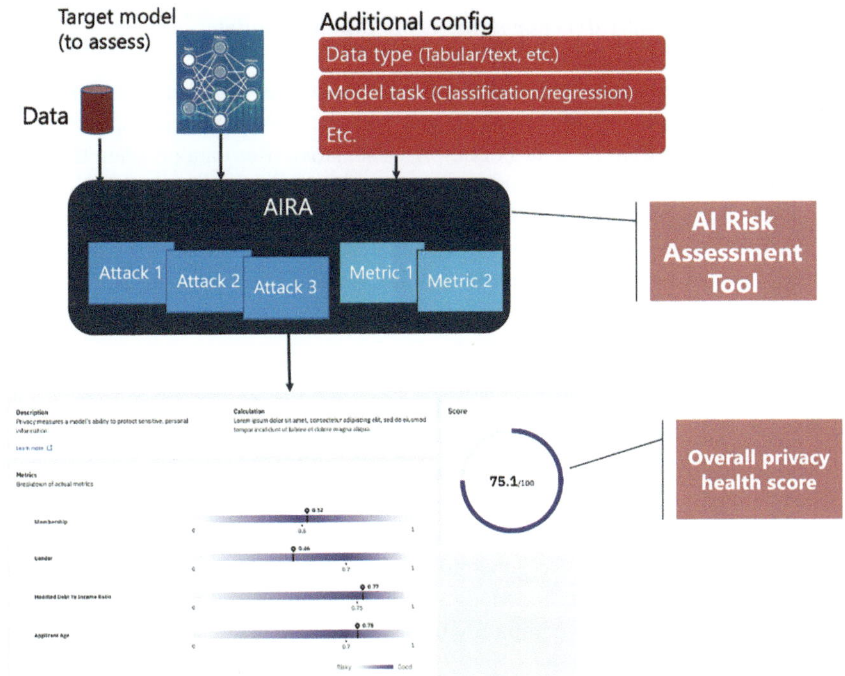

Fig. 1 AI risk assessment tool

AI Privacy Risk Assessment of Various Model Types

Let us examine the privacy risk assessment of tabular AI models, time-series AI models, and Large Language Models (LLMs).

1.8 AI Privacy Attacks on Tabular Models

There are several types of privacy attacks against ML models, including membership inference, attribute inference, model inversion, and database reconstruction. The most commonly researched and employed attack is membership inference, with dozens of papers published each year [18], and implementations available in open-source privacy assessment frameworks [10, 11]. These attacks may be performed either in a black-box manner—where only the output of the model on a particular input is known—or in a white-box manner, where internal parameters of the model are also known.

1.8.1 Membership Inference Attacks on Tabular Models

Membership inference attacks (MIA) attempt to distinguish between members, who were part of a target model's training data, and non-members. MIAs are especially relevant when the training dataset comprises sensitive user data or when the model is used to predict sensitive attributes, such as medical conditions. When such an attack succeeds, the adversary may gain insight into private information related to a person, thereby compromising their privacy. For example, by exposing the participation of an individual in the training dataset used for a model predicting the disease progression of Alzheimer's, an adversary gains knowledge about the individual having Alzheimer's disease.

MIAs have been extensively studied in the context of classification models and in the black-box setting, where the model internals are unknown to the attacker. The first MIAs were either threshold-based [19] or employed binary classifiers trained to distinguish between members and non-members based on model outputs [1]. For example, these outputs may include class probabilities or logits (for classification models), the model's loss, or possibly activations from internal layers of the model (in white-box attacks) [20]. To generate labeled (member/non-member) data to train the attack classifier without knowledge of the true member samples of the attacked model, shadow models are commonly used [1].

1.8.2 Attribute Inference and Model Inversion Attacks on Tabular Models

Attribute inference and model inversion (MI) attacks attempt to reconstruct new information about training data samples based on the trained model. Attribute inference usually refers to an attack where certain sensitive features may be inferred about individuals who participated in training a model [2, 21, 5]. Given a trained model and knowledge about some of the features of a specific person, it may be possible to deduce the value of additional, unknown features of that person.

Model inversion typically aims to reconstruct representative feature values of the training data by inverting a trained ML model [21–23]. For example, it may be possible to reconstruct what the average sample for a given class looks like. This can be considered a privacy violation if a class represents a specific person or group of people, such as in facial recognition models. MI attacks usually require white-box access to the model.

1.8.3 AI Privacy Risk Assessment of Tabular Models in Healthcare

In the healthcare domain, there are many use cases where AI models are trained on tabular data, including past clinical data, various test results, medication doses, sensor data, and more.

One such example is an ML model that uses patients' vital signs to detect sleep stages and associated disorders. These vital signs, including heart rate and respiration rate, are extracted from sensors present in wearable devices placed on the patient's body. They can be used to train different types of ML classifiers, including: Decision Tree classifier, Multi-layer Perceptron, k-Nearest Neighbors classifier, AdaBoost, Random Forest, and Logistic Regression. The result is a mapping to one of five different sleep stages.

Another example includes ML models that perform quantitative risk assessment to help determine a patient's risk for melanoma and other skin cancers based on clinical, genetic and familial history information. This includes demographic data, such as age and place of residence, medical history and historical sun exposure data. The output of the first model is a probability that indicates the risk of developing first primary melanoma. In addition, a multi-label classification model simultaneously assesses the risk of both melanoma and other skin cancers.

Both membership inference and attribute inference attacks pose a privacy threat to these ML models, since even the fact that a patient's data is part of a dataset of patients having a specific disease is already information that a patient might not want to disclose, and its leakage might result in patient harm, for example, negatively affecting the patient's employment prospects. In addition, inferring a patient's sensitive attributes directly violates their data privacy and undermines patient trust. Inference of the patient's demographic data and medical history would also go against the guidelines and recommendations described in sections Regulations Related to AI Privacy in Healthcare and AI Risk Assessment Guidelines and Frameworks. Finally, for the melanoma risk detection models, since their training data also includes family history and genetic information, the privacy risk is not only to the patient participating in the study, but also to unaware family members, which poses an even greater issue.

1.9 AI Privacy Attacks on Time-Series Models

Time-series (TS) forecasting uses historical observations to predict future values, making it indispensable in areas like healthcare, economics, and environmental science [24]. Over the years, it has progressed from classical statistical techniques, such as ARIMA and exponential smoothing [25], to state-of-the-art machine learning models like the Neural Fourier Transform (NFT) [26], TimesNet [27], and PatchTST [28].

Time-series data can be decomposed into two main components: **trend** and **seasonality**. The trend of the data represents the long-term direction of the data. For instance, in the health domain, given a series of an infant's hourly weight over two years, the overall weight increase represents the trend. In contrast, seasonality represents cycles in the data. For example, the infant's weight measurements taken in the morning are usually lower than those taken at night, producing a regular daily cycle (Fig. 2).

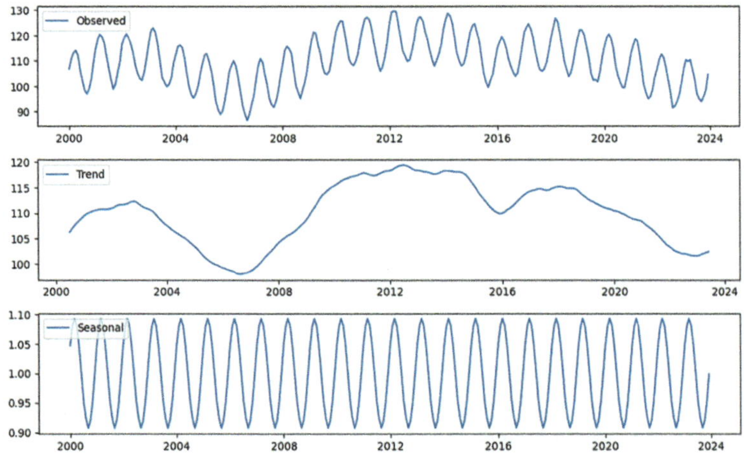

Fig. 2 Decomposition of the original time-series into its trend and seasonal components, highlighting both the long-term trajectory and recurring patterns over time

1.9.1 Membership Inference Attacks on Time-Series Models

In [29], the authors address the vulnerability of time-series forecasting models to membership inference attacks (MIAs). This is especially significant in medical contexts, where TS data often includes highly sensitive, personal medical data, such as blood-glucose levels, heart monitoring signals, or disease progression. Despite relying on highly sensitive patient data, these forecasting models have not been thoroughly examined for privacy risks, so there is a pressing need for focused research in this area.

Many TS forecasting models inherently capture trend and seasonality to improve predictive performance. In [29], the authors show that these very components can also be exploited to enhance MIAs. They introduce two novel features: one approximates the predicted time-series trend using a low-degree polynomial, while the other extracts seasonality through the Discrete Fourier Transform (DFT). Since state-of-the-art models like the Neural Fourier Transform (NFT) [26], TimesNet [27], and more explicitly incorporate trend and seasonality, they are more likely to accurately learn and reproduce these patterns if a series is included in their training data, thereby amplifying the potential for privacy leaks (Fig. 3).

The results demonstrate that embedding time-series-specific characteristics—like trend and seasonality—into the attack vector led to substantially higher risk scores

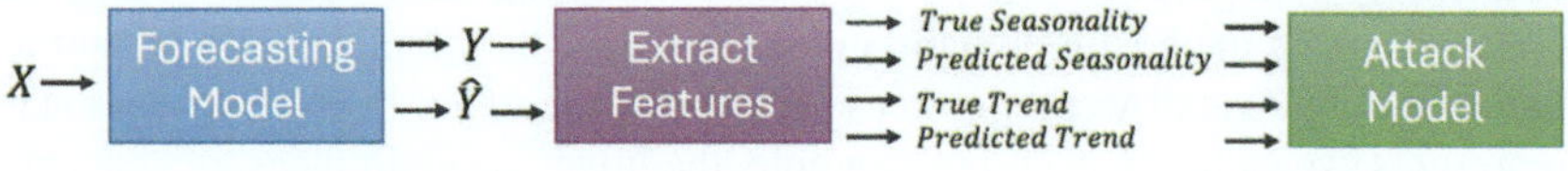

Fig. 3 Overview of the membership inference attack pipeline for time-series models

compared to conventional feature sets. In other words, constructing attack vectors around a series' unique patterns makes forecasting models far more susceptible to membership inference. By exposing this weakness, this work sets the stage for more in-depth studies of membership inference attacks on time-series forecasting.

1.9.2 AI Privacy Risk Assessment of TS Models in Healthcare

Let us examine an example model for predicting blood-glucose levels in diabetes patients. The model aims to perform personalized blood-glucose prediction for patients suffering from type 1 diabetes (PwT1D) and to reduce the effect of inter-patient variability through grouping training data according to glycemic similarities for improving personalized prediction accuracy.

The model takes as input a multi-dimensional time-series featuring: Continuous Glucose Monitoring (CGM) data, bolus insulin injections, and carbohydrate intake; and predicts the next blood-glucose values with prediction horizons of 30 and 60 min. This model is a sequence-to-sequence model based on an LSTM neural network architecture.

To reduce the setup time of a device for a new patient, the model is pre-trained on multiple patients' data in a centralized manner, and then fine-tuned on the specific patient's data once deployed on the patient's device. This both reduces the time required before the system becomes functional and decreases the computational cost of local model training.

Since this pre-trained model will be deployed on every patient's local device, it is important to ensure that it does not leak the sensitive health data of the patients used to train the model. Moreover, TS data, and especially multi-dimensional TS data, tends to be much more unique than typical tabular data, increasing even further the privacy leakage risk, and specifically the membership inference risk, as shown in [29].

1.10 AI Privacy Attacks on Large Language Models

Large Language Models are susceptible to both the same privacy attacks as classic AI models and to new attacks specific to Large Language Models.

1.10.1 Membership Inference Attacks on LLMs

Membership Inference Attacks (MIAs) aim to determine if a specific data point was part of a model's training dataset. Since they were initially introduced, MIAs have been extended to LLMs, where they exploit model overfitting and memorization tendencies to infer training membership [3].

Text embedding models (e.g., BERT, Word2Vec) encode inputs into vector representations. These models can inadvertently expose training membership, as shown in [30]. [31] extended this to include downstream NLP models that use these embeddings. In healthcare, clinical embeddings trained on patient records can expose whether a specific term or phrase was seen in training [32].

Generative LLMs, like GPT and LLaMA, are susceptible to MIAs via likelihood-based attacks, where membership is inferred by observing model confidence [33]. [3] showed that LLMs occasionally memorize training data verbatim, making MIAs particularly effective when targeting rare sequences.

Recent techniques, such as neighborhood comparison [34], generate perturbations of an input to compare loss distributions, improving attack success without requiring a reference model. Duan et al.[14], however, found that MIAs on large, diverse LLMs often fail, as models trained on vast corpora generalize well, obscuring membership signals. Document-level MIAs are an emerging risk, where complete documents (e.g., books, medical records) are used for the attack, i.e., checking if a specific document was part of the training data of an LLM [35].

MIA in LLMs remains an active research area, with text embeddings and generative models vulnerable under certain conditions. While successful MIAs are harder to perform against large, commercial models, smaller fine-tuned language models, such as those typically used in healthcare, pose significant privacy risks, making them particularly attractive targets for adversaries seeking to extract sensitive information. As research progresses, considering the tradeoff between model utility with privacy will play a crucial step for responsible LLM development.

1.10.2 Data Leakage Attacks on LLMs

Generative language models have a natural tendency to memorize training or fine-tuning samples. This tendency can be exploited by data extraction attacks to extract from the LLM partial or complete samples from the training or fine-tuning datasets, or any other dataset that the model is exposed to during inference [3, 36–38]. This is crucial for the healthcare domain as it may reveal personal medical information about individuals.

Data leakage attacks usually fall under one of two categories: attacks that utilize prior knowledge about training samples to carry out the attack, and those that do not assume such knowledge.

Attacks that utilize prior knowledge about training samples typically split the samples that are being extracted into two parts: a prefix and a suffix [3]. The prefix is used as input to the language model, aiming to elicit it to complete the suffix. The suffix is then compared with the model's actual output to determine whether leakage has occurred. Optionally, the entire dataset may be indexed to enable identification of whether generated text is from any of the training data samples. Common methods of scoring the match usually include exact matches, n-grams or other distance metrics [39, 40].

Prefix: Usually ALT or AST values higher than "two

Suffix: times the upper normal limit", is considered abnormal (in some countries, ALT or AST values of more than 100 are considered abnormal).

LLM Generated text: [INST] Thanks for posting query at Chat Doctor. Usually ALT or AST values higher than "two /[INST] at most two times the upper normal limit", is considered abnormal (in some countries, ALT or AST values of more than 100 are considered abnormal). Value of AST or ALT greater than 85 or above maybe investigated further. Alcohol ingestion and obesity are common causes of fatty liver disease. Advice :-

Fig. 4 LLM leaking the entire suffix given the prefix

Existing solutions use fixed sizes of prefixes and suffixes, based on the number of tokens in each part [3]. This token-wise split may cause the sample to be divided in a way that breaks the semantic meaning of the text and thus the model might get confused or simply not output the desired suffix. Another possible approach is to split each sample first into sentences, and then use a word-based split instead of a token-based. Since the model occasionally leaks information in subsequent sentences, it is also beneficial to "force" the model to generate a response that is longer than the target suffix. This method increases the chances of the LLM models leaking the remainder of the sentence (suffix), or even other sentences from the entire dataset. Fig. 4 demonstrates an example leakage for the mistral[48] model fine-tuned on the HealthCareMagic[49] dataset.

Common methods of scoring the match, such as considering only exact matches, or comparing long sequences of text, may miss many more subtle examples of data leakage. This method employs new lexical similarity techniques to compare the suffix to the output sentences, giving a higher score to longer leaks, allowing for a looser definition of similarity, for example, by identifying sequences that do not preserve the exact order or have some missing words.

For the second class of attacks that does not use training data, an attacker may leverage specific techniques to cause the model to "diverge" from its aligned behavior and output training data. This can be done, for example, by asking the model to repeat a word forever [41], or by using special characters [42]. This can also lead to a secondary type of PII/confidential information leakage attack if the reconstructed texts in fact contain PII or confidential information [7, 43].

Another possible use case for this kind of attack is to identify whether an LLM model was trained or fine-tuned on a dataset that is not allowed, such as cases of copyright infringement [44].

In the healthcare domain, it is common to use synthetic data to overcome privacy risks due to potential leakage of personal information [45]. However, as shown in [46], privacy leakages may occur even when synthetic data is used. Data leakage assessment can be leveraged during synthetic data generation to ensure that real data does not leak into the generated data.

[48] https://huggingface.co/mistralai/Mistral-7B-Instruct-v0.2

[49] https://huggingface.co/datasets/RafaelMPereira/HealthCareMagic-100k-Chat-Format-en

1.10.3 Prompt Leakage Attacks on LLMs

Prompt leakage [8, 9, 47], or a system prompt extraction, is a sophisticated attack vector concerning the leaking of the system instructions defined in an LLM-based application. The attack, often masked as a benign user query, tricks the system into revealing the system prompt, thus revealing the application design. This can have severe implications since the system prompt defines the application's (for example, a chatbot) capabilities, response format, and the tools it can access. It can also include personal or sensitive information in the form of a few-shot examples.

An example of a system prompt for a medical chatbot application could be:

You are a medical chatbot. You always start your answers with "Welcome to Medi-Chat." Your job is to advise on medical issues presented by clients, as if you are a family doctor. Be mindful of the advice you give, as some patients may act on it, which may cause harm.

In addition, you have access to the following tools:

Appointment Scheduler: You can use it to set up appointments for patients with real doctors. You can use it by calling appointment_scheduler(date, doctor, location).

Information Search: Use it to provide references to support your answers. You can use this tool by calling search(topic).

Prompt leakage attacks succeed due to two main factors: First, LLMs struggle to distinguish between different parts of the prompt, for example, instructions and conversation with the user. By prompting the LLM and asking it to repeat all the entire conversation, an adversary can reveal the system prompt. Second, LLMs are easy to trick: by crafting a query that redefines the LLM's instructions, an adversary can manipulate the model to extract sensitive information or even alter the behavior of the chatbot by replacing the system prompt for the current session.

In healthcare, the consequences of prompt leakage can be dire. The instructions provided to a medical chatbot define its capabilities and the style of its responses. If these instructions are leaked, adversaries can gain knowledge about the chatbot's internals, conceivably exposing vulnerabilities in the system's data protection mechanisms. An attacker could use this knowledge to craft more convincing phishing attempts or social engineering attacks.

The list of functions or tools that the chatbot can access may provide attackers with a roadmap of the chatbot's functionalities, enabling them to target specific tools for exploitation. This knowledge could be used to leak sensitive medical information or manipulate the chatbot into providing misleading or harmful advice. For instance, an attacker could attempt to trick the Appointment Scheduler function into supplying information about current appointments and patients.

Aside from the list of tools, few-shot examples, also known as In-Context Learning Examples (ICLs), are a common part of the system prompt. These are pairs of questions and answers demonstrating how the chatbot should respond to certain queries. Hence, extraction of ICLs can reveal the chatbot's response patterns and the types of queries it is designed to handle. In addition, the exposure of ICLs could also result in a patients' privacy breach and violate regulations since ICLs are usually selected from existing databases—which in the healthcare domain might be related to former (real) patients. For these reasons, it is imperative for healthcare organizations to deploy robust security and privacy measures to protect against prompt leakage and ensure the integrity and confidentiality of all the system's internal knowledge.

1.10.4 Attacks on Retrieval Augment Generation (RAG) Systems

Retrieval Augmented Generation (RAG) [48] is a technique that enriches the knowledge of Large Language Models (LLMs) with data from external databases without requiring any additional training. LLMs can thus be customized for specific needs, such as creating AI-based personal assistants or making long textual content (such as medical articles) accessible using simple text queries [49].

Prior to deploying a RAG pipeline, a retrieval database must be populated with documents. Once deployed, the RAG pipeline typically consists of two main phases: search and generation, Fig. 5. In the search phase, the retrieval database is queried to find relevant documents that match the user's query or prompt. In the generation phase, the information retrieved in the search phase is inserted into the prompt, which, together with the user prompt, is provided to the model to generate the final response.

Privacy attacks against RAG systems may aim to extract sensitive information either from the retrieval database or from the generative LLM [50]. All the attacks described in the previous sections are also relevant to RAG systems, in addition to attacks that are specific to RAG systems:

Membership Inference Attacks try to determine if a specific data sample is part of the retrieval database. This can potentially reveal sensitive or proprietary information included in the retrieval database. This kind of attack can be carried out by simply observing the final responses of the RAG system (black-box scenario), using a carefully crafted prompt intended to extract this information [51]. If the token logits are available in the model's output, a gray-box attack can be used to further increase the rate of success [17]. Another possibility is to use the semantic similarity between a target sample and the content generated by the RAG system as the

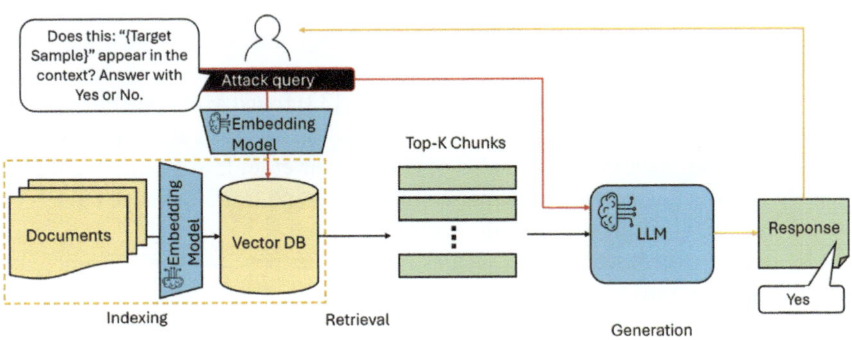

Fig. 5 Overall flow of an MIA on a RAG pipeline

membership score, based on the assumption that if a sample is present in the retrieval database, the generated content will be similar to this sample when querying the RAG system [52]. Alternatively, an Interrogation Attack [53] involves crafting natural-text queries that are answerable only with the target document's presence and measuring the correctness of the model's answers.

Data Extraction Attacks try to extract sensitive sequences of text from the retrieval database. The tendency of LLM models to repeat parts of the input prompt poses a risk of leaking information from the RAG database via the context prepended to the user prompt. One work shows that an adversary can exploit LLMs' instruction-following capabilities to easily extract text data verbatim from the datastore of RAG systems via prompt injection [54]. Attackers can also gradually extract information from the database by crafting carefully designed questions, learning from model responses, and progressively generating new queries to extract as many chunks from the knowledge base as possible [55]. This can potentially expose sensitive information present in the retrieval database, such as patients' Electronic Health Records (EHRs).

1.10.5 AI Privacy Risk Assessment of LLMs in Healthcare

One common use of LLMs in the healthcare domain is medical chatbots. These can help ease the burden on doctors and administrators in primary care. Chatbots that deliver remote health services provide patients with real-time self-support and can help them manage simple medical issues. They can improve patient experience and are promising tools for improving health care efficiency and quality [56].

In addition, we consider a language classification model that can be used in healthcare applications. This model takes as input text messages sent from a patient to their physician and tries to detect various mental disorders, such as anxiety and depression, based on these messages.

Healthcare applications of LLMs are particularly vulnerable to privacy attacks. Membership inference could reveal a patient's inclusion in a specific hospital or clinical trial dataset, which may reveal a particular disease or condition, or where and how a patient is being treated. Different leakage attacks can directly expose a patient's medical conditions, treatments, confidential interactions between patients and physicians, and many more. Jagannatha et al. [32] found MIAs on clinical LLMs to be 7% more effective than random guessing, while Zhang et al. [57] showed that synthetic health data models are also at risk. Sarkar et al. [58] further highlighted that de-identified patient notes do not eliminate MIA risks, as language models can retain implicit identifiers from training data.

Moreover, generative models, such as medical chatbots, may be vulnerable to a multitude of training, fine-tuning, and RAG data leakage attacks that directly replicate whole or partial sequences of text the model was exposed to, either during training or at inference time. It has even been found that language classification models, such as the one for detecting mental disorders based on patients' messages, can be vulnerable to text extraction attacks [59].

Performing AI risk assessment on these models can be extremely useful in making decisions such as choosing between different applicable models and model hyperparameters, or applying mitigations during the development of the model to potentially reduce the privacy risk posed. These findings also underscore the need for robust privacy defenses, such as differential privacy and data sanitization, before deploying LLMs in medical applications.

2 Conclusion

As AI advances and is becoming more embedded in healthcare solutions, so do privacy concerns and vulnerabilities increase. Subsequently, the need to safeguard and mitigate against these risks increases as well. This paper dives into AI Privacy in healthcare scenarios, from the data protection regulations and requirements, through a presentation of practical tooling for privacy risk assessment and vulnerability detection. We examine different model types, including tabular, time-series, and Large Language Models, and show in each how a targeted attack can exploit the models' behavior to expose unintended data. Our proposed framework offers an end-to-end, model agnostic, privacy evaluation which in turn enables enterprises better visibility on the privacy aspects of their AI deployments and builds their trust towards quicker adoption of AI-driven healthcare solutions.

Acknowledgments This work was performed as part of the NEMECYS project, which is co-funded by the European Union under grant agreement ID 101094323, by UK Research and Innovation (UKRI) under the UK government's Horizon Europe funding guarantee grant numbers 10065802, 10050933, and 10061304, and by the Swiss State Secretariat for Education, Research and Innovation (SERI).

References

1. R. Shokri, M. Stronati, C. Song, and V. Shmatikov, "Membership inference attacks against machine learning models," presented at the 2017 IEEE symposium on security and privacy (SP), IEEE, 2017, pp. 3–18.
2. M. Fredrikson, E. Lantz, S. Jha, S. Lin, D. Page, and T. Ristenpart, "Privacy in pharmacogenetics: An {End-to-End} case study of personalized warfarin dosing," in *23rd USENIX security symposium (USENIX Security 14)*, 2014, pp. 17–32.
3. N. Carlini *et al.*, "Extracting Training Data from Large Language Models," presented at the 30th USENIX Security Symposium (USENIX Security 21), 2021, pp. 2633–2650. Accessed: Feb. 12, 2023. [Online]. Available: https://www.usenix.org/conference/usenixsecurity21/presentation/carlini-extracting
4. Rigaki, M. and Garcia, S., 2023. A survey of privacy attacks in machine learning. ACM Computing Surveys, 56(4), pp.1–34.

5. G. Ateniese, L. V. Mancini, A. Spognardi, A. Villani, D. Vitali, and G. Felici, "Hacking smart machines with smarter ones: How to extract meaningful data from machine learning classifiers," *Int. J. Secur. Netw.*, vol. 10, no. 3, pp. 137–150, 2015.
6. M. Fredrikson, S. Jha, and T. Ristenpart, "Model inversion attacks that exploit confidence information and basic countermeasures," presented at the Proceedings of the 22nd ACM SIGSAC conference on computer and communications security, 2015, pp. 1322–1333.
7. N. Lukas, A. Salem, R. Sim, S. Tople, L. Wutschitz, and S. Zanella-Béguelin, "Analyzing Leakage of Personally Identifiable Information in Language Models," Apr. 23, 2023, *arXiv*: arXiv:2302.00539. Accessed: Aug. 14, 2023. [Online]. Available: http://arxiv.org/abs/2302.00539
8. Y. Zhang, N. Carlini, and D. Ippolito, "Effective prompt extraction from language models," *ArXiv Prepr. ArXiv230706865*, 2023.
9. B. Hui, H. Yuan, N. Gong, P. Burlina, and Y. Cao, "Pleak: Prompt leaking attacks against large language model applications," in *Proceedings of the 2024 on ACM SIGSAC Conference on Computer and Communications Security*, 2024, pp. 3600–3614.
10. M.-I. Nicolae *et al.*, "Adversarial Robustness Toolbox v1. 0.0," *ArXiv Prepr. ArXiv180701069*, 2018.
11. S. Kumar and R. Shokri, "ML Privacy Meter: Aiding regulatory compliance by quantifying the privacy risks of machine learning," in *Workshop on Hot Topics in Privacy Enhancing Technologies (HotPETs)*, 2020.
12. J. Ye, A. Maddi, S. K. Murakonda, V. Bindschaedler, and R. Shokri, "Enhanced Membership Inference Attacks against Machine Learning Models," in *Proceedings of the 2022 ACM SIGSAC Conference on Computer and Communications Security*, 2022, pp. 3093–3106.
13. P. Guldimann *et al.*, "COMPL-AI Framework: A Technical Interpretation and LLM Benchmarking Suite for the EU Artificial Intelligence Act," *ArXiv Prepr. ArXiv241007959*, 2024.
14. M. Duan *et al.*, "Do Membership Inference Attacks Work on Large Language Models?," in *Conference on Language Modeling (COLM)*, 2024.
15. H. Takahashi, "AIJack: Security and Privacy Risk Simulator for Machine Learning," *ArXiv Prepr. ArXiv231217667*, 2023.
16. A. Goldsteen, S. Shachor, and N. Raznikov, "An end-to-end framework for privacy risk assessment of AI models," in *Proceedings of the 15th ACM International Conference on Systems and Storage*, in SYSTOR '22. New York, NY, USA: Association for Computing Machinery, 2022, p. 142. https://doi.org/10.1145/3534056.3534998.
17. S. Shachor, N. Razinkov, and A. Goldsteen, "Improved Membership Inference Attacks Against Language Classification Models," Oct. 11, 2023, *arXiv*: arXiv:2310.07219. https://doi.org/10.48550/arXiv.2310.07219.
18. H. Hu, Z. Salcic, L. Sun, G. Dobbie, P. S. Yu, and X. Zhang, "Membership inference attacks on machine learning: A survey," *ACM Comput. Surv. CSUR*, vol. 54, no. 11s, pp. 1–37, 2022.
19. S. Hisamoto, M. Post, and K. Duh, "Membership inference attacks on sequence-to-sequence models: Is my data in your machine translation system?," *Trans. Assoc. Comput. Linguist.*, vol. 8, pp. 49–63, 2020.
20. M. Nasr, R. Shokri, and A. Houmansadr, "Comprehensive Privacy Analysis of Deep Learning: Passive and Active White-box Inference Attacks against Centralized and Federated Learning," in *2019 IEEE Symposium on Security and Privacy (SP)*, May 2019, pp. 739–753. 10.1109/SP.2019.00065.
21. M. Fredrikson, S. Jha, and T. Ristenpart, "Model inversion attacks that exploit confidence information and basic countermeasures," in *Proceedings of the 22nd ACM SIGSAC conference on computer and communications security*, 2015, pp. 1322–1333.
22. U. Aïvodji, S. Gambs, and T. Ther, "Gamin: An adversarial approach to black-box model inversion," *ArXiv Prepr. ArXiv190911835*, 2019.
23. Z. He, T. Zhang, and R. B. Lee, "Model inversion attacks against collaborative inference," in *Proceedings of the 35th Annual Computer Security Applications Conference*, 2019, pp. 148–162.
24. R. J. Hyndman and G. Athanasopoulos, *Forecasting: principles and practice.* OTexts, 2018.

25. G. E. Box, G. M. Jenkins, G. C. Reinsel, and G. M. Ljung, *Time series analysis: forecasting and control*. John Wiley & Sons, 2015.
26. N. Koren and K. Radinsky, "Interpretable multivariate time series forecasting using neural fourier transform," *ArXiv Prepr. ArXiv240513812*, 2024.
27. H. Wu, T. Hu, Y. Liu, H. Zhou, J. Wang, and M. Long, "Timesnet: Temporal 2d-variation modeling for general time series analysis," *ArXiv Prepr. ArXiv221002186*, 2022.
28. Y. Nie, N. H. Nguyen, P. Sinthong, and J. Kalagnanam. "A time series is worth 64 words: Long-term forecasting with transformers," arXiv Prepr. arXiv:2211.14730, 2022.
29. N. Koren, A. Goldsteen, G. Amit, and A. Farkash, "Membership Inference Attacks Against Time-Series Models," *ArXiv Prepr. ArXiv240702870*, 2024.
30. C. Song and A. Raghunathan, "Information Leakage in Embedding Models," Aug. 19, 2020, *arXiv*: arXiv:2004.00053. https://doi.org/10.48550/arXiv.2004.00053.
31. S. Mahloujifar, H. A. Inan, M. Chase, E. Ghosh, and M. Hasegawa, "Membership inference on word embedding and beyond," *ArXiv Prepr. ArXiv210611384*, 2021.
32. A. Jagannatha, B. P. S. Rawat, and H. Yu, "Membership inference attack susceptibility of clinical language models," *ArXiv Prepr. ArXiv210408305*, 2021.
33. C. Song and V. Shmatikov, "Auditing data provenance in text-generation models," in *Proceedings of the 25th ACM SIGKDD International Conference on Knowledge Discovery & Data Mining*, 2019, pp. 196–206.
34. J. Mattern, F. Mireshghallah, Z. Jin, B. Schölkopf, M. Sachan, and T. Berg-Kirkpatrick, "Membership inference attacks against language models via neighbourhood comparison," *ArXiv Prepr. ArXiv230518462*, 2023.
35. M. Meeus, S. Jain, M. Rei, and Y.-A. de Montjoye, "Did the neurons read your book? document-level membership inference for large language models," in *33rd USENIX Security Symposium (USENIX Security 24)*, 2024, pp. 2369–2385.
36. N. Carlini, D. Ippolito, M. Jagielski, K. Lee, F. Tramer, and C. Zhang, "Quantifying memorization across neural language models," in *The Eleventh International Conference on Learning Representations*, 2022.
37. J. G. Wang, J. Wang, M. Li, and S. Neel, "Pandora's White-Box: Precise Training Data Detection and Extraction in Large Language Models," *ArXiv Prepr. ArXiv240217012*, 2024.
38. X. Zhou *et al.*, "LessLeak-Bench: A First Investigation of Data Leakage in LLMs Across 83 Software Engineering Benchmarks," Feb. 10, 2025, *arXiv*: arXiv:2502.06215. https://doi.org/10.48550/arXiv.2502.06215.
39. W. Yu *et al.*, "Bag of Tricks for Training Data Extraction from Language Models," Jun. 01, 2023, *arXiv*: arXiv:2302.04460. Accessed: Aug. 14, 2023. [Online]. Available: http://arxiv.org/abs/2302.04460
40. R. Xu, Z. Wang, R.-Z. Fan, and P. Liu, "Benchmarking Benchmark Leakage in Large Language Models," Apr. 29, 2024, *arXiv*: arXiv:2404.18824. https://doi.org/10.48550/arXiv.2404.18824.
41. M. Nasr *et al.*, "Scalable Extraction of Training Data from (Production) Language Models," Nov. 28, 2023, *arXiv*: https://arxiv.org/abs/2311.17035. https://doi.org/10.48550/arXiv.2311.17035.
42. Y. Bai, G. Pei, J. Gu, Y. Yang, and X. Ma, "Special characters attack: Toward scalable training data extraction from large language models," *ArXiv Prepr. ArXiv240505990*, 2024.
43. K. K. Nakka, A. Frikha, R. Mendes, X. Jiang, and X. Zhou, "PII-Scope: A Benchmark for Training Data PII Leakage Assessment in LLMs," Oct. 09, 2024, *arXiv*: arXiv:2410.06704. https://doi.org/10.48550/arXiv.2410.06704.
44. A. Karamolegkou, J. Li, L. Zhou, and A. Søgaard, "Copyright Violations and Large Language Models," Oct. 20, 2023, *arXiv*: arXiv:2310.13771. https://doi.org/10.48550/arXiv.2310.13771.
45. V. C. Pezoulas *et al.*, "Synthetic data generation methods in healthcare: A review on open-source tools and methods," *Comput. Struct. Biotechnol. J.*, vol. 23, pp. 2892–2910, Dec. 2024, https://doi.org/10.1016/j.csbj.2024.07.005.
46. M. Meeus, L. Wutschitz, S. Zanella-Béguelin, S. Tople, and R. Shokri, "The Canary's Echo: Auditing Privacy Risks of LLM-Generated Synthetic Text," Feb. 19, 2025, *arXiv*: https://arxiv.org/abs/2502.14921. https://doi.org/10.48550/arXiv.2502.14921.

47. T. Sternak, D. Runje, D. Granoša, and C. Wang, "Automating Prompt Leakage Attacks on Large Language Models Using Agentic Approach," Feb. 18, 2025, *arXiv*: arXiv:2502.12630. https://doi.org/10.48550/arXiv.2502.12630.
48. Lewis, Patrick, Ethan Perez, Aleksandra Piktus, Fabio Petroni, Vladimir Karpukhin, Naman Goyal, Heinrich Küttler et al. "Retrieval-augmented generation for knowledge-intensive nlp tasks." Advances in neural information processing systems 33 (2020): 9459–9474.
49. Gao, Yunfan, Yun Xiong, Xinyu Gao, Kangxiang Jia, Jinliu Pan, Yuxi Bi, Yixin Dai, Jiawei Sun, Haofen Wang, and Haofen Wang. "Retrieval-augmented generation for large language models: A survey." arXiv preprint https://arxiv.org/abs/2312.10997 2, no. 1 (2023).
50. S. Zeng *et al.*, "The Good and The Bad: Exploring Privacy Issues in Retrieval-Augmented Generation (RAG)," Feb. 23, 2024, *arXiv*: https://arxiv.org/abs/2402.16893. Accessed: Mar. 13, 2024. [Online]. Available: http://arxiv.org/abs/2402.16893
51. M. Anderson, G. Amit, and A. Goldsteen, "Is my data in your retrieval database? membership inference attacks against retrieval augmented generation," *ArXiv Prepr. ArXiv240520446*, 2024.
52. Y. Li, G. Liu, C. Wang, and Y. Yang, "Generating Is Believing: Membership Inference Attacks against Retrieval-Augmented Generation," Sep. 26, 2024, *arXiv*: arXiv:2406.19234. https://doi.org/10.48550/arXiv.2406.19234.
53. A. Naseh, Y. Peng, A. Suri, H. Chaudhari, A. Oprea, and A. Houmansadr, "Riddle Me This! Stealthy Membership Inference for Retrieval-Augmented Generation," Feb. 01, 2025, *arXiv*: arXiv:2502.00306. https://doi.org/10.48550/arXiv.2502.00306.
54. Z. Qi, H. Zhang, E. Xing, S. Kakade, and H. Lakkaraju, "Follow my instruction and spill the beans: Scalable data extraction from retrieval-augmented generation systems," *ArXiv Prepr. ArXiv240217840*, 2024.
55. C. Jiang, X. Pan, G. Hong, C. Bao, and M. Yang, "RAG-Thief: Scalable Extraction of Private Data from Retrieval-Augmented Generation Applications with Agent-based Attacks," Nov. 21, 2024, *arXiv*: arXiv:2411.14110. https://doi.org/10.48550/arXiv.2411.14110.
56. M. Laymouna, Y. Ma, D. Lessard, T. Schuster, K. Engler, and B. Lebouché, "Roles, users, benefits, and limitations of chatbots in health care: rapid review," *J. Med. Internet Res.*, vol. 26, p. e56930, 2024.
57. Z. Zhang, C. Yan, and B. A. Malin, "Membership inference attacks against synthetic health data," *J. Biomed. Inform.*, vol. 125, p. 103977, Jan. 2022, https://doi.org/10.1016/j.jbi.2021.103977.
58. A. R. Sarkar, Y.-S. Chuang, N. Mohammed, and X. Jiang, "De-identification is not enough: a comparison between de-identified and synthetic clinical notes," *Sci. Rep.*, vol. 14, no. 1, p. 29669, 2024.
59. A. Elmahdy, H. A. Inan, and R. Sim, "Privacy Leakage in Text Classification: A Data Extraction Approach," Jun. 09, 2022, *arXiv*: arXiv:2206.04591. https://doi.org/10.48550/arXiv.2206.04591.

EU-CIP Data Collection and Desk Research for Critical Infrastructure Protection and Resilience

Habtamu Abie, Sandeep Pirbhulal, Emilia Gugliandolo, Rita Ugarelli, and John Soldatos

Abstract The EU-CIP project has developed a methodology for continuous data collection and desk research, aimed at consistently gathering and analyzing data from critical infrastructures. This methodology facilitates the ongoing collection, analysis, curation, extraction, and presentation of information and insights about critical infrastructure protection and resilience (CIP/CIR). In this book chapter, we present this methodology and analyze the outcomes of two questionnaire surveys conducted within the EU-CIP consortium (i) internal stakeholders feedback survey: evaluating technological gaps, needs, and innovations in CIP/CIR, and (ii) survey on applying the developed methodology for capability and technology analysis in CIP/CIR. The survey results reveal some critical capability needs identified as lacking including enhanced adaptability, reduced response times, improved detection and impact assessment, and tools for managing cascading effects between entities and across borders. The results also revealed a preliminary list of capability gaps that align with the identified needs. These include inadequate automation, insufficient control over interconnectedness, poor alignment of resilience indicators, the absence of standardized stress-testing procedures, classification issues with IoT devices, scalability challenges in mitigating DDoS attacks, and gaps in emergency management processes. Addressing the identified capabilities needs and gaps is necessary for CIP/

H. Abie (✉) · S. Pirbhulal
Norwegian Computing Center, Blindern, Oslo, Norway
e-mail: habtamu.abie@nr.no

S. Pirbhulal
e-mail: sandeep@nr.no

E. Gugliandolo
Engineering Ingegneria Informatica S.p.A, Roma, RM, Italy
e-mail: Emilia.Gugliandolo@eng.it

R. Ugarelli
SINTEF Community, Blindern, Oslo, Norway
e-mail: rita.ugarelli@sintef.no

J. Soldatos
Innov-Acts LTD, Nicosia, Cyprus
e-mail: jsoldat@innov-acts.com

J. Soldatos and H. Abie (eds.), *Technology-Enabled Resilience: Innovations in Critical Infrastructure Protection*, https://doi.org/10.1007/978-3-032-25917-2_8

CIR. These insights will contribute to refining the strategies for the EU-CIP initiative, with a focus on practical and innovative solutions for CIP/CIR.

1 Introduction

Data collection techniques encompass a range of strategies and tools aimed at gathering both quantitative and qualitative information, essential for obtaining accurate and comprehensive data. Quantitative methods utilize systematic approaches such as numerical data collection, surveys, polls, and statistical analysis to measure and identify trends. In contrast, qualitative methods focus on non-numerical data gathering through interviews, focus groups, and observations to gain deeper insights into attitudes, behaviors, and motivations. By integrating both quantitative and qualitative approaches, organizations can enhance their datasets, providing a more nuanced understanding of complex phenomena. The indispensable role of data collection in research has been underscored [1, 2]. Taherdoos (2021) provides a detailed examination of the data collection process, outlining different types of data and appropriate methods for their collection. The paper serves as an extensive resource, detailing the data collection process, classifying primary data types, examining appropriate data collection methods, assessing the pros and cons of these methods, identifying typical challenges faced during data collection, and concluding with a discussion on the ethical considerations in data collection practices. Building on this foundation, [2] have emphasized the critical importance of data collection for effectively exploring research questions. They detail both primary and secondary data collection techniques, with primary methods in behavioral sciences including observation, interviews, questionnaires, and databases. They also discuss secondary data sources such as published books and journals, as well as unpublished autobiographies and biographies. Authors provide an extensive examination of these methods, highlighting their essential role in facilitating the completion of the research process and affirming data collection as a vital research tool. Together, these studies offer a robust framework for understanding and implementing effective data collection strategies in research.

Data curation is increasingly recognized as a critical component in enhancing the quality and utility of research data across various domains [3, 4]. Data curation is defined as a series of intentional actions aimed at making research data suitable for use while ensuring they are easily discoverable and reusable [5]. This approach aligns with the FAIR principles, which underscore the importance of Findability, Accessibility, Interoperability, and Reusability. The assessment of researchers' satisfaction with data curation services and the identification of the most valued curation practices not only increase researchers' readiness to share data but also significantly enhance the data's findability, accessibility, interoperability, and reusability, as indicated in [5]. Building upon this foundational understanding, a more detailed perspective of data curation is provided by (SIKT), encompassing processes such as quality control, documentation, standardization, and organization of data. Their approach involves a structured research data repository that categorizes curation into five distinct levels, with the initial four levels adhering strictly to the FAIR principles. This structured approach ensures that data across these levels are maintained at a high standard of quality and are aligned with global best practices. In a specialized context, the critical role of data curation has been emphasized by [7] for enhancing contrastive pre-training in machine learning, improving representation quality, training efficiency, and reducing computational costs. Additionally, the All of Us Research Program (AllofUs) underscores the importance of understanding data curation for phenotypic and genomic data to maximize its research utility. Complementing these points, [9] discuss advancements in deep learning that could transform data curation techniques, thereby optimizing technological applications and correcting misconceptions. Collectively, these perspectives underscore the multifaceted role of data curation in advancing research methodologies, enhancing data quality, and ensuring compliance with international standards, thereby facilitating more effective and efficient research outcomes across various scientific disciplines.

The EU-CIP project has developed a continuous data collection and desk research methodology to enhance the protection and resilience of critical infrastructures (CIP/CIR). This methodology supports the systematic collection, analysis, curation, and presentation of data. This book chapter outlines the methodology and discusses findings from two surveys within the EU-CIP consortium: (i) an internal stakeholder feedback survey on technological gaps, needs, and innovations in CIP/CIR, and (ii) a survey assessing the methodology's application in identifying capability needs, gaps, and technology analysis. Key findings highlight critical capability needs such as enhanced adaptability, faster response times, improved detection and impact assessment, better integration of telecom and information security systems, and management of cross-border cascading effects. Identified gaps include inadequate automation, poor control over interconnectedness, misalignment of resilience indicators, lack of standardized stress-testing, issues with IoT device classification, challenges in scaling DDoS mitigation, limited AI system development, and deficiencies in emergency management. Addressing these challenges requires a multidimensional approach involving diverse stakeholders and careful consideration of standards, regulations, and policies. These insights will help refine EU-CIP strategies, focusing on practical and innovative solutions for CIP/CIR.

The rest of the paper is structured as follows. Section 2 focuses on improving the analytical capabilities in CIP/CIR. It aims to enhance the quality and impact of the evidence supporting EU initiatives by utilizing outcomes from research projects, advancements in technology, and developments in policy. Section 3 outlines the EU-CIP project's methodical approach to gathering and evaluating data on CIP/CIR. This includes continuous data collection and analysis, emphasizing cybersecurity research, the application of advanced technology in CIP/CIR, regional to EU-level CIP/CIR policies, and standards established by Standards Development Organizations (SDOs). Sections 4 and 5 detail the two surveys conducted using the developed methodology for data collection and analysis. They present the analysis and results of these surveys, detailing the research topics identified, the modality of questionnaire data collection, and the tools used for analysis, extraction, and presentation. Finally,

Sect. 6 provides a summary of the key findings, offers an outlook on potential future directions, and discusses the implications of this study.

2 Enhancing Analytical Capability in Critical Infrastructure Protection and Resilience

EU-CIP project seeks to gather and synthesize data on CIP/CIR. This effort is designed to improve the quality and effectiveness of evidence that supports EU initiatives, by leveraging results from EU security research projects, technological advances, and policy developments.

One of the main objectives of the EU-CIP project is to collect, analyze, and curate information about research, innovation, and technological development in CIP/CIR. The analysis activities of the project are aimed at providing a landscape of CIP research, innovation, and technological development activities, while at the same time producing roadmaps and foresight about future developments. These activities fall under the EU-CIP-ANALYSIS pillar of the EU-CIP project, which groups together the information collection, information curation, analysis and foresight extraction tasks.

To achieve the first project objective, analysis, the project has developed a rigorous data collection strategy guided by a set of key focus areas, ensuring comprehensive and relevant data acquisition. This systematic approach allows to address multiple aspects addressed by the EU-CIP project. The focus areas direct the research efforts, ensuring the capture of information relevant to feed the project's road maps which are sensitive documents. The focus areas comprised:

- **CIP/CIR Capability Analysis**: The research included an examination of current capabilities and identified existing gaps within the CIP/CIR landscape. This helped to pinpoint areas requiring further development or attention.
- **CIP/CIR Future Trends**: Data collected encompassed a forward-looking assessment of potential future trends in CIP/CIR, including technological advancements, regulatory changes, and evolving threat landscapes.
- **Traditional Approaches**: Cutting-edge technologies, techniques, methods, and tools have been investigated and reviewed to ensure that the data collection methods adopted were both efficient and robust, leveraging the most up-to-date approaches.
- **Policies and Standards Relevant for CIP/CIR**: Data was gathered on existing policies, regulations, and standards related to CIP/CIR.
- **Technological, Industrial, Legal, and Ethical Issues:** The analysis encompassed the CIP/CIR technological advancements, industrial practices, legal frameworks, and ethical considerations.
- **CIP/CIR Technologies (including cutting-edge tech)**: Research included an assessment of existing and emerging CIP/CIR technologies developed by CIP/

CIR EU funded projects. This analysis explored the capabilities, limitations, and potential of these technologies within the context of the project's overall goals.

- **Local Legislations Report**: Specific local and national legislations impacting CIP/CIR have been examined to identify both enabling and restrictive factors in relation to cybersecurity policies.
- **Lessons Learnt and Best Practices**: Knowledge gained from the previous projects and best practices established in the CIP/CIR field has been incorporated to refine our research and data gathering process and inform the development of recommendations.
- **Innovation and Exploitation Opportunities**: The potential for innovation and exploitation within CIP/CIR has been assessed, highlighting promising opportunities for advancement and improvement.
- **Sectorial White Papers (covering the CER sectors)**: The sector-specific white papers produced by the EU-CIP project have been analyzed to understand the unique challenges and needs within various critical infrastructure sectors. This contextual understanding informed the adopted data collection efforts, allowing for more targeted and effective data acquisition.

The focus areas represent the starting point of the developed methodology for data gathering and analysis. EU-CIP has developed a sophisticated data collection and analysis infrastructure to address knowledge fragmentation in the CIP/CIR sectors. Various methods, including interviews, questionnaires, and extensive secondary research, were employed. Regular consultations with consortium members, representing diverse stakeholders across multiple sectors, were a key component. Feedback from both internal and external stakeholders enriched the research findings. Notably, significant capability gaps such as the absence of real-time data integration frameworks and the need for enhanced early warning systems for cascading failures were identified. The analysis of the collected data, using the Context, Observation, Impact, Next Steps (COIN) framework, provided actionable insights for both policymakers and industry stakeholders. Examples of data collected, their analysis, and the results from two surveys are detailed in Sects. 4 and 5 of this chapter. The following section outlines the methodology employed for data gathering and analysis, specifically illustrating its application to the initial three focus areas.

3 Automated Desk Research Methodology in Critical Infrastructures

The EU-CIP project has developed and utilizes a comprehensive data collection methodology that integrates technological and policies learning from over 50 projects within the European Cluster for Securing Critical Infrastructures (ECSCI) [10], EU-CIP). This methodology is designed to continuously gather and analyze outcomes from research projects with a focus on CIP/CIR, cybersecurity research, advanced technology applications, as well as regional to EU-level CIP/CIR policies and

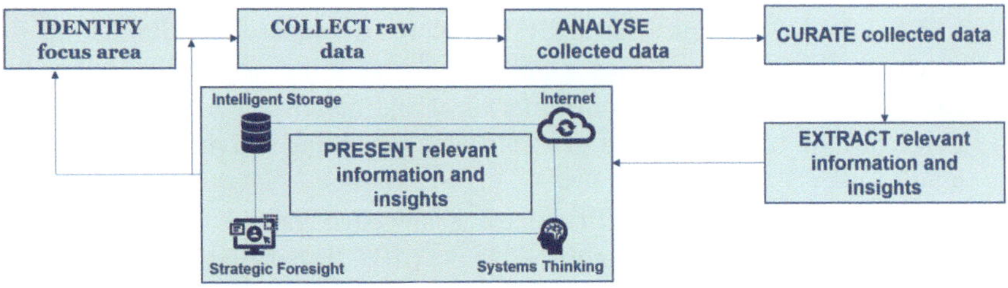

Fig. 1 EU-CIP automated data collection and desk research methodology: contributors: ENG, NRS, Innov-acts, EU-VRi [10]

standards established by Standards Development Organizations (SDOs). EU-CIP's research methodology involves continuous large-scale data monitoring and analysis from various sources, using an automated extraction pipeline that scans research, policy, and security databases. This approach was exemplified in a capability gaps analysis of critical infrastructure sectors, using data from internal consultations and the EECSCI cluster. This analysis identified key capability gaps and emerging threats, leading to the creation of sectoral reports. Sections 4 and 5 specifically detail the application of this methodology through the collection and analysis of data from two surveys, further exemplifying the methodology's practical implementation. These reports, shared via the EU-CIP Knowledge Hub, support policy development, technology planning, and investment strategies to boost resilience in critical infrastructure protection and recovery across Europe.

As illustrated in Fig. 1, the methodology employs a combination of continuous desk research, leveraging both qualitative and quantitative techniques to synthesize secondary data from a variety of relevant sources. This ongoing process ensures that the EU-CIP project stays updated with the latest developments of CIP/CIR. This provides a consistent stream of information for further detailed analysis and curation.

The comprehensive methodology is broken down into six activities, which are detailed in the subsequent sections. This structured approach ensures thorough coverage and analysis, contributing to the robustness and efficacy of the EU-CIP project.

3.1 Identify Research Topic

The first activity focuses on determining the specific information to be gathered. For CIP/CIR research topics under EU-CIP, the primary task involves selecting areas or subjects for study. These topics should be clearly defined to facilitate precise insights for decision-making, analyzing market trends, or investigating specific critical sectors. Several potential research topics have been identified for EU-CIP. Table 1 presents potential research topics for CIP/CIR and their associated benefits.

Table 1 Potential research topics for CIP/CIR

Potential research topics for CIP/CIR	Description
Policies and standards relevant to CIP/CIR	Aligns with current regulations, adheres to existing regulatory frameworks, and reflects industry best practices, thereby enhancing the relevance and applicability of the findings
CIP/CIR technologies (including cutting-edge tech)	Drives innovation, enhances security measures, improves response strategies, and supports the sustainability and resilience of critical infrastructures
Technology analysis (risk assessment, threat detection, etc.)	Identifies vulnerabilities, evaluates potential threats, and strengthens the security and resilience of systems
CIP/CIR research outcomes	Provide valuable insights into best practices, innovative solutions, and effective strategies for protecting and enhancing the resilience of critical infrastructure
Research and innovation projects	Drive advancements in technology and methodologies, fostering progress and encouraging the development of novel solutions across various fields
Sectorial white papers (covering the CER sectors)	Provide in-depth analysis, policy recommendations, and strategic insights tailored to specific industries, enhancing understanding and guiding decision-making in critical environments
Local legislations report	Offer detailed insights into regional regulatory frameworks, helping stakeholders navigate compliance challenges and adapt strategies to meet specific legal requirements
Capability needs and gaps in CIP/CIR	crucial for prioritizing resource allocation, enhancing preparedness, and strengthening resilience against potential threats to critical infrastructure
Existing developments in CIP/CIR	Represent the forefront of innovation, facilitating cutting-edge solutions and optimizing performance across various applications and industries
Technological, industrial, legal, and ethical issues	Ensuring that advancements are responsibly integrated, compliant with regulations, and aligned with societal values and norms
CIP/CIR future trends	Anticipating changes, guiding strategic planning, and ensuring that infrastructures are prepared to meet evolving threats and challenges

(continued)

Table 1 (continued)

Potential research topics for CIP/CIR	Description
Lessons learnt and best practices	Sharing knowledge, improving processes, and ensuring that past experiences inform future actions and decision-making in a constructive manner
Understanding recent trends	Provides insights into industry trends, economic dynamics, and competitive landscapes, informing strategic decisions and investment opportunities

3.2 Collect Raw Data

The second activity involves various methods of information gathering through surveys and interviews. This process is facilitated through (1) internal consultations with EU-CIP partners and various CIP/CIR stakeholders, (2) external consultations with advisory board members and ECSCI members to gather feedback on the data collection and research findings for use in developing policies, technologies, and standards, and (3) engaging open call winners in data collection to ensure their active participation and commitment.

3.3 Analyze Collected Data

The third activity employs methodologies and best practices to analyze the data gathered, including the use of the Context, Observation, Impact, and Next steps (COIN) methodology. This approach helps in deriving data-driven insights and comprehensive policy indicators for policymakers across various domains and supports continuous desk research on technologies and polices related to CIP/CIR. Various data sources such as research project outcomes, standardization bodies, white and survey papers of CIP/CIR, and trend analysis publications are utilized for this purpose. The outcomes are integrated into the EU-CIP observatory and Knowledge Hub, with validation provided by engaging EU-CIP open call winners and other stakeholders. The data is categorized into different sections like policies, gaps, needs, and technology analysis, enhancing its accessibility, curation, repurposing, and reuse by various stakeholders both within and outside the EU-CIP community.

3.4 Curate Collected Data

The fourth activity focuses on the meticulous curation of data gathered from CIP/CIR efforts. The primary goal is to ensure that data is findable, accessible, observable, distinctive, and traceable. Within EU-CIP, the data curation process includes several key steps: (i) classifying data into relevant CIP/CIR themes or topics for a systematic data overview, (ii) conducting quality control to use reliable and authoritative data sources and identify any discrepancies, (iii) representing data through graphs, charts, and videos to facilitate comprehension, and (iv) considering ethical aspects to protect privacy and prevent the disclosure of sensitive information to unauthorized individuals.

3.5 Extract Relevant Information and Insights

The fifth activity involves processing the collected CIP/CIR data to generate valuable foresight and insights. For CIP/CIR, it is crucial to unearth insights regarding existing knowledge gaps, applicable technologies to bridge these gaps, and effective research outcomes and practices. For EU-CIP, understanding the recent trends and emerging technologies in CIP/CIR through both quantitative and qualitative analysis is vital. Additionally, it is important for EU-CIP to interpret how external factors might impact CIP/CIR and identify key insights critical for safeguarding crucial sectors.

3.6 Present Relevant Information and Insights

The sixth and final activity involves how the collected data can be useful to extract in a clear and organized information and insights. While various presentation formats like dashboards, reports, or live demos are available, data visualization is particularly effective. This approach helps in identifying key patterns and results, which are crucial for analyzing CIP/CIR sectors and scenarios. The insights are shared through "Value-Added Reports, White Papers, and other Knowledge Assets" with appropriate stakeholders, providing a comprehensive overview of: (i) CIP/CIR capability analysis; (ii) cutting-edge CIP/CIR technologies, techniques, methods, and tools; (iii) outcomes, future trends, lessons learned, and best practices; (iv) opportunities and gaps in the exploitation of EU-CIP/CIR projects; and (v) areas needing standardized solutions to enhance innovation adoption and market creation.

4 Survey 1—Details, Results and Discussions

The EU-CIP Internal Stakeholders Feedback Survey: Assessing Technological Gaps, Needs, and Innovations in Critical Infrastructure Resilience builds on a comprehensive analysis of data from the EU-CIP project. This survey involved an internal consultation with consortium members, including CI operators, CIP vendors, and research organizations. It focused on identifying key capability gaps and needs, potential solutions, main trends, and innovative applications in the CIP/CIR sectors. Additionally, the survey addressed challenges in training, education, standards, and the commercialization of CIP solutions, as detailed in Chap. 2.

The survey aims to evaluate perceptions of current conditions and the impact of emerging technologies on infrastructure resilience. Initiated with a comprehensive identification of capability needs and gaps such as enhanced adaptability, quicker response times, and better system integrations, the survey revealed significant gaps in automation, interconnectedness control, and resilience indicators. Promising technologies identified include cyber-physical threat intelligence and security risk assessments, which are crucial for closing these gaps and boosting infrastructure resilience.

Despite progress in commercializing CIP solutions, challenges remain in training, education, standardization, and cross-sector collaboration. These findings will refine EU-CIP strategies, emphasizing innovative solutions for bolstering infrastructure resilience. Addressing these issues requires a multidimensional approach that integrates diverse stakeholders and adapts to emerging threats, leveraging advanced technologies to enhance resilience effectively.

4.1 Insights on Education and Training Gaps

Education, upskilling, and reskilling of CI operators' employees are crucial for deploying and operating effective CIP/CIR solutions. To address this, EU-CIP is developing training and skills development resources within the project, focusing on Innovation Management and Virtualized Solutions Validation Services. The internal stakeholders' survey seized the opportunity to gather feedback on training and skills gaps in CIP/CIR. It specifically emphasized the necessity to train stakeholders based on the criticality of end-users' information and systems access levels within a realistic training environment. Additionally, it was determined that state-of-the-art cybersecurity training curricula for security practitioners must be tailored to meet CIP/CIR requirements and industry needs. The main training topics of interest are being identified, and the training needs are being considered in the specification and implementation of the project's training and skills development activities. Table 2 outlines the key training topics for CIP/CIR programs.

4.2 Trends, Standards, and State-of-the-Art Technologies that Address Capability Gaps

Technologies and tools designed to address gaps in CIP/CIR are crucial for preventing, detecting, and mitigating threats to critical infrastructures such as cyber-attacks, physical assaults, and natural disasters. For instance, advanced cybersecurity

Table 2 Key training topics of interest for CIP/CIR programs

Training topics of interest for CIP/CIR
Cutting-edge infrastructures and technologies
Modern asymmetric threats
Different types of artificial intelligence and their role in security
Curricula that combine digital and physical security risks
Integration and interconnection of diverse systems in CI facilities, including OT/IT integration
Risk assessment and preparedness based on a good balance between risks and resilience
Applied resilience, including support for incidents caused by unexpected events
New kinds of skills relating to digitalization and high-tech society
CIP security management, leadership, and governance skills, including cyber resilience best practices and cyber diplomacy
CIP security management, leadership, and governance skills, including cyber resilience best practices and cyber diplomacy
Capacity to anticipate, prepare for, and respond to incidents and attacks
Skills for federated activities and cooperation in regional-national-EU-global federations
Cyber security skills
Hybrid security skills (e.g., leveraging outcomes of EU-HYBNET project), including hybrid impact on individuals and citizen's responsibility
Crisis communication skills, including skills against infodemic and crisis impacting, human factors, digital literacy, psychotherapy, and management of fear and mental pressures
Training on understanding interdependencies and improvement strategies for critical infrastructures

technologies aid in detecting and thwarting cyber-attacks on critical infrastructures, while physical security technologies enhance deterrence and detection capabilities. Moreover, predictive analytics and machine learning are instrumental in identifying vulnerabilities and risks, facilitating proactive mitigation efforts. The adoption of these state-of-the-art technologies enhances CIP/CIR protection, safeguarding them from the severe impacts of their disruption or destruction. Figure 2 depicts the technologies and tools currently employed in CIP/CIR, which are designed to mitigate the adverse effects resulting from potential disruptions or destruction.

The technological solutions discussed are designed to combat a range of threats, including cyber-attacks, physical intrusions, and natural disasters. These tools include advanced cybersecurity technologies to counteract cyber threats and physical security technologies to improve deterrence and detection. Additionally, predictive analytics and machine learning are becoming crucial for identifying vulnerabilities and facilitating preemptive mitigation measures. This cutting-edge technology is essential for strengthening the resilience of CI systems and preventing potentially catastrophic consequences from their compromise.

A second survey was conducted involving external stakeholders, emphasizing the importance of updating findings and gathering external perspectives. The results

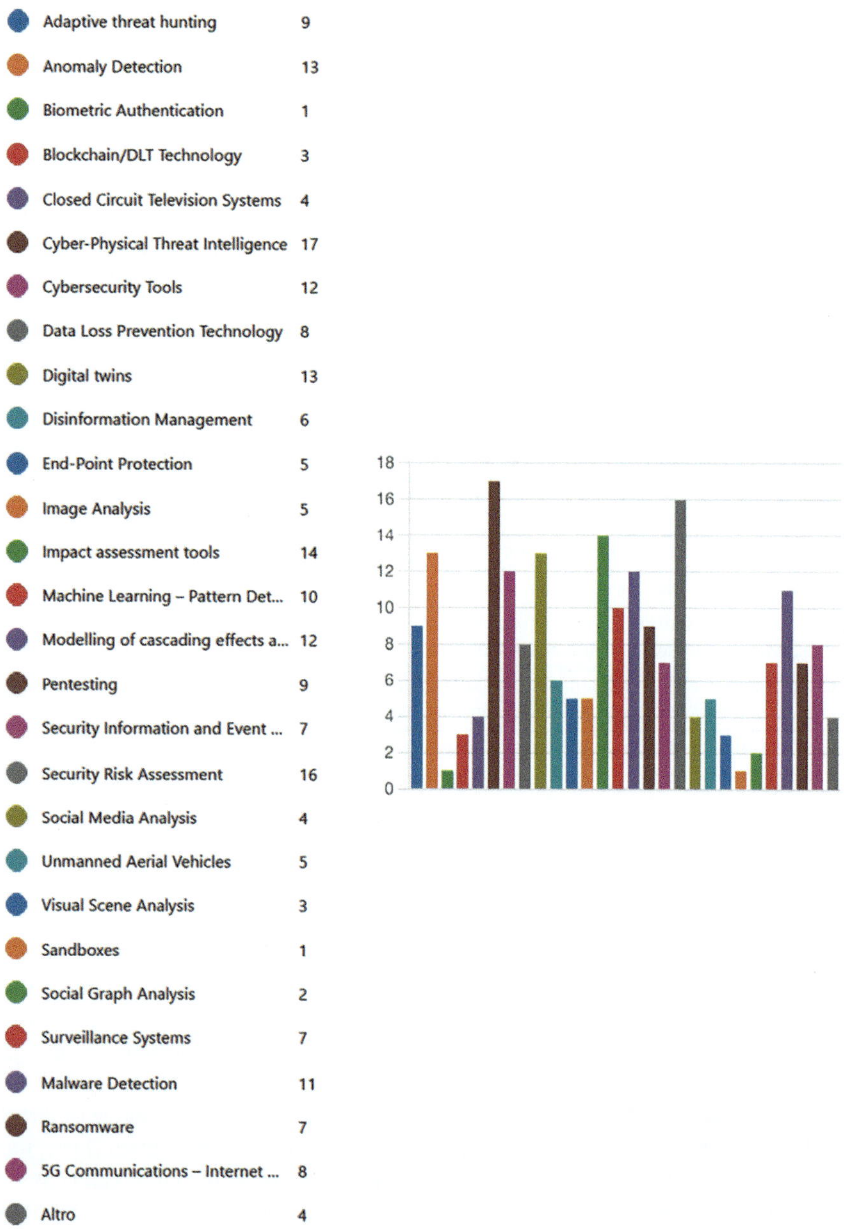

Fig. 2 Technologies and tools for CIP/CIR

(refer to Table 3) underscore the necessity of understanding and integrating advanced technologies effectively in the dynamic fields of CIP and CIR to address the identified capability gaps. Notably, some technologies were highlighted again in the second survey due to their continued relevance and importance.

The aggregated view of CIP/CIR technologies underscores the imperative for a coherent and adaptive strategy. It calls for an integration of advanced technological enablers, from AI and quantum computing to blockchain and predictive analytics, tailored to meet the evolving demands of critical infrastructure protection. Through a collaborative effort among stakeholders, these technologies can significantly contribute to bridging the identified capability gaps, fostering a secure, resilient, and sustainable critical infrastructure landscape in line with the EU-CIP vision.

The EU-CIP project's internal survey and member feedback spotlighted trends that echo these technological considerations. Figure 3 illustrates the important trends in CIP/CIR domain.

The survey results confirm the significance of the previously mentioned technologies in addressing CIP and CIR gaps. Key trends identified include cyber-physical threat intelligence, threat intelligence, and security analysis for predicting threats. The most relevant state-of-the-art technologies for bridging these gaps are:

1. Threat intelligence

Table 3 State-of-the-art technologies for CIP/CIR

State-of-the-art technologies for CIP/CIR (second survey)
Threat intelligence
Risk assessment
Impact assessment (safety and security assessment models, adaptive and anticipatory models)
Digital twin technology
Detecting anomalies
Security tools
Interdependencies modeling and cascading effects
Data analytic tools
Detection and prevention systems
Decision support systems
Blockchain technology
Quantum computing
5G network
Automation and robotics
Predictive analytics
Biometric authentication
Trusted execution environments (TEEs)
Edge computing

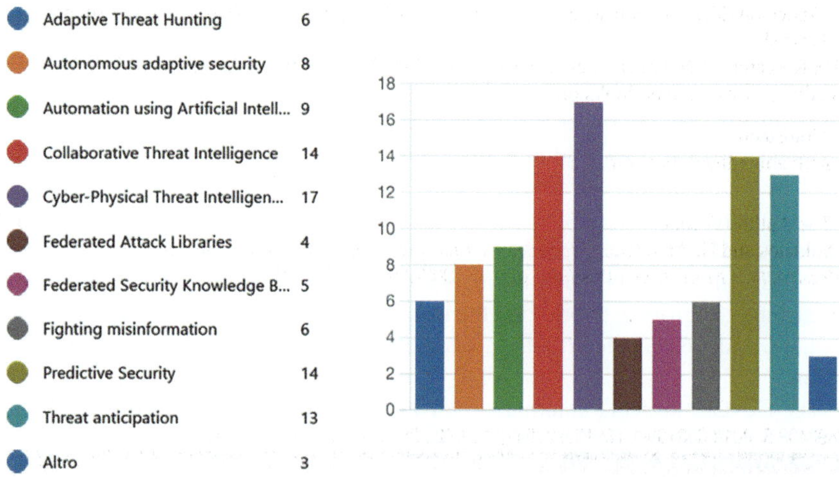

Fig. 3 Important trends IN CIP/CIR domain

2. Risk assessment
3. Impact assessment tools
4. Digital twin technology
5. Detecting anomalies
6. Cybersecurity tools
7. CIP/CIR interdependencies modeling and cascading effects.

The incorporation of above-mentioned advanced technologies can substantially contribute to addressing the identified capability gaps within the domain of CIP/CIR. Emerging technologies like AI/ML, IoT, and cybersecurity automation offer a multifaceted approach to addressing the array of capability gaps within CIP/CIR.

Critical infrastructure systems will be more prepared to handle a wide variety of emerging risks and problems by using advanced detection capabilities, expanded risk assessment tools, and greater operational flexibility. The CIP/CIR sector will make significant progress in aligning with the EU-CIP target and improving the security and resilience of infrastructure across Europe due to upcoming trends and improvements.

5 Survey 2—Results and Discussions

5.1 Capability Gaps and Needs in Critical Infrastructure

This section briefly outlines the capability needs and gaps, along with their implications, which formed the basis for this second survey and facilitated understanding. EU-CIP focuses on highlighting needs and gaps of CIP/CIR [12] so potential solutions can be developed.

The EU-CIP project has 22 members with capabilities that are not adequately supported by existing approaches. These include reduced response times, increased transparency, improved adaptation measures, improved detection capabilities, and enhanced risk and impact assessment capabilities. The list extends to supply chain security, public–private collaboration frameworks, cross-border coordination, business continuity and disaster recovery, resilience metrics and reporting, secure information exchange, and artificial intelligence integration. There are also focuses on global resilience activities, infrastructure resiliency against climate change, improved supply chain collaboration, and mapping IT and OT interdependencies. These capabilities are further detailed in [12], highlighting the areas where technological and strategic enhancements are necessary to effectively meet the evolving demands of CIP. The implications of these capability needs are far-reaching, affecting areas such as adaptability, transparency, detection, risk assessment, integration, resilience, and expertise development. Addressing these needs collectively and collaboratively is crucial for the robust protection and resilience of critical infrastructure in an increasingly interconnected and rapidly evolving threat landscape.

EU-CIP has also identified 22 preliminary capability gaps that are partly linked to previously listed capability needs. These gaps range from issues in automation and control of interconnectedness to alignment of resilience indicators and standardized stress-testing procedures. Challenges also include classification problems with IoT devices, scalability issues in mitigating security attacks and developing secure AI systems. Additional gaps concern holistic security management systems, emergency management processes, and the ability to cope with dynamically evolving threats. There are also gaps in awareness about modern CIP/CIR challenges, adaptive threat intelligence, incident response coordination, and emerging technology adoption. Rapid recovery strategies, interlaced domain integration, remote work cybersecurity, supply chain vulnerabilities, and emergency preparedness—including pandemic planning—are also notable. Furthermore, monitoring aging infrastructure, simulating crisis and cyber-physical threat intelligence, and risks associated with AI-enabled CIP tools are identified as critical areas needing attention. Each of these gaps is detailed further in [12].

The capability gaps analysis conducted by the EU-CIP Consortium has highlighted critical vulnerabilities in CIP. These gaps lead to issues such as delayed responses to cyber threats, increased risk of cascading failures, inefficient resource allocation, heightened exposure to unforeseen events, potential security breaches, service disruptions, and compromised public safety. Addressing these gaps is crucial to strengthening the resilience, security, and integrity of critical infrastructure in an increasingly complex and interconnected world. Workshops and stakeholder feedback underscore the importance of tackling these gaps by integrating systems to avoid silos, adopting emerging technologies like AI and blockchain, and developing rapid recovery strategies. There is also a recognized need to boost awareness of contemporary CIP/CIR challenges, develop adaptive threat intelligence, and ensure effective incident response coordination.

Prioritizing these gaps, especially those not adequately covered, is vital. Key focus areas such as adaptive threat intelligence, rapid recovery strategies, development and deployment of AI-based systems, and enhanced control over interconnectedness can significantly enhance CIP/CIR. By concentrating on these areas, stakeholders can allocate resources more effectively, creating a resilient, secure, and adaptable infrastructure ecosystem capable of withstanding and swiftly responding to evolving threats and challenges. Concerted efforts to address the identified capability needs and gaps through the adoption of advanced technologies, enhancing interconnectivity and transparency, and prioritizing rapid recovery and adaptive threat intelligence are essential for future-proofing Europe's critical infrastructures. Collaboration among government, industry, academia, and international partners is crucial in developing comprehensive, integrated, and flexible strategies that ensure the security and resilience of critical infrastructure systems against an increasingly complex threat landscape.

The incorporation of identified advanced technologies can substantially contribute to addressing the identified capability gaps within the domain of CIP. Also, the EU-CIP partners with sector-specific expertise have identified additional distinct requirements. During a workshop, participants envisioned the future of CIP/CIR by 2030 and highlighted several key areas, emphasizing the crucial need for a holistic and integrated approach to effectively secure critical systems:

Integration and Avoiding Silos: There was a consensus on the need to break down silos between different infrastructures. Participants stressed the interconnectedness of critical systems and the importance of addressing cascading effects in supply chains.

Trends in CIP: Emerging trends in CIP, focusing on enhanced cybersecurity measures and the adoption of advanced technologies like AI, were discussed.

Airport Challenges: Challenges related to cybersecurity at airports were highlighted, underlining the close relationship between cyber assets and physical security. Topics such as safety baselines, emerging threats like AI attacks, and real-time knowledge sharing were also mentioned.

Legislation and Proposals: Participants discussed new legislation and proposals related to CIP, including the NIS2 Resilience Critical Entities Directive and the proposed Cyber Resilience Act. The need for improved implementation was emphasized.

Water Sector Priorities: Priorities in the water sector like digitalization, cybersecurity in water treatment plants, risk management, and coordinated solutions during emergencies were included

Port Resilience: The importance of reinforcing ports as resilient infrastructures was discussed considering evolving threats, aiming to ensure the continuity of critical operations and enhance overall supply chain resilience

Re-Evaluating Vulnerability Assessments: The need to redesign risk and vulnerability assessments considering geopolitical situations and climate change impacts was highlighted, aligning with the theme of integrated resilience

5.2 Data Collection and Analysis

In this survey 2, we collected 30 responses from 25 organizations, with industry and academia comprising 56.7%, and 43.3%, respectively, as detailed in [10], EU-CIP). In this survey, we focused on recognizing the capability needs for CIP/CIR. Figure 4 depicts the capability needs with their respective percentages as per responses. From the analysis, it was noticed that the improved detection capabilities, training, reskilling, and upskilling, better integration of security tools, improved potential for assessing risks and impacts, enhanced adaptability, risk prediction and anticipation, reduced response times, ways for solving cascading effects, proactive and adaptive measures, increased transparency, better exploitation of information with 56.7%, 43.3%, 36.7%, 36.7%, 33.3%, 33.3%, 26.7%, 13.3%, 13.3%, 10%, 10%, respectively. From this survey, it was identified that the data indicates that improved detection capabilities and training are major capability needs for critical sectors.

Figure 5 illustrates the critical capability gaps for CIP/CIR with their corresponding percentages. It can be seen from the analysis that the dynamic security risks (56.7%) and lacking holistic security methods (53.3%) were main areas that need attention from industry and academia. The other notable identified capability

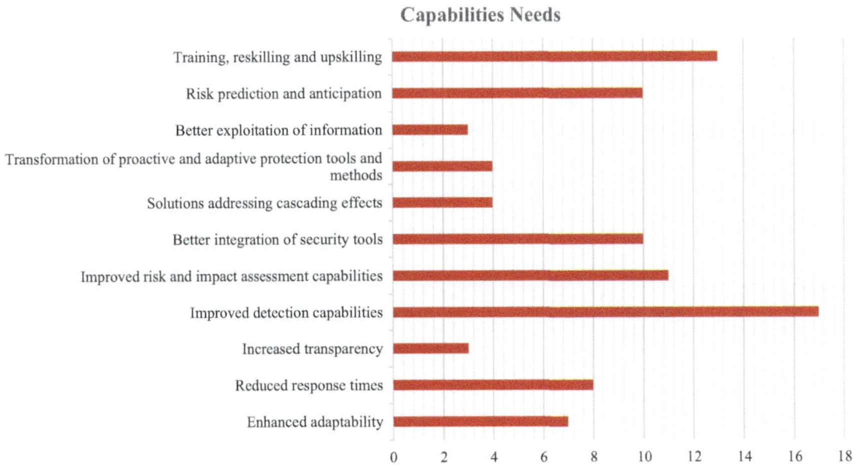

Fig. 4 Capability needs in the CIP/CIR domain

gaps were deployment of AI-based systems (40%), gaps in emergency management processes (40%), not appropriate resilience indicators aligning (13.3%), and lack of proper control of interconnectedness (13.3%). Whereas some organizations show interest in improving the following gaps with lower priority: poor automation, issues with IoT device classification, and scalability challenges from attacks, each with 10%.

Figure 6 represents the disaster phase for applying technology and tools for CIP/CIR with prevention is 50%. The other important stages were recovery, preparedness, response, and others, with 35.7%, 32.1%, 25%, and 3.6%, respectively [10]. Furthermore, the survey also collected information on which modern technologies can impact the CIP/CIR domain. It was identified that the AI-enabled threats detection and prevention systems, blockchain technology for decentralized communication, and 5G and beyond networks, quantum computing and social robotics are notable technologies.

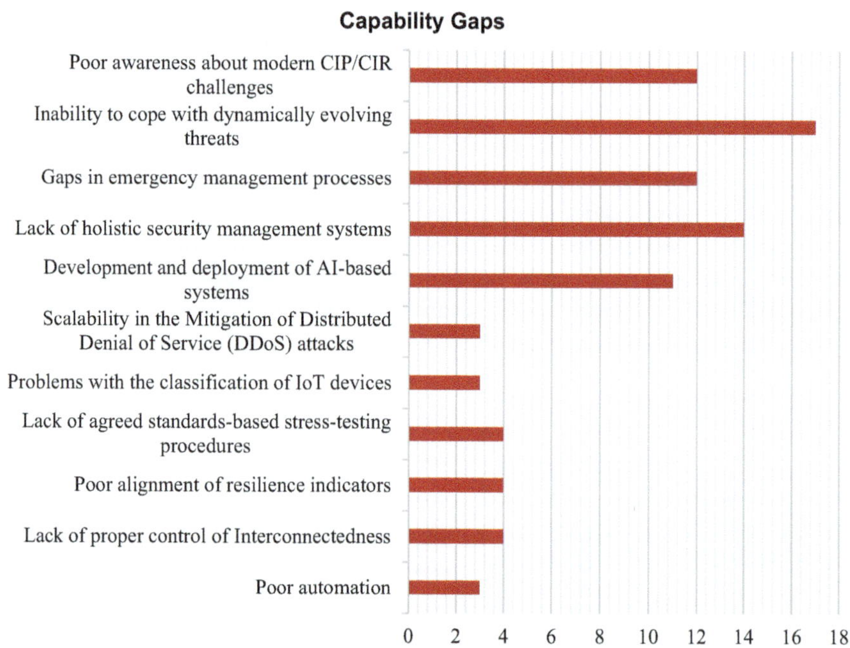

Fig. 5 Capability gaps in the CIP/CIR domain

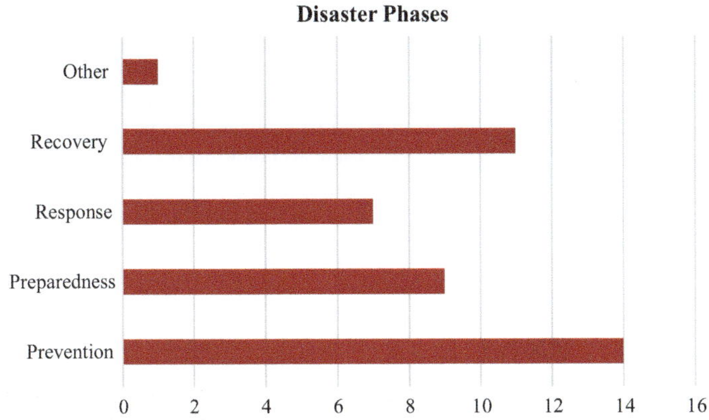

Fig. 6 Disaster phases in the CIP/CIR

6 Conclusions and Future Research Directions

We utilized the EU-CIP developed methodology to conduct two comprehensive surveys, focusing on data collection and analysis. This chapter has outlined enhancing analytical capabilities in CIP/CIR, and the analytical procedures and findings from these surveys. It has detailed the research topics identified, the modalities of questionnaire data collection, and the tools employed for analysis, extraction, and presentation. Moreover, the chapter has described the research methods used, the stages of the methodology that were implemented, and the design process of the survey questionnaires. The results revealed a preliminary list of capability gaps that align with the identified capability needs. These gaps include inadequate automation, insufficient control over interconnectedness, poor alignment of resilience indicators, the absence of standardized stress-testing procedures, classification issues with IoT devices, scalability challenges in mitigating DDoS attacks, and gaps in emergency management processes. Addressing these identified capability needs and gaps is important for CIP/CIR. These insights will contribute to refining the strategies for the EU-CIP initiative, focusing on practical and innovative solutions for CIP/CIR. Discussions with experts from the EU-CIP consortium and ECSCI have further informed and enriched this work, enhancing the overall research outcomes.

One of our future research directions includes integrating closed feedback loops with adaptive AI with the developed methodology for enhancing the automation of investigating, collecting, and merging data. Another direction involves adopting an anticipatory data curation approach that proactively manages data through continuous monitoring, forecasting, and preemptive actions to ensure its future usability and relevance.

Disclosure of Interests

The authors declare that they do not have conflict of interest.

Acknowledgements This research is conducted and funded as part of the EU-CIP project under the European Union's Horizon research and innovation program, Grant Agreement No. 101073878. Also, we would like to acknowledge that while proofreading, we used AI tools to correct grammar, punctuation, spellings, and some formulations for English language correction to give better sense and meaning. We would like to highlight that the corrected sentences by these tools were thoroughly reviewed, revised, and edited by the authors to ensure accuracy and originality.

References

1. Taherdoost H. (2021). Data Collection Methods and Tools for Research; A Step-by-Step Guide to Choose Data Collection Technique for Academic and Business Research Projects, International Journal of Academic Research in Management, 10(1): 10–38 https://elvedit.com/journals/IJARM/wp-content/uploads
2. Mazhar SA., Anjum R., Anwar AI., and Khan AA. (2021). Methods of Data Collection: A Fundamental Tool of Research. Journal of Integrated Community Health. 2021, 10(1):6–10. https://doi.org/10.24321/2319.9113.202101
3. Palathingal L., Dascalu S., Harris F., and Varol Y. (2015). A brief survey of data curation literature. Proceedings of the 30th International Conference on Computers and Their Applications, CATA 2015. 419–424.
4. Yoon A., Kim J., and Donaldson DR. (2025). Big data curation framework: curation actions and challenges. Journal of Information Science. 2025 Feb;51(1):205–223. https://doi.org/10.1177/01655515221133528
5. Marsolek W., Wright SJ., Luong H., Braxton SM., Carlson J., and Lafferty-Hess S. (2023). Understanding the value of curation: A survey of researcher perspectives of data curation services from six US institutions. PLoS ONE 18(11): e0293534. 1 Nov. 2023, https://doi.org/10.1371/journal.pone.0293534
6. SIKT Curation levels for research data, https://sikt.no/en/tjenester/arkivere-data/curation-levels-research-data
7. Desai S., Ghose D., and Chakraborty D. (2025). A Survey on Data Curation for Visual Contrastive Learning: Why Crafting Effective Positive and Negative Pairs Matters. arXiv preprint arXiv:2502.08134. 2025 Feb 12.
8. AllofUs Research Program. Data Curation Process, https://support.researchallofus.org/hc/en-us/articles/45542671110804-Data-Curation-Process
9. Thirumuruganathan S., Tang N., Ouzzani M., and Doan A. (2020). Data Curation with Deep Learning. In EDBT 2020, pp. 277–286. OpenProceedings.org.
10. Abie H. and Pirbhulal S. (2024). EU-CIP Automated Data Collection and Desk Research Methodology, Blog, March 4th, 2024, available online https://www.eucip.eu/2024/03/04/eu-cip-automated-data-collection-and-desk-research-methodology
11. EU-CIP, D3.7 Roadmaps, White Paper and Value-Added Reports v3, 2024
12. Gugliandolo E., Soldatos J., and Abie H. (2025). EU-CIP: European Knowledge Hub and Policy Testbed for Critical Infrastructure Protection. In: Gkotsis, I., Kavallieros, D., Stoianov, N., Vrochidis, S., Diagourtas, D., Akhgar, B. (eds) Paradigms on Technology Development for Security Practitioners. Security Informatics and Law Enforcement. Springer, Cham. https://doi.org/10.1007/978-3-031-62083-6_42

AI-Assisted Cybersecurity and IT Risk Management on Critical Infrastructures: Lessons Learned from Financial Sector Use Cases

Ramon Martin de Pozuelo, Martí Fabregat, Konstantina Tripodi, Elias Dakos, Amaia Gil, Lander Segurola, Gianfranco Caputo, Nicola Durante, and Cristian Raul Vintila

Abstract This chapter presents the integration of artificial intelligence (AI) applied to cybersecurity practices in the financial sector, focusing on the pilots' use cases involved in the European research projects AI4CYBER and ATLANTIS. These projects provide novel approaches to risk identification, threat detection, and incident

R. M. de Pozuelo (✉) · M. Fabregat
CaixaBank S.A. Sabino Arana, Barcelona, Spain
e-mail: rmartindepozuelo@caixabank.com

M. Fabregat
e-mail: mfabregatip@caixabanktech.com

K. Tripodi
JRC Capital Management, Berlin, Germany
e-mail: ktripodi@jrconline.com

E. Dakos
Athens Technology Center (ATC), Athens, Greece
e-mail: edakos@atc.gr

A. Gil · L. Segurola
Vicomtech Foundation, Basque Research and Technology Alliance (BRTA), Donostia-San Sebastián, Spain
e-mail: agil@vicomtech.com

L. Segurola
e-mail: lsegurola@vicomtech.com

G. Caputo
Links Foundation, Torino, TO, Italy
e-mail: Gianfranco.caputo@linksfoundation.com

N. Durante
Engineering—Ingegneria Informatica Spa (ENG), Roma, Italy
e-mail: nicola.durante@eng.it

C. R. Vintila
Siemens SRL, Bucuresti, Romania
e-mail: Cristian.vintila@siemens.com

J. Soldatos and H. Abie (eds.), *Technology-Enabled Resilience: Innovations in Critical Infrastructure Protection*, https://doi.org/10.1007/978-3-032-25917-2_9

response through AI-driven technologies. The document explores how the financial entities apply the developed tools across various use cases to enhance protection against insider threats, automate vulnerability management, detect GNSS spoofing attacks, and manage systemic risks such as disinformation. A technical yet sector-oriented overview is provided, highlighting methodologies, lessons learned, and the potential of AI for strengthening security and resilience in the financial domain.

Keywords Resilience · Cybersecurity · Financial sector

1 Introduction

The financial sector is a high-value target for cybercriminals due to its critical role in the global economy and its vast amount of sensitive data. As attack surfaces grow with digitalization and interconnectivity, traditional reactive defense mechanisms are no longer sufficient. Proactive, intelligent, and automated approaches are essential to maintain operational continuity and protect against sophisticated cyber threats.

AI has emerged as a critical enabler of innovation in cybersecurity. It offers predictive capabilities, autonomous detection of anomalies, and faster response to incidents. Two of the most prominent initiatives in this field are AI4CYBER and ATLANTIS, both funded by the European Union under the Horizon Europe program.

AI4CYBER is focused on the development of AI-enabled components to reinforce cybersecurity in critical infrastructures. ATLANTIS, on the other hand, provides a systemic approach to managing risks and threats that go beyond traditional cyberattacks, such as disinformation campaigns and hybrid threats.

CaixaBank (CXB), one of Spain's largest financial institutions, and JRC Capital Management Consultancy & Research GmbH (JRC), a financial institution regulated by BaFin and supervised by the Deutsche Bundesbank specializing in quantitative investment strategies in highly liquid markets such as Forex, are the main actors and leader of the financial sector use cases for projects that push the boundaries of cybersecurity through AI.

This chapter details how these companies leverage the tools and methodologies developed in these projects to protect financial operations. It presents real use cases and subcases tested in controlled and near-operational environments. By focusing on identity protection, vulnerability management, situational awareness, and systemic risk mitigation, these initiatives provide a blueprint for the future of cybersecurity in banking.

2 AI4CYBER–Artificial Intelligence for Next Generation Cybersecurity

2.1 Project Overview

Artificial intelligence (AI) has lately proved to be a coin with two sides. On the one hand, it can be leveraged as a powerful defensive mechanism to improve system preparedness and response against cyber incidents and attacks, and on the other hand, it can be a formidable weapon attackers can use to damage, compromise, or manipulate systems. While an increasing number of examples of defensive AI are flourishing in the market and in the literature [1], fewer examples are documented on real attacks that incorporate AI. The recent news on deepfakes, voice imitations, phishing, stegonets, etc., that use AI are just some examples of the AI-fueled attack tsunami we may be facing soon. However, while we adapt our cybersecurity solutions against AI-powered attacks, we need to find ways to better interpret the results of the attacks, ensuring no harm has been caused to European citizens. When confronting defensive AI to offensive AI, in order to really understand the effectiveness of defenses and evolve them in the right direction, it is necessary to find means to enable the interpretability of the rationale for the decisions taken by the defensive AI and the conclusions it arrives to. Furthermore, it is necessary to keep privacy, equity, ethics, and security while processing, exchanging, storing the required big data when training, testing, and operating the AI models. In this scenario, the view of AI4CYBER[1] is presented as follows in regards with the main challenges it intends to address.

Therefore, the main aim of AI4CYBER is to establish an ecosystem framework of AI-based services for supporting critical system developers and operators to efficiently manage system robustness, resilience, and appropriate response in the face of advanced and AI-powered cyberattacks [2].

2.2 Financial Sector Use Case

CXB is currently leveraging a variety of identity and access management (IAM) and privileged access management (PAM) tools to manage the security of its applications, services, and internal systems. This approach ensures that different types of users—primarily employees—have secure access to the bank's digital infrastructure. CXB's security infrastructure is structured around monitoring and controlling user identity, their permissions for accessing various tools, and tracking user actions. This is facilitated by a set of commercial and self-developed tools that provide static and dynamic application security testing, vulnerability management, and penetration testing analysis.

[1] https://ai4cyber.eu/.

However, with increasing remote work and the growing involvement of third-party service providers in accessing CXB's applications, maintaining a secure and controlled access environment has become more complex. These additional external actors accessing internal environments necessitate an enhanced security approach to mitigate potential risks. Remote work has further complicated the access control landscape, as employees now require secure, controlled, and reliable access to the bank's internal applications from home, necessitating a robust identity and access management system that is adaptable to changing security dynamics.

CXB's existing Security Information and Event Management (SIEM) system plays a critical role by collecting logs from network devices and applications. This system aggregates data that is continuously analyzed to detect potential threats or attacks. To improve incident response, CXB has implemented Security Orchestration, Automation, and Response (SOAR), a tool designed to automate responses when alerts or suspicious activity are identified. These capabilities are essential in detecting potential security incidents across the bank's internal systems, but there are still areas where improvements can be made. Specifically, CXB aims to enhance the automated detection of anomalies—such as suspicious user behavior or abnormal network traffic—which could signal security vulnerabilities or attacks.

The core objective of CXB's cybersecurity strategy is to offer highly resilient and secure banking services. For this reason, the bank must ensure that both its infrastructure and its applications are protected from cyber threats. CXB is focused on maintaining strong defenses not only at the infrastructure level but also at the application level, as its digital services are integral to its operations. To bolster its efforts, CXB has been deploying tools aimed at securing its applications, identifying vulnerabilities, and responding proactively to any potential risks.

In addition to traditional security measures, CXB is investigating the use of Artificial Intelligence (AI) to enhance its cybersecurity strategy, particularly in the areas of vulnerability detection and mitigation. The bank is exploring AI-based solutions that can perform more advanced vulnerability analysis at the code level. These AI tools will be integrated into CXB software development lifecycle, particularly within the Continuous Integration/Continuous Deployment (CI/CD) pipeline, to automatically assess newly committed code for potential vulnerabilities. When issues are detected, the AI-driven system will provide detailed reports in formats such as JSON, XML, or CSV, outlining the vulnerabilities and recommending possible fixes. This integration will help prevent the deployment of vulnerable or misconfigured code, which is especially critical in the highly regulated and security-sensitive banking sector.

Furthermore, CXB is looking into the potential of AI to enhance its threat detection and response capabilities. By integrating AI models with its existing SIEM system, CXB aims to leverage machine learning and advanced anomaly detection techniques to identify sophisticated attack patterns, such as zero-day exploits or advanced persistent threats. These AI systems will not only help detect vulnerabilities but also assist in identifying abnormal user behaviors that might indicate an insider threat or a compromised account. Through machine learning and pattern recognition, AI can provide CXB with the ability to assess potential risks and respond to emerging threats faster and more effectively more accurately.

The integration of AI in vulnerability detection and threat analysis is expected to reduce the need for human intervention in routine security tasks, thereby increasing the efficiency and effectiveness of the bank's cybersecurity operations. By automating aspects of vulnerability management and threat detection, CXB can proactively identify and address security issues before they can escalate into significant threats. This proactive approach ensures that the bank's infrastructure, applications, and services are consistently safeguarded against the evolving threat landscape.

To conclude, CXB, with the available information and the current global situation with AI, has defined the following use cases to better secure its financial operations:

- AI-based vulnerability detection and mitigation at code level
- Advanced attack for data exfiltration
- Anomaly detection for an insider threat.

2.2.1 AI-Based Vulnerability Detection and Mitigation at Code Level

As the name of the use case implies, the main objective of this use case is to thoroughly inspect the code that needs to be deployed in critical systems such as that of the banking sector and recommend mitigation for the vulnerabilities identified. In the financial sector, where sensitive customer data and financial assets are on the line, any security lapse can lead to significant breaches, financial losses, and reputational damage.

Proactively inspecting and securing the code is not merely a technical necessity but a strategic imperative to not let primary attack vectors to cybercriminals that will exploit them to gain unauthorized access to banking systems.

In order to detect and mitigate vulnerabilities, AI4CYBER offers AI4VULN [3] and AI4FIX [4] respectively. These tools automatically assess the code after each change to generate a list of the vulnerabilities found and recommend mitigations which can be handled either manually or semi-automatically with a notification with the recommended code.

AI4VULN is a solution to automatic identification and verification of vulnerabilities and weaknesses in the code with much higher accuracy rate than existing vulnerability analysis solutions thanks to applying symbolic execution and the use of AI to support scalability.

On the other hand, AI4FIX is a smart automated testing of software and weaknesses correction service based on AI knowledge, capable to correct even those "warning level" problems that are dangerous and may lead to exploitable security vulnerabilities in the future. By removing the basic root causes of vulnerabilities, the AI4FIX service can reduce the number of actual exploitable weaknesses and increase system robustness.

AI4FIX is designed to enhance software development processes through the power of AI. This framework combines various tools and technologies to provide comprehensive code analysis, vulnerability detection, and automated bug fixing capabilities. By leveraging static code analysis, symbolic execution, and so much more, AI4FIX aims to improve code quality, identify potential security risks, and streamline the development workflow.

To be able to test and measure their capabilities, CXB has deployed these components in Azure Cloud, where the code can be inspected in a controlled and highly secure environment. Being able to test these use case with real data and in a near-production environment grants CXB a perfect opportunity to assess the results, measure KPIs and evaluate the viability of AI4CYBER's components.

2.2.2 Advanced Attack for Data Exfiltration

The primary objective is to effectively detect and mitigate advanced persistent threats (APTs) specifically designed to exfiltrate sensitive data from critical banking systems. In the banking sector, the stakes are incredibly high. A successful data exfiltration can compromise customer information, financial assets, and the overall integrity of the institution, leading to substantial financial losses and irreparable damage to its reputation.

Defending against these sophisticated attacks requires more than just traditional security measures. APTs are characterized by their stealth, persistence, and ability to evade conventional detection mechanisms. They represent a critical attack vector that attackers will inevitably try to exploit to gain unauthorized access. Proactive and intelligent detection and response capabilities are therefore paramount to safeguard sensitive data and maintain the overall security posture of the bank.

To address this challenge, AI4CYBER offers a suite of interconnected components, primarily AI4TRIAGE and AI4SOAR [5], which orchestrate to detect anomalous behavior, automate incident response, and mitigate the impact of advanced attacks. These tools work synergistically to provide enhanced detection accuracy, faster response times, and a more resilient security posture.

AI4TRIAGE acts as an intelligent security analyst, automatically analyzing security events (logs) to identify root causes, prioritize alerts, and provide enriched context. Using machine learning, it can distinguish subtle indicators of compromise that may be missed by conventional methods, enabling earlier detection of APT activity. AI4TRIAGE identifies patterns and anomalies in network traffic, system logs, and user activity indicative of malicious activity. It leverages trained models based on past attacks and threat intelligence that recognizes deviation and flags suspicious activities. Its algorithms help prioritize vulnerabilities, focusing on the most critical issues which enhances efficiency in a threat environment.

AI4SOAR, on the other hand, is the automated incident response engine. It takes the prioritized and enriched information from AI4TRIAGE and automatically initiates predefined mitigation actions based on playbooks. These playbooks can include a wide range of responses, such as isolating infected systems, blocking malicious network traffic, disabling compromised user accounts, and triggering forensic investigations. It has the capacity to detect inconsistencies in the playbooks, checking the different nodes to avoid issues during execution. By automating these responses,

AI4SOAR significantly reduces response times, minimizes the impact of attacks, and frees up security analysts to focus on more complex and strategic tasks.

Furthermore, the use case leverages AI4CTI to continuously refine and improve the AI models and SOAR playbooks. By incorporating the latest open-source from the Internet, vendors, and internal security data, AI4CTI ensures that the detection and response capabilities remain effective against evolving threats and emerging attack techniques.

TRUST4AI [6] helps to ensure transparency, fairness, and security of the AI models being implemented. It has an explainability to understand why a model is making a particular decision, evaluations models to ensure no bias, and ensures that the implemented AI models are secure from adversarial attacks.

For effective collaboration and information sharing that allows the threat intelligence community to share cyber threat intelligence, AI4COLLAB [7] anonymizes data and ensures compliance to the GDPR when doing so.

To thoroughly validate the capabilities of these components, CXB can deploy them within a secure and controlled environment in Azure Cloud. This allows them to test the use case with realistic data and scenarios, assess the results against predefined KPIs, and critically evaluate the viability of AI4CYBER's components in a near-production environment. Such rigorous testing ensures that the bank can confidently deploy these solutions and realize their full potential in protecting its critical assets.

2.2.3 Anomaly Detection for an Insider Threat

Insider threats represent a significant and growing risk to critical infrastructure sectors, including banking. Cybersecurity and Infrastructure Security Agency (CISA) defines an insider threat as "the potential for an insider to use their authorized access or understanding of an organization to harm that organization." This can range from unintentional data leaks to malicious acts of sabotage or data theft.

In the financial sector, the potential consequences of an insider threat are severe. Trusted employees or contractors can leverage their access to sensitive systems and data to cause significant financial losses, damage the bank's reputation, and compromise the confidentiality of customer information. Traditional perimeter security measures are often ineffective against insider threats because they originate from within the trusted zone.

To combat this danger, AI4CYBER provides a sophisticated solution centered around AI for Federated Intrusion Detection System (AI4FIDS) [8] coupled with AI4TRIAGE to automatically detect anomalous behavior indicative of insider threat activity. This system constantly monitors user activity and system logs, identifying deviations from established baselines and triggering alerts when suspicious patterns emerge. The output of these systems feeds other important systems involved in this use case, such as TRUST4AI, AI4SOAR, AI4COLLAB, necessary for the use case to be complete.

AI4FIDS employs federated learning (FL) techniques with a comprehensive approach to intrusion detection to ensure data privacy and collaboration between multiple systems. The system includes four main Intrusion Detection Systems (IDS) that leverage Federated Learning (FL):

- The Network Federated Intrusion Detection System (N-FIDS), which monitors TCP/IP and application-layer flow statistics.
- The Logs Federated Intrusion Detection System (L-FIDS), which analyzes system log data.
- The Operational Federated Intrusion Detection System (O-FIDS), which receives aggregated operational data.
- The Visual Federated Intrusion Detection System (V-FIDS), which processes network traffic data via binary visual representations and CNNs.

These systems can detect abnormal patterns such as unusual login timestamping and user geolocation anomalies, detecting and analyzing logs and traffic data to detect potential insider threat attacks. Each one of the IDS is supported by modules that are responsible for the data collection, preprocessing, detection process through the trained federated model, as well as generating security events. For the implementation, maintenance, and model training of the IDS's, AI4FIDS's training module is also important in the federated training procedure.

AI4TRIAGE then processes these alerts, applying advanced machine learning techniques to identify the root cause, prioritize the incident based on its potential impact, and enrich the alert with relevant contextual information. This allows security analysts to quickly understand the nature of the threat and make informed decisions about the appropriate response. AI4TRIAGE is also used to generate additional information about the events taking place in Active Directory as well as authentication logs from the IAM platform.

When a high-priority insider threat is detected and triaged, the AI4SOAR system automatically initiates predefined response actions, as outlined in playbooks. In this case, there will be actions such as an investigation of current activities, notification to the human operators for the analysis of actions and depending on the risk level isolation will occur. Playbooks provide a coordinated effort for high impact issues and threats. These may include locking the affected user account, restricting access to sensitive systems, and initiating a forensic investigation to gather evidence and assess the extent of the damage.

Furthermore, the AI4CYBER components are designed to promote collaboration and information sharing. Should the security professional want to communicate with the threat intelligence community, AI4COLLAB enables secured, anonymous communication that will allow for an improved threat awareness. Additionally, the TRUST4AI tool is of great importance for ensuring the constant reliability, fairness, and security of the models.

In a banking environment, where access to critical systems often involves the SWIFT network, and user data is stored and accounted for using systems like Active Directory, these use case benefits provide a clear demonstration of its efficacy in anomaly detection.

By adopting the AI4CYBER components and leveraging their powerful anomaly detection capabilities, CXB can significantly enhance its ability to detect and mitigate insider threats, protecting sensitive data and maintaining the integrity of its operations. Deploying this use case inside of the CXB premises would be a step toward ensuring the viability of AI4CYBER, and in doing so could validate the viability of these components.

3 ATLANTIS–Improved Resilience of Critical Infrastructure Against Large-Scale Transnational and Systemic Risks

3.1 Project Overview

ATLANTIS[2] project addresses the escalating threat landscape facing Essential Critical Infrastructures (ECIs) in Europe. Recognizing the increasing frequency and sophistication of cyber-physical attacks, the interconnected nature of ECIs, and the potential for widespread disruption or cascading failures, ATLANTIS aims to bolster the resilience of these vital systems through innovative AI-powered solutions, enhanced cooperation among stakeholders, and improved awareness of systemic risks.

The project focuses on four strategic goals:

1. Enhancing awareness of large-scale vulnerabilities and long-term systemic risks.
2. Improving the systemic resilience of ECIs through adaptive and customizable security measures and tools.
3. Fostering effective cooperation among CI operators and governmental security stakeholders while preserving CI autonomy and sovereignty.
4. Delivering an open technological framework providing ECIs with AI-based solutions for increased awareness, capability, and cooperation in managing systemic threats.

Central to ATLANTIS is the creation of an open, AI-driven technological framework tailored specifically for ECIs. This innovative framework will facilitate the integration of existing and future security technologies, data feeds, and decision-support services enabling a holistic approach to ECI security. By leveraging cutting-edge technologies like Federated Machine Learning (FML), Explainable AI (XAI), and 5G/multi-constellation satellite systems, ATLANTIS will provide ECIs with the necessary tools and insights to proactively identify, assess, mitigate, and respond to evolving cyber-physical threats, ultimately enhancing their operational and systemic business continuity. The project is committed to using open-source and/or EU-based technologies whenever possible, contributing to the strategic autonomy of ECIs and

[2] https://www.atlantis-horizon.eu/.

the European Union in critical service provision. Through real-world demonstrations and focused skills improvement initiatives, ATLANTIS will ensure the practical application, ethical considerations, legal compliance, and societal acceptance of its innovative solutions.

3.2 Financial Sector Use Case

The financial sector, a cornerstone of global economic stability, navigates an increasingly perilous digital landscape. The very modernization that drives efficiency and innovation—the interconnected infrastructures, the ubiquitous online banking access, the automated data centers—simultaneously expands the threat surface, inviting sophisticated cyber-physical attacks. Cybercriminals, nation-state actors, and malicious insiders alike seek to exploit vulnerabilities, underscoring the urgent need for heightened and strategically integrated security measures.

The core problem lies within the defensive architecture of many financial institutions. Despite significant investments, security measures frequently exist in a fragmented state, compartmentalized and reactive, proving insufficient against the sophisticated, asymmetric nature of modern threats. The absence of real-time visibility hinders accurate risk assessment, impeding the orchestration of effective responses.

A web of interconnected data centers and critical service providers creates a complex web of dependencies, often harboring unknown risks that existing solutions are incapable of accurately assessing. Simply put, financial institutions must move beyond best-effort approaches and embrace a holistic, proactive security strategy.

The impact of this vulnerability is far-reaching. A successful attack against a major bank or payment processor transcends mere data breaches; it can trigger systemic instability, erode consumer trust, and ultimately damage the overall economic outlook. Regulatory pressures further intensify the need for robust defenses. Financial institutions face stringent requirements to safeguard customer data, maintain stability, and prevent illicit activities. Data breaches, service disruptions, and compliance failures carry the threat of crippling financial penalties, irreparable reputational damage, and costly legal liabilities.

Responding to this critical imperative, Large-Scale Pilot #3 (LSP#3), led by CXB in collaboration with key partners like JRC, is designed to validate, integrate, and markedly improve cyber defenses within the financial sector. LSP#3 recognizes that the challenge demands both real-time response capabilities and more reliable cyber threat detection mechanisms.

The bedrock of this initiative is a comprehensive framework built upon ATLANTIS technologies, aimed at bolstering cybersecurity resilience across a broad range of threats, from malicious data breaches to full-scale bank network disruptions.

Key facets of this framework include:

1. Identifying and mitigating influence operations aimed at swaying institutional decision-making, ensuring strategic actions are based on reliable intelligence.

2. Uncovering system cyber interdependencies and accelerating the responses necessary to contain breaches and minimize sensitive data exposure.
3. Validating Position, Navigation and Timing (PNT) services independently of GPS, mitigating the risks associated with spoofing attacks and ensuring regulatory compliance.
4. Preparing for and mitigating disruptions caused by attacks against critical service providers and essential infrastructure.

The ATLANTIS framework [9] integrates a formidable suite of technologies to realize this protective vision. Services that analyze social media, identify threats, discover risks and provide organized interfaces are designed with a unified and collaborative perspective. This collective approach surpasses disparate defenses—a proactive defense that can anticipate and neutralize threats before they can take root. The specific aims of LSP#3 include:

- Cultivating a robust security ecosystem that leverages a broad range of security options to enhance resilience in the face of constantly evolving threats.
- Shrinking the window of vulnerability by automating tasks and accelerating the detection and response lifecycle.
- Preventing the proliferation of cyberattacks by containing incidents and strengthening the digital ecosystem as a whole.
- Establishing a collaborative security model that fosters information sharing among financial institutions, promoting a more resilient financial ecosystem.

In a world where cyberthreats are relentless and ever-changing, the financial sector must embrace innovation to maintain stability and trust. LSP#3, through its emphasis on real-time response, proactive defense, and collaborative engagement, offers a path toward a more secure future for finance, ensuring that institutions can confidently serve their customers and contribute to a thriving global economy.

3.2.1 Tackling Disinformation Campaigns Affecting Financial Businesses

In today's digital financial ecosystem, disinformation campaigns have become a serious cybersecurity risk, capable of disrupting market stability, influencing investor decisions, and damaging the reputation of financial institutions, like JRC. Malicious actors, including cybercriminals and state-sponsored entities, exploit online platforms, fake news websites, and deepfake technologies to spread misleading information. These campaigns can manipulate stock prices, erode consumer confidence, and even trigger regulatory scrutiny. To counteract these threats, financial institutions, like JRC, are increasingly turning to AI-driven cybersecurity and IT risk management solutions.

One of the most effective ways to combat financial disinformation is through Natural Language Processing (NLP). AI-powered NLP models analyze vast amounts of online content, scanning social media, news websites, and financial reports to detect misleading narratives. These systems use sentiment analysis and pattern recognition to identify anomalies that could indicate the spread of false or manipulative financial information. By monitoring digital platforms in real time, JRC can detect disinformation before it gains traction.

Truly Media plays a critical role in this verification workflow. It enables the identification of forged documents, manipulated social media posts, and synthetic multimedia—including deepfakes and altered recordings. With forensic analysis capabilities, this tool flags inconsistencies such as metadata manipulation or unnatural visual/audio patterns. Verified evidence can then be disseminated to media partners and regulators, discrediting false claims and helping restore public confidence in the institution's integrity.

Another critical approach is machine learning-based fact-checking. AI-driven tools can cross-reference news articles, social media claims, and market reports with verified sources, assessing their credibility. By leveraging large datasets of historical financial information, machine learning models can recognize inconsistencies in reports and issue real-time alerts. This technology helps JRC to protect investors and decision-makers from being misled by false narratives designed to manipulate market behavior.

The rise of deepfake and synthetic media has further complicated the disinformation landscape. Malicious actors can create convincing videos or audio clips that appear to show executives, regulators, or financial analysts making damaging statements. To counter this, AI tools like Truly Media[3] and specialized deepfake detection systems analyze multimedia content for irregularities—such as unnatural facial movements, voice mismatches, or suspicious origin metadata. Financial institutions leveraging such tools are able to stop fraudulent content from going viral and influencing sentiment. This also enables institutions to time their factual responses with key sentiment shifts, thereby stabilizing fund performance and avoiding long-term valuation disruptions.

At the strategic level, SAFER [10] offers advanced risk forecasting and regulatory modeling. It predicts how investors, media, and oversight bodies are likely to react to specific narratives. Using scenario simulations and behavioral models, SAFER enables institutions like JRC to engage in preemptive mitigation—whether through investor communications, regulatory engagement, or public disclosure—before a crisis escalates.

Additionally, behavioral analysis and threat intelligence play a key role in mitigating disinformation threats. AI-powered cybersecurity platforms track suspicious online activity, identifying coordinated efforts by bots and fake accounts to amplify false narratives. By integrating global threat intelligence feeds, these systems can detect emerging disinformation campaigns and provide early warnings to financial institutions, allowing them to implement countermeasures before widespread damage occurs.

The effectiveness of these tools is reflected in real-world outcomes. In coordinated responses to emerging disinformation, institutions have successfully restored brand

[3] https://www.truly.media/.

reputation, stabilized fund performance, and averted investor panic. For example, by deploying such platforms in a high-risk scenario, a firm was able to expose and disprove false claims, issue verified counter-evidence, avoid major redemptions, and maintain its standing with regulators and institutional clients.

Importantly, this proactive, AI-supported approach has helped preserve investor trust. Through direct outreach, real-time transparency dashboards, and data-driven reassurances, stakeholders were convinced of the institution's capability and intent to manage disinformation threats effectively. What began as a reputational crisis became a demonstration of digital resilience and operational maturity. Furthermore, archived records from the detection and response process now support legal and compliance departments in the future regulatory or forensic scenarios.

3.2.2 Bank Network Intrusion and Sensitive Information Leak

Modern financial institutions like CXB face a relentless barrage of cyberattacks, targeting everything from financial transactions and customer data to core banking systems. These attacks are not only becoming more frequent but also more sophisticated, increasing the potential for significant financial losses, reputational damage, and regulatory penalties. The interconnected nature of banking systems means that a breach in one area can quickly spread throughout the entire infrastructure, causing cascading failures and widespread disruption.

Therefore, the ability to quickly detect, analyze, and respond to network intrusions is critical for maintaining business continuity and protecting sensitive assets. Traditional security measures, while necessary, are often insufficient to address the dynamic and evolving threat landscape. A more proactive and integrated approach is needed, one that leverages real-time threat intelligence, advanced risk assessment techniques, and automated mitigation capabilities.

This Use Case addresses these challenges head-on by implementing a real-time risk reporting mechanism focused on CXB's IT network. The objective is to utilize Cyber Threat Intelligence extracted for events, system vulnerabilities, and external network threats to enable proactive risk management strategies, a secure response for analysis, and automation from the security plan. If any event occurs, responses depend upon many levels of systems. This system hopes to address that by combining intelligence, AI, and security.

The ATLANTIS framework works to create a well-rounded approach, combining various tools to provide a real-time protection against cyber threats, identifying cascading effects that could jeopardize the business continuity of the financial institution. With the combination of different components and models inside the ATLANTIS framework and the security tools owned by CXB, we are able to make the organization more reliable, secure and resilient to any cyber-physical attacks. Key components for this use case include: EBRA [11], SAFER and RRIM [12].

EBRA, standing for Entropy-Based Risk Assessment, is a cutting-edge analytical tool designed to provide dynamic and granular insights into the security posture of network infrastructures. Within the realm of cybersecurity, EBRA leverages principles of information theory, particularly the concept of entropy, to quantify the uncertainty and potential risk associated with different nodes and connections within a network. It goes beyond traditional risk assessment methods by incorporating machine learning techniques and network topology analysis, offering a truly comprehensive and adaptive approach to threat detection.

Threat Intelligence is a proactive cybersecurity approach that transforms raw data into actionable insights, empowering organizations to anticipate, identify, and mitigate threats before they inflict damage. This involves a structured lifecycle of planning, collection, processing, analysis, and dissemination, leveraging various types of intelligence (strategic, tactical, and technical) to improve threat identification, incident response, and overall security posture, fostering a culture of proactive cybersecurity preparedness.

SAFER, the Situational Awareness Framework for Emergency and Resilience, is a holistic cybersecurity framework designed to elevate organizational preparedness and responsiveness. By emphasizing complete situational comprehension, dynamic real-time analysis, and clear communication, SAFER enables organizations to anticipate, manage, and recover from cyber incidents effectively. The data is always iterative. SAFER relies on key components like a comprehensive risk list, asset modeling, and real-time data feeds to provide a 360-degree view of the security landscape. This empowers security teams to make informed decisions and take proactive steps to mitigate threats, with machine learning enhancing accuracy and performance in a constant framework.

RRIM, or Risk Reduction and Incident Mitigation, is a post-crisis management framework focused on providing actionable recommendations for mitigating short-to-medium term risks stemming from cyber-physical incidents. Its goal is to achieve higher integration for rapid and accurate responses, improving real-time operations and ensuring increased transparency and system credibility. The RRIM framework leverages Explainable AI (XAI) to enhance understanding and build trust, fostering an environment for rapid improvements in both security and operational quality. By promoting strong AI-human collaboration and enabling constant adaptation, it delivers an integrated system that allows organizations the ability improve decision-making in the most optimal way [12].

In Use Case 3.2, the ATLANTIS framework orchestrates a multi-layered defense against cyberattacks targeting CXB's banking infrastructure. EBRA and Threat Intelligence, in coordination with SIEM's security logs, act as the initial detection and assessment team to identify and map unusual traffic patterns and potential threats coming for the outside. Upon threat identification, all the important data goes to the team that will evaluate the different routes it has for the incident to be solved. The whole purpose rests to improve quick real-time implementation, that helps understand security operation.

Afterward, SAFER builds a comprehensive understanding of the incident, drawing on EBRA's network insights and interdependencies to enrich the context with other relevant parts within the organization. Then the information is presented to the DSS for quick decision for the operator.

Finally, RRIM analyzes potential mitigation actions for security assessment and gives quick council to the operator, enhancing the AI-human interaction for swift mitigation and response actions that can later be integrated with SOAR solutions or security playbooks.

3.2.3 Data Center Time (PNT) Disruption

In this scenario, the vulnerability of financial systems to GNSS disruptions takes center stage. As precise timing is essential for financial transactions, data integrity and regulatory compliance, GPS signals can be interrupted with cyberattacks like spoofing or jamming and can cause the system to suffer data loss or data corruption. This use case seeks to address this weakness, focusing on identifying these types of attacks (i.e., spoofing or jamming) to the GNSS while detecting possible cascading effects.

The current system relies heavily on GPS time synchronization and while it has some backup solutions like internal timing reference systems, being able to quickly detect the attacks on PNT services will improve a quick responsive analysis of systems as well of resilience for the banking services. By doing all this framework, the financial institutions can provide stronger systems of safety and protection. Providing for better metrics to react when any security events take place also helps improving analysis, since all systems come together.

ATLANTIS framework leverages an AI-powered detection tool that utilizes AI and machine learning models to analyze incoming GNSS signals. It is trained to recognize normal signal patterns and detect anomalies that indicate jamming or spoofing attacks [13]. These anomalies trigger alerts within the ATLANTIS framework providing interference classification, distinguishing between different types of attacks and potential false positives. This classification contributes to subsequent decision-making within the framework.

After the model detects an attack in the GNSS, SAFER aggregates the information from the AI-powered tool and other sources to help analysts quickly grasp the situation. Visualizations like network topology maps would allow them to understand dependencies of time-critical systems on GNSS and the potential scope of a disruption. SAFER's role is to provide a high-level overview of the unfolding incident, focusing on real-time updates and clear communication. Based on SAFER's overview, the DSS automatically recommends a course of action. This might include switching to a backup PNT source or implementing other mitigation strategies. When multiple solutions are available and the DSS helps narrow down the list to facilitate quick decision-making by the system operators, this helps provide faster alternatives, enhancing quick responses.

Finally, RRIM plays a crucial role in optimizing the response during, and after. In the response phase, RRIM analyzes the risks associated with the different mitigation options recommended by DSS. After the event takes place, now the analysis and evaluation of what happened takes its main focus and all the information from how the event was handled is given to the AI models for further analysis. Recommendations from RRIM help operators to improve and adapt to new issues and threats, improving and strengthening a system from future issues.

In essence, the ATLANTIS Framework, in conjunction with LINKS' solution, combines real-time detection, risk assessment, decision support, automated mitigation, and post-incident analysis to ensure rapid response and business continuity in the face of GNSS disruptions. It empowers CXB to maintain reliable PNT services, even under attack, safeguarding critical operations and protecting against potentially severe financial and operational consequences.

3.2.4 Banking Business Continuity and Critical Providers

Financial institutions rely on a network of critical providers, such as communication and electricity companies, to ensure continuous operation. These providers are essential for maintaining core banking functions, serving both internal processes and communications with customer-facing services and external partners. Any disruption in these services can have significant repercussions, affecting financial transactions, customer service, regulatory reporting, and overall business continuity.

Therefore, rapid detection, analysis, and response to critical provider interruptions are crucial. In Use Case 3.4, the focus shifts to physical dependencies and how their disruption can trigger cascading failures within CXB's infrastructure. Using internal data combined with real-time assessment from SAFER creates a quick visualization and risk scoring that is then analyzed in RRIM for potential improvements. The ATLANTIS framework empowers CXB to maintain robust business procedures under said events.

SAFER provides important information to the operator of the potential issues that are likely to occur, which helps improve quick decision-making. Then, RRIM uses Explainable AI (XAI) to help operators by providing new data, insights, to quickly adapt and fix any outage or network failure in the shortest amount of time.

By leveraging these tools, the framework not only improves CXB's ability to identify and respond quickly but also enhances transparency and accountability, which provides for a more robust system, better data collection methods, and overall improve the system.

4 Lessons Learned

Both AI4CYBER and ATLANTIS projects have yielded valuable lessons for enhancing cybersecurity resilience in the financial sector, particularly in the face of evolving threats. These lessons can be categorized into several key areas:

1. **The Power of Proactive, AI-Driven Security**: Both projects reinforce the necessity of moving beyond traditional, reactive cybersecurity measures. AI's ability to analyze vast datasets, detect anomalies, and predict potential threats in real-time enables a proactive security posture, crucial for mitigating increasingly sophisticated attacks [14]. Specifically, AI-powered solutions have demonstrated effectiveness in:

    - **Vulnerability Detection and Mitigation** [3]: AI4VULN and AI4FIX from AI4CYBER automate the identification and remediation of code-level vulnerabilities, crucial for preventing exploitation by attackers and ensuring CI/CD pipeline security within CXB's development processes.
    - **Advanced Threat Detection:** AI4TRIAGE and AI4SOAR from AI4CYBER, along with AI-powered GNSS spoofing detection model solution integrated into ATLANTIS, provide advanced threat detection capabilities, identifying complex attack patterns, and initiating automated responses.
    - **Insider Threat Detection:** AI4FIDS from AI4CYBER offers a federated learning approach to insider threat detection, enhancing data privacy while enabling collaborative anomaly detection across multiple systems.
    - **Disinformation Countermeasures:** NLP, fact-checking, and deepfake detection tools, as highlighted in ATLANTIS, demonstrate the power of AI in combating disinformation campaigns that can negatively impact financial markets. Moreover, regulatory compliance frameworks increasingly emphasize AI-assisted risk management. Disinformation campaigns that manipulate financial markets can violate securities laws, prompting regulatory investigations. AI-powered compliance tools help institutions monitor and report disinformation incidents, ensuring transparency and adherence to financial regulations.

2. **The Importance of Systemic Resilience**: ATLANTIS underscores the interconnectedness of critical infrastructure and the necessity of systemic resilience. By mapping interdependencies and considering cascading effects, financial institutions can better anticipate the broader impact of localized disruptions, whether from cyberattacks, GNSS spoofing, or critical provider outages. This systemic approach enables more comprehensive risk assessment and informed decision-making during incident response.
3. **The Value of Explainable AI (XAI)**: The use of XAI, as demonstrated by the TRUST4AI component and in ATLANTIS' framework, is crucial for building trust and transparency in AI-driven security solutions. Understanding the rationale behind AI's decisions enables human analysts to validate alerts, refine models, and ensure fairness and accountability in automated responses.
4. **Collaboration and Information Sharing**: AI4COLLAB in AI4CYBER promotes the importance of secure, anonymous data sharing for enhanced threat intelligence. Similarly, ATLANTIS emphasizes the need for cooperation among different stakeholders to enable a coordinated response to large-scale systemic threats. This collaboration is crucial for building a robust and resilient defense against increasingly sophisticated attacks.

5. **Integration and Automation**: Both projects demonstrate the value of integrating AI solutions with existing security tools and workflows. Automating routine tasks and responses enables faster detection, accelerated mitigation, and frees up security analysts to focus on more complex and strategic activities. AI4SOAR's automated playbooks, triggered by AI4TRIAGE's prioritization of alerts, exemplifies the value of this automated workflow and helps reduce response times.

By applying these lessons, the financial sector can continue to evolve its cybersecurity practices, adopting a proactive, systemic, AI-driven approach that will be critical for achieving enhanced resilience against the ever-growing threat landscape.

5 Conclusions

This chapter highlights the transformative potential of AI in cybersecurity and IT risk management within the financial sector. By leveraging AI-powered solutions, financial institutions can adopt a proactive, rather than reactive, security posture. AI enables real-time threat detection, automated incident response, and enhanced risk assessment capabilities essential for mitigating increasingly complex attacks. The use of XAI builds trust and transparency in these solutions, fostering collaboration between AI systems and human analysts. CXB's active participation in the AI4CYBER and ATLANTIS projects demonstrates the practical application of these AI-driven technologies in addressing critical security challenges, providing valuable insights and lessons for the broader financial industry. As cyber threats evolve and the sophistication level increases, AI will be even more important in the future to strengthen security and enable operational resilience for financial services with a global reach.

Acknowledgements This work has received funding from the European Union's Horizon Europe research and innovation program under grant agreements No 101070450 (AI4CYBER) and No 101073909 (ATLANTIS).

References

1. Nour, S.M. and Said, S.A. (2024). Harnessing the Power of AI for Effective Cybersecurity Defense. https://doi.org/10.1109/icci61671.2024.10485059.
2. Iturbe, E., Rios, E., Rego, A. and Toledo, N. (2023). Artificial Intelligence for next generation cybersecurity: The AI4CYBER framework. https://doi.org/10.1145/3600160.3605051.
3. Eberhardt, G. and Milánkovich, Á. (2024). VulnGPT: Enhancing Source Code Vulnerability Detection Using AutoGPT and Adaptive Supervision Strategies. [online] pp. 450–454. https://doi.org/10.1109/dcoss-iot61029.2024.00072.
4. Jasz, J., Hegedus, P., Milankovich, A. and Ferenc, R. (2022). An End-to-End Framework for Repairing Potentially Vulnerable Source Code. 2022 IEEE 22nd International Working

Conference on Source Code Analysis and Manipulation (SCAM), [online] pp. 242–247. https://doi.org/10.1109/scam55253.2022.00034.

5. Bouaziz, A., Nguyen, M.-D., Valdés, V., Cavalli, A. and Mallouli, W. (2023). Study on Adversarial Attacks Techniques, Learning Methods and Countermeasures: Application to Anomaly Detection. https://doi.org/10.5220/0012125100003538.
6. Pawlicki, M., Pawlicka, A., Śrutek, M., Kozik, R. and Choraś, M. (2023). Interpreting Intrusions - The Role of Explainability in AI-Based Intrusion Detection Systems. Lecture Notes in Networks and Systems, pp. 45–53. https://doi.org/10.1007/978-3-031-41630-9_5.
7. Dalamagkas, C., Asimopoulos, D., Radoglou-Grammatikis, P., Maropoulos, N., Lagkas, T., Argyriou, V., Sargsyan, G. and Sarigiannidis, P. (2024). AI4COLLAB: An AI-based Threat Information Sharing Platform. [online] pp. 783–788. https://doi.org/10.1109/csr61664.2024.10679429.
8. Radoglou-Grammatikis, P., Bouzinis, P.S Ioannis, Makris Thomas, Lagkas Vasileios, Argyriou Georgios Th., Papadopoulos Panagiotis, Fouliras George, Seritan Panagiotis, Sarigiannidis AI4FIDS: Multimodal Federated Intrusion Detection IEEE Transactions on Emerging Topics in Computing 1-15 10.1109/TETC.2025.3562346
9. Voulkidis, A. (2024). The CCI-SAAM Framework. [online] Available at: https://www.atlantis-horizon.eu/wp-content/uploads/2024/06/ATLANTIS-The-CCI-SAAM-Framework.pdf [Accessed 10 Jul. 2025].
10. Stira, C., Puglisi, I. and Giunta, G. (2024). The Awareness and Comprehension Framework. [online] Available at: https://www.atlantis-horizon.eu/wp-content/uploads/2024/10/ATLANTIS-06-_The-Awareness-and-Comprehension-Framework_1.pdf
11. Segurola-Gil, L. and Gil-Lerchundi, A. (2024). Entropy-Based Risk Assessment for Communication Networks. [online] Available at: https://www.atlantis-horizon.eu/wp-content/uploads/2024/06/ATLANTIS-Entropy-Based-Risk-Assessment-for-Communication-Networks.pdf
12. Dinu, V., Plesa, A., Jarca, A., Vintila, R.C., Nechifor, C.-S.N. and Ilie, I. (2024). Enhancing Risk Reduction and Incident Mitigation Through Automated Explainable Recommendations and XAI. [online] Available at: https://www.atlantis-horizon.eu/wp-content/uploads/2024/06/ATLANTIS-Enhancing-Risk-Reduction-and-Incident-Mitigation-through-XAI.pdf.
13. Caputo, G. (2024). AI-Based GNSS Jamming and Spoofing Detection and Classification. [online] Available at: https://www.atlantis-horizon.eu/wp-content/uploads/2024/06/ATLANTIS-AI-Based-GNSS-Jamming-and-Spoofing-Detection-and-Classification.pdf.
14. Karampasi, A., Radoglou-Grammatikis, P., Pawlicki, M., Choraś, R., Martin de Pozuelo, R., Sarigiannidis, P., Puchalski, D., Pawlicka, A., Kozik, R. and Choraś, M. (2024). Towards Transparent AI-Powered Cybersecurity in Financial Systems: The Deployment of Federated Learning and Explainable AI in the CaixaBank pilot. 2022 IEEE International Conference on Data Mining Workshops (ICDMW), pp. 270–277. https://doi.org/10.1109/icdmw65004.2024.00041.

AI-Enhanced Security Tools with Virtual Reality for Ensuring Digital Resilience

Shareeful Islam, Spyridon Papastergiou, Armend Duzha, Kostas Drakonakis, Grigorios Chrysos, Sotiris Ioannidis, Iordanis Xanthopoulos, George Spanoudakis, Marc-Oliver Pahl, Haralambos Mouratidis, Michael Kühr, and Mohammad Hamad

Abstract The cybersecurity challenges facing today's digital infrastructure make it critical to improve the disruption preparedness and resilience of these systems.

S. Islam
Anglia Ruskin Univerity, Cambridge, UK
e-mail: shareeful.islam@aru.ac.uk

S. Papastergiou (✉)
R&D Maggioli, Athens, Greece
e-mail: spyros.papastergiou@maggioli.gr

A. Duzha
R&D Maggioli, Santarcangelo di Romagna, Italy
e-mail: armend.duzha@maggioli.it

K. Drakonakis · G. Chrysos · S. Ioannidis
Technical University of Crete, Crete, Greece
e-mail: gxrysos@tuc.gr

S. Ioannidis
e-mail: sotiris@ece.tuc.gr

I. Xanthopoulos · G. Spanoudakis
Sphynx Technology Solutions AG, Zug, Switzerland
e-mail: i.xanthopoulos@sphynx.ch

G. Spanoudakis
e-mail: spanoudakis@sphynx.ch

M.-O. Pahl
IMT Atlantique, Rennes, France
e-mail: marc-oliver.pahl@imt-atlantique.fr

H. Mouratidis
Security Labs Consulting, Cork, Ireland
e-mail: haris@securitylabs.ie

M. Kühr · M. Hamad
Computer Engineering, Technical University of Munich, Munich, Germany
e-mail: michael.kuehr@tum.de

J. Soldatos and H. Abie (eds.), *Technology-Enabled Resilience: Innovations in Critical Infrastructure Protection*, https://doi.org/10.1007/978-3-032-25917-2_10

Resilience can only be achieved through (i) having a clear and comprehensive understanding of cyber incidents and their cascading effects, including potential disruptions, and (ii) providing a system with dynamic recovery and response capabilities. The CyberSecDome project aims to ensure the continuity of operations even in adverse circumstances of complex and heterogeneous systems in various business domains. This chapter outlines the AI-empowered security tools developed by CyberSecDome, including intrusion detection and prediction, dynamic risk assessment, dynamic and adaptive incident response, and VR-based interactive and collaborative user interface.

Keywords Artificial intelligence · Digital infrastructure · Resilience · Risk assessment · Incident handling and response · Virtual reality

1 Introduction

The adoption of AI for cybersecurity solutions is now widely considered due to the capabilities for processing a vast amount of security data to identify and predict security patterns, threats, and vulnerabilities for informed security decision-making. Specifically, AI models are adopted to enhance specific security services, including risk assessment, incident handling, and response, which certainly improve the security posture and resilience of service delivery within the organization. Cyber threats have become increasingly sophisticated, employing new and continuously evolving methods, while digital infrastructure has grown more complex due to varying levels of cyber and physical interdependencies. It is necessary to understand the possible security scenario that can pose a potential risk for major disruption to the business. In this context, virtual reality can effectively support the visualization of security scenarios by creating context-specific attack and defense environments, thus enhancing the understanding of attacks and their mitigation strategies. Despite wider recognition of considering AI in cybersecurity, there is a need to visualize the possible attack scenarios based on the security information identified and predicted from the AI models. Hence, the outputs produced by AI-enhanced security tools need to be consolidated and visually represented for a better understanding of the context. This allows providing more insight about the threat and justification for choosing security controls. This chapter presents a comprehensive overview of the AI-enhanced security tools based on the technical outcomes of the CyberSecDome[1] project. The focus is to consider the adoption of AI in different security services as a part of the project

M. Hamad
e-mail: mohammad.hamad@tum.de

[1] https://cybersecdome.eu/.

activities. Moreover, building on the insights gained from the project's advancements, the chapter concludes with a set of concrete recommendations and best practices that can be leveraged by cybersecurity practitioners or contribute toward related standardization efforts.

2 State-of-the-Art Analysis

This section provides a comprehensive overview of recent advancements in incident detection and prediction tools, focusing on innovative AI methodologies and hardware technologies that enhance detection effectiveness. Foudhaili et al. [6] introduced a framework for deploying an intelligent intrusion detection system in edge environments, demonstrating superior performance and energy efficiency with minimal hardware overhead compared to existing solutions. Zeng et al. [16] developed a hardware/software co-designed IDS tailored for IoT edge devices, achieving high accuracy, low latency, and low power consumption on a compact FPGA platform—key attributes for scalable IoT deployment. Pigasus [17] presents an FPGA-accelerated IDS capable of 100 Gbps throughput on a single server by offloading TCP reassembly and pattern matching, achieving low latency, high accuracy, and a 38-fold improvement in power efficiency with only five CPU cores. Köylü et al. [13] conducted a survey on machine learning applications in hardware security, categorizing approaches addressing threats such as side-channel attacks, hardware Trojans, and counterfeit detection, while also emphasizing vulnerabilities inherent to ML models in hardware and the necessity for robust defensive mechanisms. Additionally, [15] reviews publicly available datasets for AI-based IDS research, assessing features, attack types, and labeling accuracy, while addressing challenges including outdated content, lack of standardization, and limited realism, advocating for improved dataset design and standardization practices. Last, prior studies [7] demonstrate that time-based traffic features effectively enhance intrusion prediction systems (IPS) within encrypted and VPN environments, enabling threat detection solely through timing information without payload inspection, thereby supporting real-time, privacy-preserving, and efficient threat prediction via lightweight machine learning models.

3 Intrusion Detection and Prediction

The Intrusion Detection and Prediction System (IDPS) is a comprehensive and high-performance cybersecurity solution that integrates parallel analytical frameworks for the detection and prediction of intrusion attempts. Specifically, the system is engineered to conduct continuous, real-time monitoring and AI-driven analysis of dynamic network traffic via a specialized architecture, enabling both intrusion detection and the anticipation of emerging threats. The IDPS processes incoming network

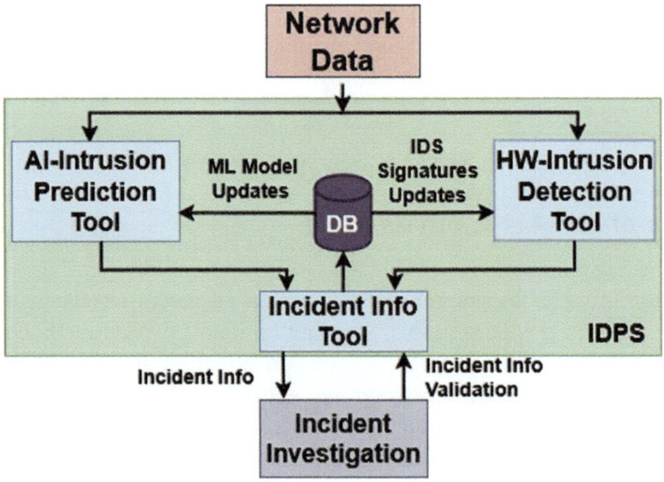

Fig. 1 Architecture of intrusion detection and prediction tool

packets in a real-time streaming way through two independent and complementary modules: the Intrusion Detection System (IDS) and the Intrusion Prediction System (IPS). Each module operates autonomously, employing distinct methodologies to analyze network traffic and identify potential security incidents. The IDS focuses on signature-based and pattern-matching techniques, while the IPS utilizes predictive models to recognize anomalous or malicious behavior. The alerts generated by both components are subsequently aggregated and forwarded to downstream elements of the CybeSecDome framework, where they are validated, further investigated, and used to notify end users.

As shown in Fig. 1 the IDPS system comprises the following core modules:

- Hardware-based Intrusion Detection System (HW-IDS): This component is deployed on a high-performance hardware acceleration platform, i.e., a Field-Programmable Gate Array (FPGA), and is optimized for low-latency, real-time evaluation of network traffic for intrusion attempts.
- AI-based Intrusion Prediction System (AI-IPS): This module integrates advanced artificial intelligence techniques, employing machine learning (ML) models to identify patterns indicative of malicious behavior. It automates the development and optimization of ML models and also facilitates their retraining to ensure they remain up to date.

The HW-IDS tool, Fig. 2, is a real-time cybersecurity threat detection tool implemented on a high-performance reconfigurable computing platform. Its principal objective is the continuous monitoring of network traffic to identify potential attack signatures and issue corresponding security alerts. The detection methodology is based on an efficient framework [4] that maps predefined security signatures into a compact internal data structure, enabling rapid evaluation of each incoming network

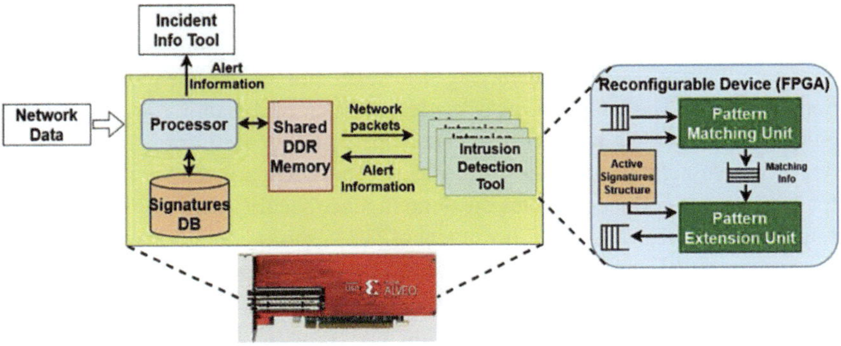

Fig. 2 Hardware architecture of intrusion detection tool

packet. The HW-IDS processing pipeline operates in a real-time, streaming way, exploiting the parallelism and concurrency offered by reconfigurable hardware.

The AI-IPS tool employs advanced machine learning (ML) methodologies to detect potential cyber threats by analyzing historical network traffic data. The predictive process is initiated through the development of ML models trained on labeled datasets. The IPS component utilizes the eXtreme Gradient Boosting (XGBoost) [2] classifier—a high-performance, open-source library implementing gradient-boosted decision trees. XGBoost is widely recognized for its efficiency, scalability, and predictive accuracy, making it a standard in both academic research and industrial applications. The IPS XGBoost model is trained via the federated client in each CyberSecDome node. The training dataset comprises aggregated network traffic samples, organized into flows, and default hyperparameters are applied uniformly across all clients to ensure consistency. Upon completion of the training phase, the XGBoost model is deployed for inference. The IPS predictor processes raw network samples in real time, dynamically aggregates them into flows, and generates alerts for the Security Information and Event Management (SIEM) system whenever malicious behavior is detected.

4 Dynamic Risk Assessment

Cybersecurity risk management plays a critical role in managing risks within an organizational context. However, there is a limited focus on the dynamic context of the organization by the existing risk management practices. The Dynamic Risk Assessment (DRA) focuses on analyzing the key dynamic parameters of the organizational infrastructure and security context, such as vulnerability exploitation and asset dependencies, that influence or quantify the risk level. DRA considers vulnerability exploitation because a large number of vulnerabilities are published, but not all are relevant in a specific system context [10]. Therefore, it is necessary to focus on relevant vulnerabilities with exploitation potential. DRA leverages the capability of an AI model to identify vulnerability exploitation by correlating security data from open intelligence sources and linking it to related assets within specific organizational contexts. The adoption of AI models to enhance dynamic risk analysis relies on specific AI models and security-related open intelligence information containing threat and vulnerability-related information. The DRA utilizes a hybrid AI model for this purpose and incorporates open intelligence security data from various open-source repositories. The workflow model of DRA is triggered once an asset is registered in the DRA service, making it necessary to calculate the risk level for each asset. The methodology used for the risk score primarily considers the exploitation level of vulnerabilities associated with the affected asset. Based on the exploitation level of a specific vulnerability, the risk for the asset is calculated.

The DRA employs a hybrid model of AI, shown in Fig. 3, to effectively conduct vulnerability assessment and prioritization in complicated systems. This hybrid AI forms part and parcel of the DRA framework, given that it allows for an accurate assessment and prioritization of a vulnerability based on dynamically computed risk through enhanced feature selection and predictive modeling. The hybrid model starts with a linear regression that embeds the Extra Trees Regressor to emphasize the most important features of the dataset. This ensures that only the most influential attributes are considered, hence reducing noise and enhancing the efficiency of the model. After selecting the best features, a neural network is trained on how to contextualize and prioritize vulnerabilities based on the extracted insights. This approach brings together the interpretability of linear regression with the flexibility and predictive power of neural networks to provide a robust, scalable solution for the identification and mitigation of high-priority risks in dynamic environments. A brief overview and rationale of the hybrid model are added below:

- Extra Trees Regressor is chosen for its efficiency in handling big data with high-dimensional feature space, and it finds a perfect application in feature selection for dynamic risk analysis. Unlike the traditional tree-based models, it is an ensemble learning method, reducing overfitting by adding extra randomness to the construction of each tree. Instead of searching for the best split at each node, it selects a random subset of features and then splits based on random thresholds, yielding diverse and uncorrelated trees [11]. Therefore, it is employed by DRA for key feature selection of vulnerability datasets.
- Linear regression is a simple statistical method that models the relationship between one or more independent variables and a dependent variable. It works under the assumption that such a relationship can be modeled as a linear equation, where a dependent variable is a weighted sum of independent variables plus a bias. The major goal of Linear Regression is to find the best possible weights (or coefficients) that minimize the error between predicted values and the actual data points.
- Neural networks are chosen because they can model complex nonlinear relationships in data, making them ideal for applications such as dynamic risk analysis,

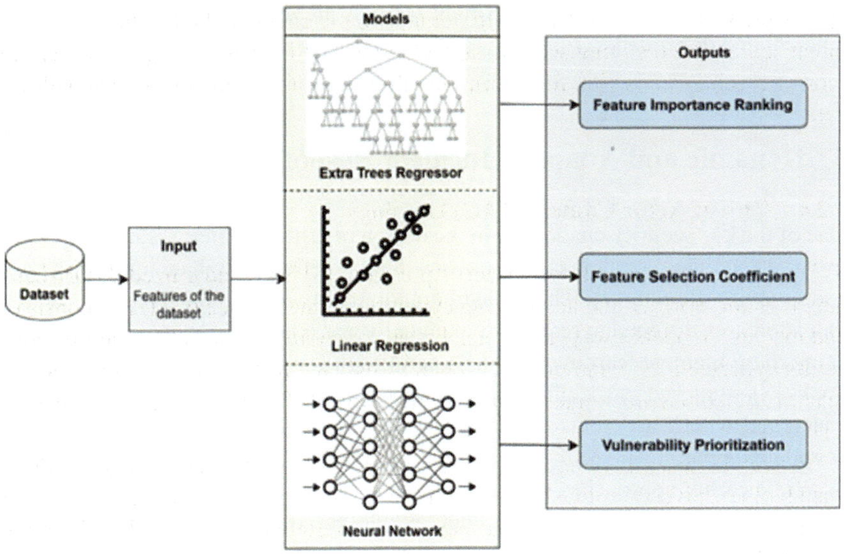

Fig. 3 Hybrid AI model for dynamic risk assessment

whereby interactions between features can be pretty complex and multidimensional. Neural networks differ from traditional models in that they do exceptionally well at picking up patterns and dependencies implicitly hidden, which allows them to prioritize vulnerabilities more effectively. The model begins with an input layer, where each neuron represents a feature or variable from the input data [12]. Information from the input layer is passed to one or more hidden layers, where each neuron receives inputs from all neurons in the previous layer, applies a mathematical transformation using a weight, bias, and an activation function, and passes the result to the next layer. These are scalable and flexible enough to process big, high-dimensional data with the ability to adapt to various problem domains.

Figure 3 presents the hybrid model, where Extra Trees Regressor is employed to perform feature selection from the chosen data set. The execution of this step begins with data preprocessing, where data cleaning involves handling missing values and normalizing the data. The Extra Trees Regressor is then trained on this dataset to calculate feature importance. Based on these importance scores, features are ranked, and those with high importance scores are chosen for further analysis and for building the models. This way, less important features are filtered out. Additionally, linear regression is used to extract the coefficients of the selected features. This is useful for establishing the correlation between the features and the target variable, providing clear insight into how each feature affects the EPSS score obtained. Applying both the Extra Trees Regressor for feature selection and linear regression for coefficient analysis ensures that the most important features are used, making the model more efficient and dependable.

5 Dynamic and Adaptive Incident Response

One of the key security challenges in the design of critical systems is ensuring adaptive and intelligent response mechanisms that can effectively handle unpredictable and evolving threats [9]. The Dynamic and Adaptive Incident Response (DAIR) tool addresses the need to respond dynamically and adaptively to incidents by leveraging reinforcement learning (RL) to generate optimal response strategies—or playbooks [3]—tailored to specific cyber incidents. RL is a subfield of machine learning concerned with how agents ought to take actions in an environment to maximize cumulative reward.

While earlier approaches have explored non-AI-based response systems [8], and others have applied AI techniques such as rule-based logic or supervised learning for cyber incident mitigation [1], DAIR is applying RL to enable context-aware and continuously evolving strategies that adapt to diverse and complex threat scenarios. It integrates seamlessly with other components, particularly IDPS, to deliver intelligent, adaptable, and collaborative cybersecurity defense. The DAIR is composed of two primary modules:

- AI-based Optimal Response Decision Module: This module generates playbooks based on normalized incident data by the incident investigation component.
- Policy Quality Estimator: It evaluates the effectiveness of the selected responses and updates the AI model accordingly.

The architecture is designed to support modular AI model integration, allowing different RL algorithms to operate on a shared data abstraction layer. Communication with external components is facilitated via REST APIs. Figure 4 shows the architectural overview of DAIR.

DAIR receives incident data from the IDPS, which includes information such as IP addresses, ports, used communication protocols, attack types, and attack severity. This data is normalized and used as input for the RL algorithms. The selected playbook is then returned to the IDPS module and visualized in the VR interface for Security Operations Center (SOC) analysts to review and execute. DAIR employs Q-learning as a variant of RL to learn optimal response strategies. Q-learning uses a tabular approach (Q-table) to map state-action pairs to expected rewards. The Q-table is trained using feedback from the SOC, which evaluates the effectiveness of proposed playbooks. The reward is a binary value specifying the success of the proposed response. This feedback loop ensures that DAIR continuously adapts to the evolving threat landscape and organizational context.

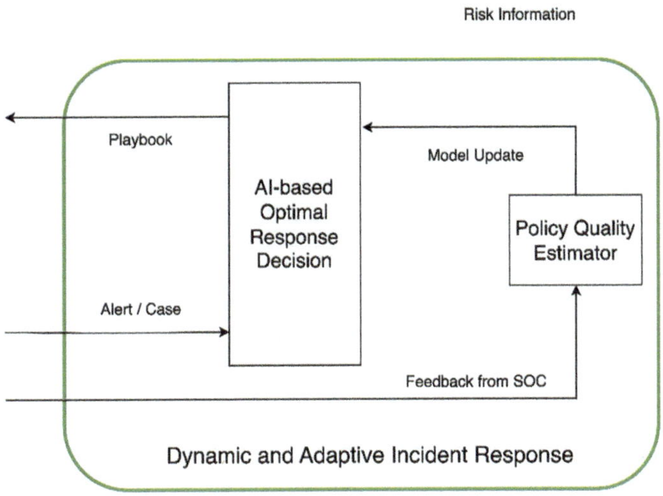

Fig. 4 Architecture of dynamic and adaptive incident response

6 Extended Reality Interfaces for Immersive Cybersecurity Analytics

Cybersecurity operations involve analyzing large volumes of heterogeneous, time-sensitive data. Analysts must detect threats, assess risks, and coordinate responses, often across distributed infrastructures and teams. Conventional tools—such as Security Information and Event Management (SIEM) systems, log analyzers, and dashboards—are typically presented on 2D screens and offer limited contextual integration. These interfaces require users to mentally synthesize information from separate sources, which can impair situational awareness and slow down decision-making. Extended Reality (XR) technologies—spanning Virtual Reality (VR), Augmented Reality (AR), and Mixed Reality (MR)—offer an emerging paradigm that enables immersive, spatial representations of cybersecurity data. While XR is increasingly used for training and simulation, its application in professional, real-time cybersecurity operations remains rare.

Current SOC environments rely on multiple monitors, flat visualizations, and isolated tools that require extensive cognitive effort to interpret in complex scenarios. Analysts switch between different systems and interfaces to form a comprehensive picture of a situation. Remote collaboration is also limited, typically relying on video calls and screen sharing, which further fragments information flow. These limitations become critical in high-stakes environments such as smart cities, critical infrastructure, or national-scale cyber defense, where a fast and coherent understanding of the threat landscape is essential. Thus, there is a huge need to transform this landscape by developing next-generation interfaces for cybersecurity. We leverage XR to integrate diverse data streams into a shared, immersive environment. Our approach enhances situational awareness, facilitates collaboration, and allows analysts to interact with their tools in new and more intuitive ways. By embedding cybersecurity data directly into spatial representations of the protected environment, we enable analysts to see, explore, and understand threats in context.

Visual analytics has long been a focus in cybersecurity research. Early systems focused on 2D graph visualization, anomaly detection via heatmaps, and timeline-based event tracking. While effective for static data or historical analysis, these interfaces often fall short in real-time monitoring or large-scale infrastructure management. XR has been explored in cybersecurity education and training, where simulations of attack scenarios can improve user preparedness. Some research has investigated 3D visualization of network topologies or attack information [14]. However, professional applications of XR remain limited, often due to concerns over integration complexity, usability, and real-time performance. Our work builds on these initial efforts by focusing not on education or simulation, but on real time, professional use. We combine immersive visualization, synchronized data interaction, collaborative spaces, and tool integration into one operational system designed for analysts and incident responders. Our approach centers on a digital twin of the monitored infrastructure, such as a smart city, energy plant, or transport hub. This 3D map serves as a base layer for overlaying cybersecurity-relevant information such as threats, active alerts, anomalies, and risk levels.

Threat indicators are spatially linked to the actual assets they affect—e.g., a vulnerable router in a network zone or a compromised IoT device in a smart building. Visual encodings such as color gradients, glyphs, or animations make it easy to grasp severity, attack vectors, and propagation patterns. The analyst can zoom, filter, and query the environment, supporting both overview and detail-on-demand analysis.

To support multi-perspective analysis, we integrate synchronized views within the XR space. Users can examine multiple visualizations—such as timeline-based event sequences, topological maps, asset dependency trees, or log stream summaries—simultaneously. These views are linked: selecting a compromised server in one view will highlight its representation across all others. This allows analysts to correlate findings across dimensions without leaving the immersive workspace.

Importantly, we embed virtual screens within the XR environment that provide direct access to native tools—such as SIEM dashboards, log viewers, or terminal sessions. This hybrid approach ensures operational continuity and allows analysts to combine traditional and immersive workflows without disruption.

Collaboration is supported through multi-user functionality. Multiple analysts can inhabit the same virtual environment as avatars, share observations, annotate objects, and jointly investigate incidents. This capability is especially valuable in geographically distributed response teams, where a shared virtual space can replace fragmented communication channels.

Contributions: Our approach makes the following key contributions to the field of cybersecurity analytics:

1. Immersive Situational Awareness: We embed real-time cybersecurity data directly into 3D representations of critical infrastructure, providing analysts with intuitive spatial context.
2. Synchronized Multiview Analytics: Our system supports multiple coordinated visualizations within the same XR environment, enabling multifaceted exploration of complex incidents.
3. Hybrid Interaction Model: We integrate native cybersecurity tools via virtual screens in XR, maintaining familiarity and operational reliability while enhancing spatial interaction.
4. Collaborative Decision Spaces: We enable multi-user virtual environments that support real time, spatial collaboration among distributed analysts and incident responders.
5. Domain-Specific Pilots: Through the CyberSecDome project, we validate our system across multiple use cases, including smart city infrastructure, cloud security, and public administration, ensuring practical relevance and real-world impact.

7 Key Takeaways and Best Practices

Based on the comprehensive description of the aforementioned AI-enhanced components, a set of key insights and takeaways has been extrapolated. These can serve as concrete recommendations and best practices for cybersecurity practitioners (e.g., CSIRTs, SOC operators) willing to replicate or adopt such approaches, as well as for standardization and policy-making bodies aiming toward standardizing key aspects in the area of digital infrastructure protection and resilience. We detail our recommendations in the following.

Intrusion detection components should ideally:

- Be mapped on reconfigurable hardware platforms and integrate HPC frameworks, so as to address high-volume network traffic challenges and enable real-time analysis and timely threat mitigation.
- Showcase scalability through autonomous, remotely operable nodes that effectively collaborate to secure expansive digital infrastructures by exchanging relevant information (e.g., signatures).
- Be adaptive to evolving threats by promptly incorporating new validated attack signatures as detected by other collaborating components, such as AI-enhanced intrusion prediction systems.

AI and ML-enhanced Intrusion prediction approaches are recommended to:

- Incorporate data preprocessing methods to generate additional data variables originating from the captured network traffic, to further improve the models' performance.
- Use non-black-box models to foster better results' interpretation and explainability.

- Establish periodic retraining routines to keep models updated based on new benign and malicious network traffic.

An effective risk assessment approach should:

- Take into consideration system context and asset dependencies, alongside vulnerability exploitation (e.g., EPSS scores [5]), so as to compute accurate and relevant risk scores.
- Utilize advanced neural networks to identify non-discernible, implicit patterns and dependencies and achieve effective vulnerability prioritization.

An effective incident response component should:

- Showcase adaptivity by automatically selecting context-aware responses to incidents, enabling faster and more accurate reactions without relying on static rules.
- Leverage advanced reinforcement learning approaches (e.g., Q-Learning) to adapt response strategies over time, improving resilience against evolving threats based on real SOC feedback.

Extended Reality interfaces are recommended to:

- Incorporate real-time threat overlays over infrastructure maps to provide additional insights into where incidents occur, how they spread and what assets are at risk, offering enhanced situational awareness and prompt response capabilities.
- Combine multiple, linked visualizations (e.g., network graphs, timelines, asset trees) to enable different analytical perspectives on the same incident and support better-informed decisions.
- Integrate native security tools (e.g., SIEM, log analyzers) via virtual screens to ensure continuity of workflows, allow for gradual transition to immersive systems and retain the reliability of proven toolsets.

8 Conclusion

AI-enhanced security solutions are now widely considered to enhance the security and resilience of digital infrastructure. However, adoption of AI for cybersecurity needs to be systematic based on specific security services or security functionalities. Moreover, it is also necessary to select an appropriate AI model along with a relevant dataset to train the model so that the security service can benefit from AI. In this context, this chapter presented AI-enhanced security services considering three main security services, including Intrusion Detection and Prediction, Dynamic Risk Assessment, and Dynamic and Adaptive Incident Response. Each service adopts a specific AI model to support its functionalities, such as eXtreme Gradient Boosting (XGBoost) for IDPS, linear regression, and neural networks for DRA, and Q-learning for DAIR. Moreover, the chapter also presents how visual analytics, such as heatmaps and timeline-based event tracking, can effectively support the creation of cybersecurity situational awareness. Finally, we also provide the best practice guidelines for adopting AI capabilities in cybersecurity services. We believe this contribution advocates further adoption of AI in a systemic and effective manner.

Acknowledgements This work was supported by the European Union's Horizon Europe project, CyberSecDome-An innovative virtual reality-based intrusion detection, incident investigation, and response approach to improve the resilience, security, privacy, and accountability of complex and heterogeneous digital systems and infrastructures, funded under grant agreement No. 101120779.

References

1. Gauhar Ali, Sajid Shah, and Mohammed ElAffendi. Enhancing cybersecurity incident response: Ai-driven optimization for strengthened advanced persistent threat detection. Results in Engineering, 25:104078, 2025.
2. Tianqi Chen and Carlos Guestrin. Xgboost: A scalable tree boosting system. CoRR, abs/1603.02754, 2016.
3. Cybersecurity and Infrastructure Security Agency (CISA). Cybersecurity incident & vulnerability response playbooks. Technical report, November 2021. U.S. Department of Homeland Security.
4. Dimitris Deyannis, Eva Papadogiannaki, Grigorios Chrysos, Konstantinos Georgopoulos, and Sotiris Ioannidis. The diversification and enhancement of an ids scheme for the cybersecurity needs of modern supply chains. Electronics, 11(13):1944, 2022.
5. FIRST.org. Exploit prediction scoring system (epss). https://www.first.org/epss/, 2024. Accessed: 2025-07-03.
6. Wadid Foudhaili, Anouar Nechi, Celine Thermann, Mohammad Al Johmani, Rainer Buchty, Mladen Berekovic, and Saleh Mulhem. Reconfigurable edge hardware for intelligent ids: Systematic approach. In International Symposium on Applied Reconfigurable Computing, pages 48–62. Springer, 2024.
7. Gerard Drapper Gil, Arash Habibi Lashkari, Mohammad Mamun, and Ali A Ghorbani. Characterization of encrypted and vpn tra"c using time-related features. In Proceedings of the 2nd international conference on information systems security and privacy (ICISSP 2016), pages 407–414. SciTePress Setúbal, Portugal, 2016.
8. Mohammad Hamad, Andreas Finkenzeller, Michael Kühr, Andrew Roberts, Olaf Maennel, Vassilis Prevelakis, and Sebastian Steinhorst. React: Autonomous intrusion response system for intelligent vehicles. Computers & Security, 145:104008, 2024.
9. Mohammad Hamad and Sebastian Steinhorst. Security challenges in autonomous systems design. In International Conference on Computational Technologies and Electronics, pages 142–154. Springer, 2023.
10. Shareeful Islam, Abdulrazaq Abba, Umar Ismail, Haralambos Mouratidis, and Spyridon Papastergiou. Vulnerability prediction for secure healthcare supply chain service delivery. Integrated Computer-Aided Engineering, 29(4):389–409, 2022.
11. Sadiqa Jafari and Yung-Cheol Byun. E"cient state of charge estimation in electric vehicles batteries based on the extra tree regressor: A data-driven approach. Heliyon, 10(4), 2024.
12. Nancy Kanwisher, Meenakshi Khosla, and Katharina Dobs. Using artificial neural networks to ask 'why'questions of minds and brains. Trends in Neurosciences, 46(3):240–254, 2023.
13. Troya Çağıl Köylü, Cezar Rodolfo Wedig Reinbrecht, Anteneh Gebregiorgis, Said Hamdioui, and Mottaqiallah Taouil. A survey on machine learning in hardware security. ACM Journal on Emerging Technologies in Computing Systems, 19(2):1–37, 2023.

14. A. Oline and D. Reiners. Exploring three-dimensional visualization for intrusion detection. In IEEE Workshop on Visualization for Computer Security, 2005. (VizSEC 05)., pages 113–120, 2005.
15. Daniela Pinto, Ivone Amorim, Eva Maia, and Isabel Praça. A review on intrusion detection datasets: tools, processes, and features. Computer Networks, page 111177, 2025.
16. Qingyu Zeng and Yuko Hara-Azumi. Hardware/software codesign of real-time intrusion detection system for internet of things devices. IEEE Internet of Things Journal, 2024.
17. Zhipeng Zhao, Hugo Sadok, Nirav Atre, James C Hoe, Vyas Sekar, and Justine Sherry. Achieving 100gbps intrusion prevention on a single server. In 14th USENIX Symposium on Operating Systems Design and Implementation (OSDI 20), pages 1083–1100, 2020.

Machine Unlearning for Cybersecure AI in Critical Infrastructures

Efi Kafali, Christoforos N. Spartalis, Theodoros Semertzidis, Charalampos Z. Patrikakis, and Petros Daras

Abstract The increasing integration of artificial intelligence (AI) systems in critical n frastructures (CIs) is raising concerns regarding security, privacy, and safety. In such high-risk environments, any potential compromises of AI systems should be addressed promptly. However, discarding a compromised AI system and building another one from scratch is inefficient in the modern AI landscape, mainly due to the significant time and cost required. This underscores the need for model correction methods, that mitigate vulnerabilities during the AI system's lifecycle without necessitating complete replacement. In this chapter, we emphasize the challenges that highlight the importance of such corrective measures and their potential for mitigating privacy and security risks in AI systems deployed in CIs. By addressing these aspects and their limitations, the chapter provides a comprehensive understanding of how to enhance resilience, response capabilities, and reliability of AI systems operating in CIs.

Keywords Machine unlearning · Computer vision · Critical infrastructures

E. Kafali · C. N. Spartalis · T. Semertzidis (✉) · P. Daras
CERTH, Thessaloniki, Greece
e-mail: theosem@iti.gr

E. Kafali
e-mail: ekaf@iti.gr

C. N. Spartalis
e-mail: c.spartalis@iti.gr

P. Daras
e-mail: daras@iti.gr

C. Z. Patrikakis
Department of Electrical and Electronics Engineering, University of West Attica, Egaleo, Athens, Greece
e-mail: bpatr@uniwa.gr

J. Soldatos and H. Abie (eds.), *Technology-Enabled Resilience: Innovations in Critical Infrastructure Protection*, https://doi.org/10.1007/978-3-032-25917-2_11

1 Introduction

Machine unlearning (MU) has emerged as a promising research area in AI and machine learning (ML), aiming to remove the influence of specific data from trained models [1]. Un like traditional ML, which focuses on acquiring and retaining knowledge from training data, MU operates as a reverse process, securing that a model behaves as if it never encountered particular data points.

Recent regulations such as the general data protection regulation (GDPR) [2] and the California Consumer Privacy Act (CCPA)[1] have heightened the need for MU, particularly in enforcing the "right to be forgotten" (RTBF) [3]. MU is a valuable tool for privacy protection, as it allows the targeted removal of sensitive information from AI models when individuals choose to withdraw their data. By securing that personal data are not only deleted from storage but also erased from the model's internal representations, MU mitigates the risk of unintended data retention and enforces compliance with evolving privacy laws.

Moreover, beyond privacy concerns, MU is increasingly recognized as a critical tool for AI security and safety [1]. Specifically, MU can play an important role in mitigating AI-specific threats, including adversarial attacks, data poisoning, and backdoor manipulations [1]. Large language models (LLMs) and deep learning (DL) systems are susceptible to these threats, as they learn from vast datasets that may include biased, harmful, or sensitive in formation [4]. For example, an LLM trained on unverified internet data might internalize and reproduce security-sensitive information or unethical content, while an image recognition system may be compromised through poisonous samples that manipulate its decision-making [5]. Unlearning techniques enable AI developers to downscale or even erase such problematic knowledge without requiring full-model retraining, reducing the risk of model exploitation.

Additionally, the increasing demand for auditing AI systems for robustness, fairness, and accountability is further underscoring the importance of MU for AI fairness and governance [5, 6]. As models are scrutinized for biases, vulnerabilities, and potential security risks, the ability to selectively remove harmful or problematic learned patterns becomes essential. In this context, MU can be critical in industry and CIs, where AI-driven systems must operate reliably under strict regulatory and security constraints, especially in sectors such as finance, healthcare, and energy, where security vulnerabilities can have far-reaching consequences and cascading effects.

In this chapter, we present an analysis of MU techniques and their potential as an effective response mechanism against current threats in CIs. We begin this analysis by discussing representative state-of-the-art MU techniques. Next, we define a threat model for MU, outlining the specific threats where MU can serve as an effective countermeasure. In addition, we examine the system architecture design and identify the key components that AI CI systems should include to effectively support and optimize the use of MU. In order to demonstrate its impact in the real world, we present examples in which MU has been successfully applied in real-world applications, and a case study demonstrating the effectiveness of MU in mitigating

[1] https://oag.ca.gov/privacy/ccpa.

a data poisoning attack within a healthcare CI. Finally, we conclude the chapter, summarizing the insights gained and highlighting future research directions.

2 Machine Unlearning Concept and Techniques

Machine unlearning is the process of removing the influence of specific data points from a previously trained ML model. In typical ML, a model is trained on a dataset and learns patterns from all the data it is given. However, in specific cases, such as privacy concerns, erroneous data, or regulatory requirements, it may become necessary to eliminate the influence of specific data points after the model has already been trained.

To achieve the latter, MU involves two principal concepts: the retain set and the forget set [7]. The retain set consists of the data points that should remain in the model, as they continue to be relevant for the model's performance. The forget set contains the data points whose influence needs to be removed. The goal of unlearning is to adjust the model such that it "forgets" the data in the forget set, while retaining the knowledge from the retain set. This process allows the model to continue making accurate predictions without retaining the influence of the removed data. The main concept of MU, as described in this paragraph, is shown in Fig. 1.

Current MU methodologies are broadly categorized into exact and approximate unlearning [8]. Exact unlearning involves retraining the model from scratch without the data to be forgotten, securing complete removal, but incurring significant computational overhead. On the other hand, approximate unlearning leverages model updates, influence tracking, and fine-tuning techniques to efficiently reduce the influence of unwanted data without full retraining. In the context of cybersecurity, these techniques can enable AI systems to adapt to new security threats by forgetting compromised or adversarially crafted information without compromising the integrity of the model.

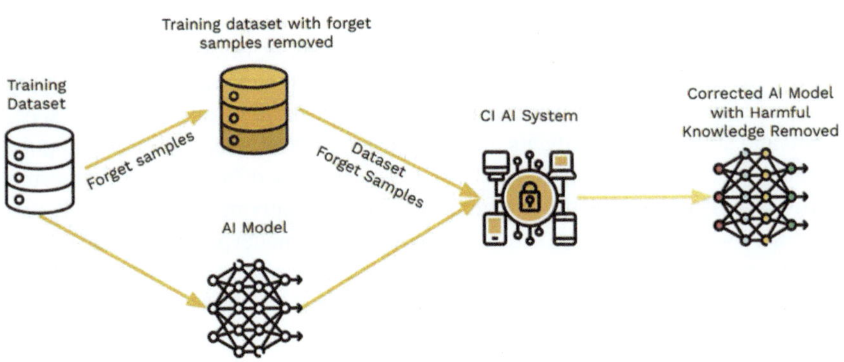

Fig. 1 The concept of machine unlearning for cybersecure AI systems for CIs

2.1 Exact Unlearning Related Works

In the context of exact unlearning, sharding, isolation, slicing, and aggregation (SISA) [9] is an advanced framework designed to improve the efficiency of MU by structuring model training in a way that minimizes the computational burden of removing specific data points. Instead of retraining the entire model, SISA partitions the dataset into multiple independent shards, each of which is further divided into slices. This hierarchical structure enables precise and localized updates when data needs to be forgotten. When a request for unlearning is made, SISA firstly determines the exact shard and slice in which the target data resides. This process allows the system to isolate the affected portion of the model while keeping the rest of the network untouched. The unlearning operation proceeds as follows:

1. **Locating the Target Data**: The framework identifies the shard containing the data to be removed. Within this shard, it identifies the slice where the data were last introduced, securing that only the affected portion of the model undergoes modification, thus preventing unnecessary updates to unrelated data.
2. **Rolling Back to a Prior State**: Instead of applying complex parameter updates across the entire model, SISA reverts the corresponding slice to the state it was in before the data were introduced. This rollback is responsible for erasing the influence of the unwanted data while preserving the remaining learned information.
3. **Selective Retraining**: Once the rollback is complete, the affected shard undergoes incremental retraining using only the remaining valid data. This step secures that the model regains stability and maintains predictive accuracy without requiring a full reprocessing of the dataset. Since only one shard is retrained, the computational cost is significantly reduced compared to model-wide retraining.
4. **Aggregation Layer Update**: After the updated shard is re-integrated, the final step involves adjusting the aggregation layer, which combines the outputs from all shards. This step secures that the system's decision-making remains coherent and unaffected by the removal of a specific data point.

The structured nature of SISA, as shown in Fig. 2, offers several benefits for unlearning in ML models. Firstly, SISA significantly reduces the computational overhead associated with full-model retraining by limiting retraining to a specific shard and slice. Additionally, its modular design improves scalability by supporting parallel processing, which is ideal for large-scale AI deployments. Furthermore, SISA establishes precise and targeted data removal, applying unlearning only where necessary while minimizing disruptions to the overall model performance [10]. These features make SISA a valuable tool for privacy-preserving and secure AI applications, especially in scenarios that demand compliance-driven data deletion or continuous adaptation to evolving threats.

Several adaptations of SISA have been developed for specific use cases, particularly in enhancing exact unlearning efficiency across different ML schemes. In the context of ensemble learning, DaRE Forest [11] extends SISA principles to

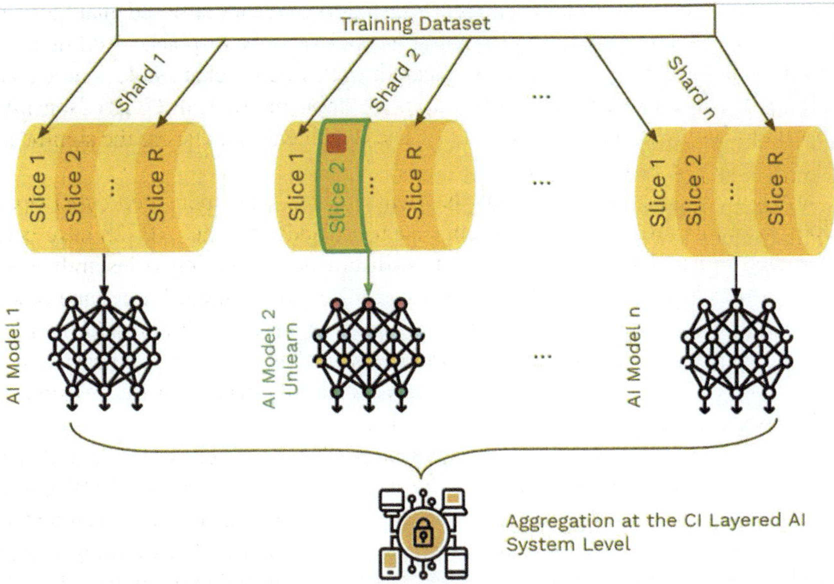

Fig. 2 SISA [9] Unlearning-based architecture supporting data and knowledge deletion in CIs

random forests by structuring decision trees into random and greedy nodes. This separation allows for efficient unlearning by limiting updates to only necessary subtrees while maintaining robust model performance. Similarly, HedgeCut [12] optimizes unlearning in extremely randomized trees by precomputing subtree variants for different split decisions, enabling low-latency updates when data need to be removed. These approaches improve the adaptability of tree-based models, thus they are well-suited for real-time applications where rapid data removal is required.

For more complex data structures such as graphs and federated learning, specialized adaptations have been introduced. GraphEraser [13] applies SISA principles to Graph Neural Networks (GNNs) by implementing a balanced graph partitioning strategy and an optimized aggregation mechanism that secures minimal impact when removing specific nodes. In federated learning, KNOT [14] leverages a clustered aggregation mechanism that confines unlearning operations to specific client groups, reducing retraining overhead while maintaining overall model consistency. These advancements demonstrate the versatility of SISA framework in addressing unlearning challenges across diverse ML architectures.

2.2 Approximate Unlearning Related Works

Approximate unlearning provides an efficient alternative to exact unlearning. Approximate methods reduce computational and storage costs by minimizing the influence of removed data rather than fully eliminating it, enabling in this way their use large-scale and real-time applications.

Approximate methods such as those proposed by Guo et al. [15] adjust model parameters based on the influence of removed data points, avoiding full retraining and its complexities. Storage efficiency is another advantage, as demonstrated by Sekhari et al. [16], who utilize stored statistics instead of retaining entire datasets.

A principal approach in approximate unlearning is re-optimization. Golatkar et al. [17] introduce a selective forgetting mechanism using a Forgetting Lagrangian to adjust model weights post-training, ensuring minimal accuracy loss. Their follow-up work [18] refines this technique by leveraging Neural Tangent Kernels (NTK), focusing on final activations to achieve more precise forgetting. In mixed-privacy settings, Golatkar et al. [19] propose a method that isolates core model weights from user-specific weights, allowing selective updates for efficient unlearning.

Another promising direction is gradient-based unlearning. DeltaGrad [20] leverages stored gradients to update models incrementally, avoiding costly retraining. Similarly, FedRecover [21] applies historical gradient information to correct poisoned federated models, while descent-to-delete [22] fine-tunes models using limited gradient steps with added noise to obscure traces of removed data. BAERASER [23] extends this to backdoor removal, applying gradient ascent to erase adversarial triggers while preserving model accuracy.

Beyond gradient updates, model pruning has been explored for efficient unlearning. Wang et al. [24] use TF-IDF-based pruning to remove class-specific features in CNNs, effectively forgetting targeted categories. For linear models, Izzo et al. [25] propose projective residual update (PRU), projecting parameter updates onto a low-dimensional subspace to efficiently remove data influence.

In addition to model-centric approaches that directly modify weights, typically requiring white-box access, data-centric methods achieve unlearning through indirect weight updates by modifying the model's input or output. For instance, some approaches apply adversarial attacks to the input data [26, 27], while others increase entropy in the output distribution [28, 29] to enhance privacy.

These examples represent a subset of the works in approximate unlearning. The benefits of this approach, including improved efficiency and reduced computational overhead, have been extensively discussed in the literature. However, several challenges persist, such as the potential for accumulated approximation errors and the reliance on convexity assumptions.

3 Threat Landscape and Machine Unlearning as a Response

Although MU has gained attention in the context of security and privacy, existing research has yet to define a clear threat model, leaving open questions about which specific threats MU can effectively defend against. In this section, we examine the

Table 1 Threat actors in the context of privacy, security, and safety risks mitigated by MU

Threat actor	Description
Malicious data contributors	Inject poisoned data into training sets, causing models to learn incorrect or adversarial behaviors
Privacy violators	Exploit privacy attacks, such as model inversion and membership inference at tacks to extract sensitive training data
Insider threats	Internal actors introducing bias, misinformation, or unauthorized data into training pipelines
Unintentional data collectors	Entities that scrape public data but unintentionally learn from sensitive, personal, or copyrighted information

evolving threat landscape and explore the extent to which MU can mitigate various risks. Additionally, we map the advantages of MU approaches to the specific challenges and constraints faced by CIs, demonstrating the potential of MU-based corrective measures to enhance CI resilience and response capabilities against evolving threats.

3.1 *Threat Model for Unlearning-Based Defenses*

3.1.1 Threat Actors

MU is designed to mitigate privacy, security, and safety risks introduced by a wide range of threat actors, who may act with malicious intent or inadvertently compromise AI model integrity. Some adversaries, such as malicious data contributors and privacy violators, deliberately manipulate data or models to exploit vulnerabilities or extract sensitive information. Others, such as unintentional data collectors, may not intend harm but still pose significant challenges by unintentionally incorporating sensitive information into AI systems. Understanding these diverse threat actors is important for developing robust MU-based defense strategies. The primary threat actors we identify in this context are listed in Table 1.

3.1.2 Threat Vectors and Attack Surfaces

MU serves as a defense mechanism against security and privacy threats that arise throughout the AI lifecycle. In this section, we detail attack vectors that MU can mitigate based on existing research, highlighting their impact on AI models and the advantages MU offers.

Data Poisoning Attacks. Data poisoning involves injecting malicious data into a model's training set to manipulate its behavior. Attackers can introduce mislabeled examples (label flipping with or without triggers) or craft data that strategically influences decision bound aries (gradient manipulation). These attacks degrade model reliability, potentially causing targeted misclassifications or systemic failures.

MU provides a response path by enabling the removal of compromised training data with out requiring full retraining. For example, the Corrective Machine Unlearning approach [30] focuses on mitigating the impact of both label flipping and label flipping with poisoning in models, even when only a small subset of the manipulated data is identified. This method successfully removes the detrimental effects of backdoor attacks (i.e., a type of data poisoning attack that causes the model to behave incorrectly on specific trigger inputs), such as those caused by feature-and-label manipulations [31] and label-only manipulations [32], using techniques based on Selective Synaptic Dampening [33] to target the identified corrupted samples, without the need for full retraining.

Privacy Attacks: Model Inversion and Membership Inference. Privacy-violating attacks exploit learned representations to extract sensitive information from a trained model. Model inversion attempts to reconstruct training data based on model outputs, while membership inference determines whether a specific data point was part of the training set. These two are the most prominent privacy attacks discussed in the MU literature. For a more comprehensive list of privacy attacks, please refer to Chap. 7.

MU can counter these risks by removing specific user data, weakening the ability of attackers to infer private information. For instance, a recent approach to zero-shot machine unlearning [34] addresses privacy risks by removing sensitive data without requiring access to the original data. This is achieved by modifying the model's input using error-minimizing noise and gated knowledge transfer, which together defend against model inversion and membership inference attacks, demonstrating its robustness in protecting private information.

Another approach [28] focuses on the model's output, aiming to eliminate sample-specific information learned by the model. It addresses the memorization capacity of deep neural networks, which increases privacy risks by producing overconfident predictions that facilitate membership inference attacks. To mitigate this, the method smooths the output distributions of specific user data, reducing overconfidence up to an information-theoretic bound. This ensures that only sample-specific information associated with the targeted user data is unlearned while preserving the model's utility on the remaining data.

Concept Drift Exploitation. AI models continuously update their understanding of data distributions. Attackers can exploit concept drift to gradually introduce harmful patterns that go unnoticed during training, causing models to adapt in unintended ways.

MU can enable corrective measures by supporting the removal of outdated or adversarially influenced data, helping models recover from drift-based attacks. In addressing concept drift exploitation, [35] introduces an unsupervised, model-agnostic method based on the logic of unlearning to counteract concept drift using autoencoders. By training an autoencoder on the data distribution before concept drift, this method applies unsupervised transfer learning to "undo" the drift, restoring the original distribution. This approach unlearns the impact of concept drift without retraining models, offering an efficient solution for adapting to distribution shifts while maintaining model performance.

Bias and Out-of-Distribution (OOD) Data. Biases in training data can lead to discriminatory or skewed AI decisions, while OOD data can cause models to behave unpredictably when encountering unseen distributions.

MU offers a mechanism to remove biased or non-representative data, improving fairness and robustness. For example, the Fast Model Debiasing (FMD) framework [36] removes biases from trained models using a small counterfactual dataset and unlearning mechanism. It reduces biases in models trained on datasets such as CelebA and achieves competitive performance with lower computational costs compared to traditional methods requiring retraining or bias labeling. Similarly, the Forsaken method [37] addresses OOD unlearning by using neuron masking to efficiently "unlearn" memorized samples. Forsaken avoids the need for full retraining, focusing instead on removing the negative impact of unexpected training data, such as privacy risks from OOD data, while maintaining model utility with minimal computational cost.

Copyrighted or Unauthorized Data Usage. AI models trained on publicly scraped data may inadvertently incorporate copyrighted, proprietary, or confidential information. Legal and ethical concerns, such as those outlined in GDPR and other regulatory frameworks, necessitate the ability to erase specific data points.

MU can enable AI developers to comply with takedown requests and regulatory obligetions efficiently. For instance, a recent study introduces a novel approach called Stable Sequential Unlearning (SSU) [38] to address the challenge of unlearning copyrighted content from LLMs in response to takedown requests. SSU works by identifying and removing weight updates associated with copyrighted material, using techniques including random labeling loss and targeted weight adjustments. This sequential unlearning method preserves the model's general purpose capabilities while effectively mitigating the risks of copyright infringement.

3.2 *Unlearning as a Response Mechanism in CIs*

As highlighted in the previous subsection, MU provides a structured response mechanism that allows for the selective removal of harmful, biased, or unauthorized data without the need for full-model retraining or access to the entire dataset. This capability is especially valuable in critical environments such as CIs, where maintaining operational continuity and compliance with security standards is of critical importance.

In the context of CIs, MU can offer several advantages that can significantly improve system resilience, security, and adaptability. These benefits are especially critical given the unique constraints faced by CIs, such as limited access to complete datasets, the challenge of updating outdated systems, and the need for rapid response

Table 2 Mapping MU advantages to CI constraints

Advantage of MU	Constraint in CIs	How MU addresses the constraint
Rapid mitigation of data poisoning	CIs require uninterrupted operation, making full model retraining impractical	Selective unlearning removes only compromised data, minimizing downtime, and ensuring availability
Availability of training data	The full training dataset may not be available for retraining, hindering mitigation of CI vulnerabilities	Some MU methods require only a small portion of the dataset, making unlearning feasible despite limited access
Post-attack recovery	Recovery methods must not destabilize critical decision-making	Targeted forgetting erases only harmful knowledge while preserving essential functionality
Regulatory compliance and privacy protection	CIs are required to comply with regulations mandating full traceability of model and data modifications, as well as the protection of data privacy	MU frameworks support logged modifications, ensuring AI decisions remain auditable and accountable
Adaptability against evolving threats	Continuously adapting to new threats is resource-intensive for CIs	MU offers flexible, efficient responses without requiring full system reconstruction
Handling outdated systems and data	Legacy systems are costly to upgrade and may contain outdated data that are hard to replace	MU enables forgetting obsolete knowledge without full retraining or redesign

to emerging threats. In this section, we identify these advantages and illustrate how MU addresses these specific constraints.

This detailed mapping of MU's advantages to CI constraints is presented in Table 2.

4 System Architecture for Hosting Machine Unlearning in CIs

As evidenced through the literature analysis, implementing MU in CI systems requires a well-structured approach that integrates MU capabilities while at the same time maintaining essential properties such as availability, scalability, and security. Given the operational constraints typical of CI environments, such architectures must allow for efficient data removal, computational feasibility, and minimal service disruption. This section outlines the key technical components and architectural considerations required to support MU in CI contexts.

4.1 Core Components for MU Integration in CIs

To enable MU, a CI system must be equipped to handle either exact or approximate unlearning, depending on the application's precision and performance requirements. One foundational element is a **Data Provenance and Logging Layer**, which systematically records the flow and usage of data throughout the training pipeline. This layer is essential for tracing specific data points and identifying those that must be removed.

Complementing this is a **Data Indexing and Sharding Module**, which partitions the training data into smaller, manageable subsets. This design facilitates localized updates, especially under architectures based on SISA principles (see Sect. 2.1), making unlearning more efficient. For approximate unlearning strategies, it is also important to include mechanisms for **Influence Tracking or Gradient Storage**, which estimate the impact of specific data on model parameters and enable targeted adjustments.

A robust **Model Checkpointing System** is another crucial component, providing periodic snapshots of the model state to allow rollback or selective retraining when necessary. An **Unlearning Execution Component** itself can serve as the core engine of the MU process, implementing operations such as gradient reversal, loss function adjustments, neuron pruning, or partition-level retraining to effectively remove learned information.

Ensuring these operations do not interfere with normal CI workflows requires a **Computational Resource Orchestrator** that dynamically allocates resources between ongoing inference, retraining, and unlearning tasks. Finally, a **Verification and Auditing Layer** must validate that unlearning has been successfully applied. This is typically achieved through post-unlearning evaluations such as divergence analysis, ensuring compliance with privacy and regulatory standards.

Together, these components form a blueprint for integrating MU into CI environments, balancing technical efficacy with the rigorous demands of critical system operations.

4.2 Challenges and Technical Considerations

Several technical challenges must be addressed when designing a CI system with MU capabilities. One major concern is minimizing unlearning latency, as data removal must be performed efficiently without introducing excessive delays, particularly in real-time CI applications. An other critical issue is obtaining model stability, as frequent unlearning operations may destabilize model parameters. Additionally, robust verification techniques are required to maintain the integrity of model predictions after unlearning is applied. Furthermore, many CIs handle large volumes of data streams, necessitating MU mechanisms that can process continuous data updates without requiring full-model retraining. Finally, adversarial resilience presents further challenges. The MU pipeline itself must be secured to prevent adversarial unlearning attacks, where malicious actors may attempt to degrade model performance through excessive unlearning requests. Addressing these challenges effectively is important to secure that MU remains a feasible solution for secure and resilient AI deployment in CI environments.

At this point, it is important to acknowledge that while MU offers a powerful defense mechanism, it also introduces potential vulnerabilities, as is common with any novel defense. The process of selectively removing data from a model, though effective in mitigating certain threats, can itself become an attack surface. Adversaries might exploit weaknesses in the unlearning process, leading to system failures or unpredictable behavior. However, exploring these risks and how MU might be targeted is beyond the scope of this chapter. Our focus here is on the benefits and applications of MU in enhancing security and privacy, rather than the threats it might introduce. For readers interested in exploring the potential attack surface that MU might create, we refer them to the works of [39, 40].

4.3 Selecting the Right MU Strategy for a CI Environment

The choice between exact and approximate unlearning depends on the specific operational and security requirements of a given CI system. Exact unlearning, such as SISA-based approaches, is best suited for high-security applications where regulatory compliance mandates the full removal of specific data points. This approach can support complete erasure of sensitive or compromised information, making it ideal for environments where data integrity is paramount. On the other hand, approximate unlearning is preferable for performance-critical applications where computational efficiency is the primary concern. This MU approach allows models to reduce the influence of unlearned data while avoiding the high costs associated with full retraining. It is particularly useful in CI sectors where maintaining system uptime and responsiveness is essential.

5 Real-World Examples and a Case Study

5.1 Examples of MU Applied to the Real World

This section presents several representative examples where MU has been successfully applied, demonstrating its practical value in real-world scenarios. From large-scale tech companies to financial institutions, the integration of MU showcases its potential in addressing challenges related to data privacy, security, and model robustness. We explore these applications in detail, providing insight into how MU is reshaping the landscape of AI development and deployment.

Harry Potter Removed from META's Llama2. A notable example of MU was demonstrated by Meta in their work on removing the Harry Potter books from the Llama2-7b model [41]. Since LLMs are trained on vast internet corpora, they often absorb copyrighted content, raising ethical and legal concerns. To address this, the authors developed a method to unlearn specific data, in this case, Harry Potter, without retraining the entire model, which would be prohibitively expensive (Llama2 required over 184,000 GPU-hours). Instead, they fine-tuned the model for just one hour on a single GPU to effectively erase related content.

Their three-step approach begins by identifying tokens strongly associated with the target data using a reinforced model. Then, unique expressions are replaced with generic alternatives, and new labels are generated to reflect predictions from a hypothetical model not trained on Harry Potter. Finally, the model is fine-tuned on these labels, removing the targeted content while maintaining performance on standard benchmarks.

While the model forgot detailed content, it retained some surface-level knowledge (e.g., from Wikipedia), showing that MU may not fully eliminate well-known cultural references. Still, this work marks significant progress toward ethical and adaptable language models, showcasing MU as a promising post-training tool for customizing LLM behavior.

Google Hosting the First MU Challenge Promoting the RTBF. In an effort to further advance the field of MU, Google organized the first MU Challenge in 2021[2] [42], aimed at promoting the RTBF in ML systems. It encouraged researchers and developers to create methods that allow ML models to forget specific data points without retraining from scratch, thus addressing privacy concerns and ensuring data removal is possible without significant performance degradation. The challenge highlighted the growing importance of privacy-preserving AI systems and set the stage for further advancements in the field of MU.

JPMorgan Chase Adapts UT Austin's MU Framework to Secure Customer Privacy. Researchers from UT Austin, in collaboration with JPMorgan Chase, developed a MU algorithm for image-to-image (I2I) generative models [43]. This framework enables models to efficiently forget specific data, protecting sensitive information without retraining. It ensures minimal impact on performance for retained data and doesn't require access to it, supporting compliance with data retention policies. The approach was validated on large-scale datasets, including ImageNet1K and Places-365.

JPMorgan Chase applied this MU technique to improve privacy protections, particularly for handling sensitive consumer data. For instance, when a customer departs, the model can forget any personal images or records associated with them, safeguarding private information and aligning with regulatory standards.[3] The framework has broad potential across industries where privacy is critical. By removing

[2] https://research.google/blog/announcing-the-first-machine-unlearning-challenge/.

[3] https://thedailytexan.com/2024/02/20/ut-and-jpmorgan-chase-researchers-develop-unlearning-algorithm-for-ai/.

sensitive content without full retraining, it lowers computational costs and enhances real-time responsiveness. For data-heavy enterprises such as JPMorgan Chase, MU ensures customer privacy is upheld throughout the AI lifecycle—meeting legal standards while preserving model performance.

5.2 Case Study: Responding to a Training Data Poisoning Attack with MU in a Healthcare CI

In this case study, we explore the application of MU to protect a diagnostic model used in healthcare, specifically one designed to analyze X-ray images, from a poisoning attack. To simulate a real-world scenario, we assume a healthcare CI where a diagnostic model is deployed to prioritize treatment decisions for COVID-19 patients, reflecting the high-stakes conditions hospitals faced during the pandemic. As a baseline, we first train a clean ResNet-18 model using the COVID-PNEUMONIA-NORMAL public X-ray dataset.[4] This model serves as the baseline for evaluating the impact of data poisoning attacks and the effectiveness of unlearning-based mitigation strategies.

An attacker could have several motives for conducting a data poisoning attack against a healthcare CI. One possible goal is to degrade the reliability of the diagnostic model, leading to misdiagnoses that downplay serious conditions such as COVID-19 or pneumonia. This could be exploited for financial gain, such as fraudulent insurance claims, or to disrupt healthcare operations by misprioritizing critical patients. Another potential motivation is to deliberately harm the reputation of the healthcare CI, causing a loss of trust among patients and stakeholders. Additionally, in targeted attacks, an adversary could manipulate outcomes for specific individuals, ensuring they are misclassified and denied appropriate medical intervention.

To simulate this risk, we assume that an attacker has gained unauthorized access to the healthcare CI and can manipulate the training dataset and diagnostic model. We employ a neural Trojan attack, applying patch triggers as in [31]. Using this method, we poison 20% of the training dataset, introducing targeted mislabeling as shown in Table 3. Example generated poisonous samples are shown in Fig. 3. The compromised dataset is then used to train the diagnostic model, resulting in a poisoned baseline that exhibits manipulated decision-making, potentially leading to widespread misdiagnoses and disruptions within the healthcare CI.

To counter this attack, we employ a MU approach based on the Bad Teacher method [44]. In this context, we assume that we know the entire forget set (all poisonous samples have been identified by other security components of the CI). The idea is to apply the unlearning framework to reduce the influence of the poisoned samples while retaining the integrity of the clean data. For a more efficient approach, we use only 30% of the retain set and the full forget set to unlearn the poisoned model.

[4] https://www.kaggle.com/datasets/prashant268/chest-xray-covid19-pneumonia?resource=download.

Table 3 Label flipping strategy used in the poisoning attack, where all labels are altered toward the target class "normal"

Ground truth label	Flipped label
COVID	Normal
Normal	Normal
Pneumonia	Normal

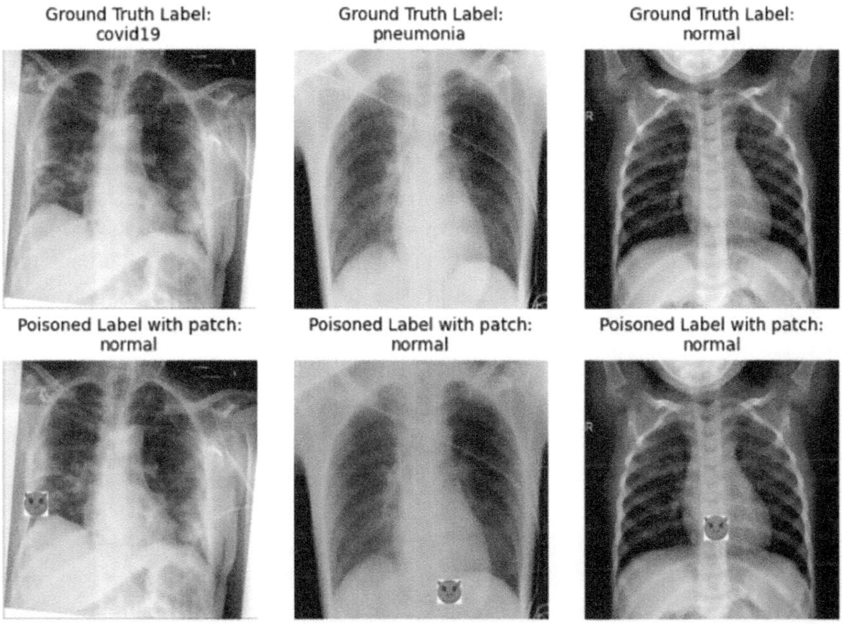

Fig. 3 Example generated poisonous samples used in the attack

We compare this approach with two other responses to poisoned training data: fine-tuning the model exclusively on the clean data and retraining the diagnostic model from scratch using only clean data. The results, as depicted in Table 4, show that applying unlearning significantly reduces the success rate of the attack, with a minor accuracy drop on clean samples while using less than half of the training dataset.

Beyond its effectiveness in mitigating data poisoning attacks, MU addresses a core challenge in healthcare CIs: the limited availability of training data. In many healthcare environments, full access to training datasets is often restricted due to stringent privacy regulations, institutional policies, data-sharing agreements, or even fragmented data storage across different healthcare providers. Additionally, ethical concerns and patient consent requirements may prevent the aggregation of datasets for model retraining.

The Bad Teacher MU approach enables targeted unlearning using only a small fraction of the clean dataset (we use only 30% of the retain set, while the full retain set corresponds to 80% of the original dataset). The forget set consists of all the

Table 4 Experimental results comparing different RN18 models in terms of clean accuracy, attack success rate (ASR), and percentage of original training data used

Model	Clean Acc.	ASR	(%) of data used
Clean	1.00	–	100
Poisoned with patch triggers	0.96	1.00	100
Finetuned only on clean samples	0.98	0.86	80
Retrained from scratch only on clean samples	0.98	0.18	80
Unlearned (100% forget + 30% retain set)	0.94	0.08	44

malicious data injected during the attack (20% of full training dataset), which, while challenging to isolate, can be identified through detection techniques. The resulting total number of samples used is just 44% of the original training dataset. In contrast, retraining from scratch or finetuning would require full access to the entire retain set, which may not always be feasible. As such, the Bad Teacher MU approach offers a more efficient and practical solution, enabling healthcare institutions to protect sensitive information, preserve AI diagnostic integrity, and ensure compliance with data privacy and regulatory standards.

6 Conclusions and Future Directions

This chapter presents an overview of MU techniques and their potential to enhance the security, privacy, and adaptability of AI-driven CIs. It begins with a survey of key state-of-the-art MU approaches, followed by the introduction of a threat model outlining scenarios where MU serves as an effective countermeasure. The discussion then continues with system architecture considerations, highlighting essential components for effectively integrating MU into AI CI systems. To illustrate practical relevance, the chapter presents real-world examples and a healthcare case study demonstrating MU's effectiveness in mitigating data poisoning attacks.

While MU presents promising applications in AI governance, privacy preservation, and security resilience, challenges remain in developing robust, reliable, and scalable unlearning mechanisms. Researchers and practitioners should focus their research on identifying and counteracting new vulnerabilities that unlearning may introduce, aiming to prevent unintended consequences such as over-forgetting or knowledge reconstruction, and to maintain model accuracy post-unlearning.

Acknowledgements This work was partially supported by the EU funded project ATLANTIS (Grant Agreement Number 101073909).

References

1. Z. Liu, H. Ye, C. Chen, Y. Zheng, and K.-Y. Lam, "Threats, attacks, and defenses in machine unlearning: A survey," IEEE Open Journal of the Computer Society, pp. 1–13, 2025.
2. European Parliament and Council of the European Union, "Regulation (EU) 2016/679 of the European Parliament and of the Council."
3. H. Zhang, T. Nakamura, T. Isohara, and K. Sakurai, "A review on machine unlearning," SN Computer Science, vol. 4, no. 4, p. 337, 2023.
4. J. Chen, Z. Deng, K. Zheng, Y. Yan, S. Liu, P. Wu, P. Jiang, J. Liu, and X. Hu, "Safeeraser: Enhancing safety in multimodal large language models through multimodal machine unlearning," arXiv preprint arXiv:2502.12520, 2025.
5. T. Shaik, X. Tao, H. Xie, L. Li, X. Zhu, and Q. Li, "Exploring the landscape of machine unlearning: A comprehensive survey and taxonomy," IEEE Transactions on Neural Networks and Learning Systems, 2024.
6. A. Boer, L. de Beer, and F. van Praat, "Algorithm assurance: Auditing applications of artificial intelligence," Advanced Digital Auditing, vol. 149, 2023.
7. M. H. Huang, L. G. Foo, and J. Liu, "Learning to unlearn for robust machine unlearning," in European Conference on Computer Vision, pp. 202–219, Springer, 2024.
8. C. Fan, J. Liu, A. Hero, and S. Liu, "Challenging forgets: Unveiling the worst-case forget sets in machine unlearning," in European Conference on Computer Vision, pp. 278–297, Springer, 2024.
9. L. Bourtoule, V. Chandrasekaran, C. A. Choquette-Choo, H. Jia, A. Travers, B. Zhang, D. Lie, and N. Papernot, "Machine unlearning," in 2021 IEEE symposium on security and privacy (SP), pp. 141–159, IEEE, 2021.
10. J. Xu, Z. Wu, C. Wang, and X. Jia, "Machine unlearning: Solutions and challenges," IEEE Transactions on Emerging Topics in Computational Intelligence, 2024.
11. J. Brophy and D. Lowd, "Machine unlearning for random forests," in International Conference on Machine Learning, pp. 1092–1104, PMLR, 2021.
12. S. Schelter, S. Grafberger, and T. Dunning, "Hedgecut: Maintaining randomised trees for low-latency machine unlearning," in Proceedings of the 2021 International Conference on Management of Data, pp. 1545–1557, 2021.
13. M. Chen, Z. Zhang, T. Wang, M. Backes, M. Humbert, and Y. Zhang, "Graph unlearning," in Proceedings of the 2022 ACM SIGSAC conference on computer and communications security, pp. 499–513, 2022.
14. N. Su and B. Li, "Asynchronous federated unlearning," in IEEE INFOCOM 2023-IEEE conference on computer communications, pp. 1–10, IEEE, 2023.
15. C. Guo, T. Goldstein, A. Hannun, and L. Va n Der Maaten, "Certified data removal from machine learning models," in Proceedings of the 37th International Conference on Machine Learning, pp. 3832–3842, 2020.
16. A. Sekhari, J. Acharya, G. Kamath, and A. T. Suresh, "Remember what you want to forget: Algorithms for machine unlearning," Advances in Neural Information Processing Systems, vol. 34, pp. 18075–18086, 2021.
17. A. Golatkar, A. Achille, and S. Soatto, "Eternal sunshine of the spotless net: Selective forgetting in deep networks," in Proceedings of the IEEE/CVF conference on computer vision and pattern recognition, pp. 9304–9312, 2020.
18. A. Golatkar, A. Achille, and S. Soatto, "Forgetting outside the box: Scrubbing deep networks of information accessible from input-output observations," in Computer Vision–ECCV 2020: 16th European Conference, Glasgow, UK, August 23–28, 2020, Proceedings, Part XXIX 16, pp. 383–398, Springer, 2020.
19. A. Golatkar, A. Achille, A. Ravichandran, M. Polito, and S. Soatto, "Mixed-privacy forgetting in deep networks," in Proceedings of the IEEE/CVF conference on computer vision and pattern recognition, pp. 792–801, 2021.
20. Y. Wu, E. Dobriban, and S. Davidson, "Deltagrad: Rapid retraining of machine learning models," in International Conference on Machine Learning, pp. 10355–10366, PMLR, 2020.

21. X. Cao, J. Jia, Z. Zhang, and N. Z. Gong, "Fedrecover: Recovering from poisoning attacks in federated learning using historical information," in 2023 IEEE Symposium on Security and Privacy (SP), pp. 1366–1383, IEEE, 2023.
22. S. Neel, A. Roth, and S. Sharifi-Malvajerdi, "Descent-to-delete: Gradient-based methods for machine unlearning," in Algorithmic Learning Theory, pp. 931–962, PMLR, 2021.
23. Y. Liu, M. Fan, C. Chen, X. Liu, Z. Ma, L. Wang, and J. Ma, "Backdoor defense with machine unlearning," in IEEE INFOCOM 2022-IEEE conference on computer communications, pp. 280–289, IEEE, 2022.
24. J. Wang, S. Guo, X. Xie, and H. Qi, "Federated unlearning via class-discriminative pruning," in Proceedings of the ACM web conference 2022, pp. 622–632, 2022.
25. Z. Izzo, M. A. Smart, K. Chaudhuri, and J. Zou, "Approximate data deletion from machine learning models," in International Conference on Artificial Intelligence and Statistics, pp. 2008–2016, PMLR, 2021.
26. M. Chen, W. Gao, G. Liu, K. Peng, and C. Wang, "Boundary unlearning: Rapid forgetting of deep networks via shifting the decision boundary," in Proceedings of the IEEE/CVF Conference on Computer Vision and Pattern Recognition, pp. 7766–7775, 2023.
27. S. Cha, S. Cho, D. Hwang, H. Lee, T. Moon, and M. Lee, "Learning to unlearn: Instance-wise unlearning for pre-trained classifiers," in Proceedings of the AAAI conference on artificial intelligence, vol. 38, pp. 11186–11194, 2024.
28. C. N. Spartalis, T. Semertzidis, E. Gavves, and P. Daras, "LoTUS: Large-Scale Machine Unlearning with a Taste of Uncertainty," in Proceedings of the IEEE/CVF Conference on Computer Vision and Pattern Recognition, 2025.
29. L. Graves, V. Nagisetty, and V. Ganesh, "Amnesiac Machine Learning," Proceedings of the AAAI Conference on Artificial Intelligence, vol. 35, pp. 11516–11524, May 2021. Number: 13.
30. S. Goel, A. Prabhu, P. Torr, P. Kumaraguru, and A. Sanyal, "Corrective machine unlearning," Transactions on Machine Learning Research, 2024.
31. T. Gu, B. Dolan-Gavitt, and S. Garg, "Badnets: Identifying vulnerabilities in the machine learning model supply chain," arXiv preprint arXiv:1708.06733, 2017.
32. S. Goel, A. Prabhu, A. Sanyal, S.-N. Lim, P. Torr, and P. Kumaraguru, "Towards adversarial evaluations for inexact machine unlearning," arXiv preprint arXiv:2201.06640, 2022.
33. J. Foster, S. Schoepf, and A. Brintrup, "Fast machine unlearning without retraining through selective synaptic dampening," in Proceedings of the AAAI conference on artificial intelligence, vol. 38, pp. 12043–12051, 2024.
34. V. S. Chundawat, A. K. Tarun, M. Mandal, and M. Kankanhalli, "Zero-shot machine unlearning," IEEE Transactions on Information Forensics and Security, vol. 18, pp. 2345–2354, 2023.
35. A. Artelt, K. Malialis, C. G. Panayiotou, M. M. Polycarpou, and B. Hammer, "Unsupervised unlearning of concept drift with autoencoders," in 2023 IEEE Symposium Series on Computational Intelligence (SSCI), pp. 703–710, IEEE, 2023.
36. R. Chen, J. Yang, H. Xiong, J. Bai, T. Hu, J. Hao, Y. Feng, J. T. Zhou, J. Wu, and Z. Liu, "Fast model debias with machine unlearning," Advances in Neural Information Processing Systems, vol. 36, pp. 14516–14539, 2023.
37. Z. Ma, Y. Liu, X. Liu, J. Liu, J. Ma, and K. Ren, "Learn to forget: Machine unlearning via neuron masking," IEEE Transactions on Dependable and Secure Computing, vol. 20, no. 4, pp. 3194–3207, 2022.
38. G. Dou, Z. Liu, Q. Lyu, K. Ding, and E. Wong, "Avoiding copyright infringement via large language model unlearning," arXiv preprint arXiv:2406.10952, 2024.
39. L. Milner, "Threat models to machine unlearning," Preprints, September 2024.
40. M. Chen, Z. Zhang, T. Wang, M. Backes, M. Humbert, and Y. Zhang, "When machine unlearning jeopardizes privacy," in Proceedings of the 2021 ACM SIGSAC conference on computer and communications security, pp. 896–911, 2021.
41. R. Eldan and M. Russinovich, "Who's harry potter? approximate unlearning in llms," arXiv preprint arXiv:2310.02238, 2023.

42. E. Triantafillou, P. Kairouz, F. Pedregosa, J. Hayes, M. Kurmanji, K. Zhao, V. Dumoulin, J. C. S. J. Júnior, I. Mitliagkas, J. Wan, L. Sun-Hosoya, S. Escalera, G. K. Dziugaite, P. Triantafillou, and I. Guyon, "Are we making progress in unlearning? Findings from the first neurips unlearning competition," CoRR, vol. abs/2406.09073, 2024.
43. G. Li, H. Hsu, C.-F. Chen, and R. Marculescu, "Machine unlearning for image-to-image generative models," arXiv preprint arXiv:2402.00351, 2024.
44. V. S. Chundawat, A. K. Tarun, M. Mandal, and M. Kankanhalli, "Can bad teaching induce forgetting? unlearning in deep networks using an incompetent teacher," in Proceedings of the AAAI Conference on Artificial Intelligence, vol. 37, pp. 7210–7217, 2023.

AI-Assisted Orchestration and Automation for Business Continuity, Incident Response, and Information Exchange: The PHOENI2X Project

Kostantinos Lampropoulos, Kostantinos Fysarakis, Vasileios Mavroeidis, and Chrysanthos Chrysanthou

Abstract As our society becomes more reliant on digital systems and technologies, cybersecurity attacks continue to increase, threatening everyday lives and public welfare. To be able to address these challenges, NIS2 Directive requires from EU Member States and their designated Operators of Essential Services (OES) to implement a minimum set of capabilities and ensure coordination and cooperation across borders. However, this is only the first step toward full European cyber resilience. Preparedness, shared situational awareness, and coordinated incident response are also essential for effective crisis management. This chapter presents PHOENI2X, (https://phoeni2x.eu/) a project designed to meet these needs by creating and delivering a cyber resilience framework (CRF) providing artificial intelligence (AI)-assisted orchestration, automation and response capabilities for business continuity and recovery, incident response, and information exchange, tailored to the needs of OES and of the EU Member State (MS) National Authorities entrusted with cyber-security.

Keywords Cybersecurity · AI-Assisted orchestration · Situational awareness · Incident response · Business continuity

---

K. Lampropoulos (✉)
Department of Electrical and Computer Engineering, University of Patras, Patras, Greece
e-mail: klamprop@ece.upatras.gr

K. Fysarakis · V. Mavroeidis
Department of Informatics, University of Oslo, Oslo, Norway
e-mail: konstaf@ifi.uio.no

V. Mavroeidis
e-mail: vasileim@ifi.uio.no

C. Chrysanthou
Sphynx Analytics Limited, Nicosia, Cyprus
e-mail: c.chrysanthou@sphynx.ch

J. Soldatos and H. Abie (eds.), *Technology-Enabled Resilience: Innovations in Critical Infrastructure Protection*, https://doi.org/10.1007/978-3-032-25917-2_12

1 Introduction

As our societies and economies become increasingly dependent on digital infrastructures, the frequency, diversity, and impact of cyber-security incidents also rise. ENISA's Threat Landscape reports from 2020 and after, demonstrate how cybersecurity risks are becoming harder to identify and assess. They also reveal an increase in sophisticated and targeted attacks in critical domains (e.g., energy, transport, healthcare), and a growing prevalence of hybrid threats, i.e., threats combining both the cyber and the physical domains.

This led to the introduction of the NIS Directive (2016/1148) [1], which now forces EU Member States (MS) and their OESs to have a minimum baseline set of capabilities while providing cross-border coordination and cooperation. However, we still have a long way to go toward European cyber resilience. Dealing with systemic and complex cyber risks remains a significant challenge. The problem is also exacerbated by the increased motivation (often combining financial, nation-state interests, nationalistic, and political motives) and sophistication of, often state-sponsored, threat actors, who carry out growing targeted and persistent attacks on high-value data. Their target is to exploit the vulnerabilities of the ever-increasing interconnectedness of various systems and networks [2].

In this context, the aspects of preparedness, shared situational awareness, and coordinated incident response play a crucial role in cybersecurity resilience. They are not just the main drivers of effective crisis management but guide strategic and political decisions to effectively tackle threats that go beyond the digital world. Essential enablers of these aspects include (a) automation and orchestration mechanisms for business continuity and incident response; (b) actionable and pertinent Cyber Threat Intelligence (CTI) and threat landscape information, and interoperable, standardized alerting and information exchange mechanisms for building a shared situational awareness, as well as; (c) access to sophisticated early warning systems and structured, realistic trainings (e.g., through hand-on, realistic Cyber Ranges) for preparedness.

The rest of this chapter is organized as follows. Section 2 presents the approach and architecture of the PHOENI2X project. Sections 3–6 analyze the key components that drive its innovation addressing relevant research challenges, and Sect. 7 concludes the chapter discussing open issues and future work.

2 The PHOENI2X Architecture

To address the critical cybersecurity challenges of the EU, PHOENI2X introduces the PHOENI2X Cyber Resilience Centre (PHOENI2X CRC). This is not just a system installed inside OES's premises but a part of a wider cybersecurity ecosystem that adopts a conceptual blueprint that promotes the establishment of interoperable Cyber Security Operations Centres and combines capacity for Shared Situational

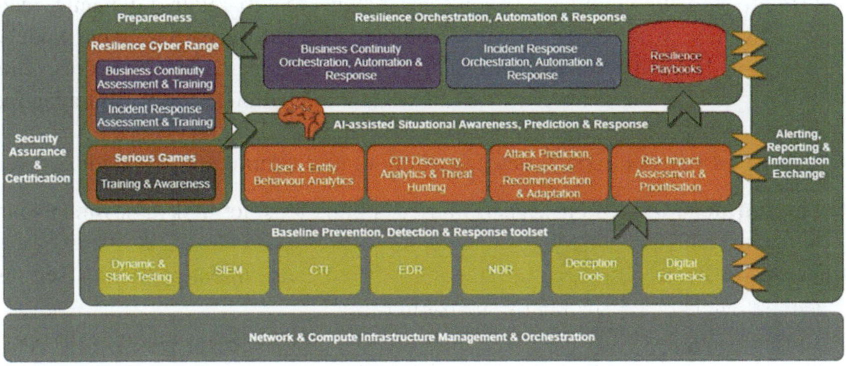

Fig. 1 PHOENI2X high-level architecture

Awareness, Coordinated Response, and Joint Preparedness [3]. PHOENI2X CRCs integrate capacities for:

- Enhanced Situational Awareness: Achieved through AI-driven Prediction, Prevention, Detection, and Response capabilities, alongside prioritization based on business risk impact assessments.
- Resilience Automation, Orchestration, and Response (ROAR): Encompasses both proactive and reactive mechanisms, ensuring Business Continuity, Recovery efforts, and Cyber and Physical Incident Response (IR).
- Increased Preparedness: Supported through the deployment of pertinent Serious Games and realistic Resilience Crisis Response (RCR) Assessment and Training programs.
- Timely and Actionable Information Exchange: Emphasizes the importance of communication between OES, National Authorities, and EU stakeholders, leveraging standardized and interoperable alerting and reporting processes.

A high-level architectural view of the PHOENI2X CRCs is provided in Fig. 1. This architecture is organized in horizontal and vertical layers that intersect exchanging and processing information from multiple sources and of different scopes.

The main blocks of the PHOENI2X architecture (starting from the bottom) are the following.

Baseline prevention, detection, and response toolset

This block includes a holistic toolset of open-source solutions and systems, covering areas like: dynamic and static testing (e.g., OpenVAS[1]); security information and event management (SIEM; e.g., Wazuh[2]), CTI (e.g., OpenCTI[3]); endpoint detection

[1] Https://www.openvas.org/

[2] Https://wazuh.com/

[3] Https://filigran.io/platforms/opencti/

and response (EDR; e.g., OSSEC[4]); network detection and response (NDR; e.g., Suricata[5]); deception tools (e.g., Dionaea[6]), and forensic tools (e.g., GRR[7]). This block is usually formed by security tools that are often already installed by the OESs for prevention, detection, and response.

AI-assisted situational awareness, prediction, and response

Next is the block where PHOENI2X integrates and combines different AI technologies focused on AI-assisted Situational Awareness, Prediction, and Response. Such technologies include:

- User and Entity Behavior Analytics: This component uses ML-based analytics for the creation of a behavioral baseline for users and entities (e.g., systems, devices), identifying any deviations to provide early warnings for security events.
- CTI Discovery, Analytics, and Threat Hunting: This component utilizes AI (namely NLP and Sentiment Analysis, supported by Heuristic Analytics) for the discovery, extraction, ingestion, correlation, and contextualization of (MISP and STIX) CTI indicators, providing an up-to-date view of the threat landscape, also facilitating proactive threat hunting.
- Attack Prediction, Response Recommendation, and Adaptation: This component leverages attack categorization to identify the scope and impact of potential attacks. It offers attack prediction through training ad hoc behavioral models for each system, and attack response and adaptation, suggesting proactive actions and mitigation strategies.
- Risk Impact Assessment and Prioritization: This component is based on a near-real-time risk assessment engine (through predefined machine-readable model rules: a qualitative model based on DEXi9 and a quantitative one to be developed within PHOENI2X), thus facilitating response prioritization.

Resilience orchestration, automation, and response (ROAR)

Being the core response automation and orchestration enabler of PHOENI2X, the ROAR block will support: (i) the specification of Resilience Playbooks (RPs) (ii) the execution and orchestration of the RPs. Regarding the latter, the execution and orchestration of multiple RPs in parallel is also needed, along with monitoring the progress/status of their execution, and the collection of the evidence generated during said executions. RPs will provide a structured, machine-processable encoding of a sequence of actions comprising the organization's Business Continuity, Recovery, and Incident Response operations. Such actions represent a fundamental activity (e.g., add a rule to a firewall) allowing organizations to specify and automate the execution, monitor the progress and assess the effectiveness of all their internal protection, recovery, and response-related processes.

---

[4] Https://www.ossec.net/

[5] Https://suricata.io/

[6] https://github.com/DinoTools/dionaea.

[7] https://www.grr-response.com/

Preparedness

Aiming at increased Preparedness, PHOENI2X offers realistic scenario assessment and training capabilities, based on a Resilience Cyber Range (RCR) and Serious Games. The former supports OES preparedness through (a) the assessment of defined RPs in a realistic environment that emulates involved cyber systems and their interactions and (b) hands-on training to OES employees in various topics like monitoring and protection, recovery, incident response, etc. On the other hand, the latter (Serious Games) are used for training that specifically targets the humans (e.g. employees of OESs). Their capabilities will support the RCR, by improving the levels of Training and Awareness of employees and technicians in different cyber-attacks, threat elicitation, thus improving organizational security robustness depending on the human factors (e.g., social engineering, phishing etc.).

Alerting, reporting, and information exchange

This subsystem runs vertically across all layers and targets to address the common need for standardized and coordinated cyber-security notification in every-day practice and, during a cyber and/or operative incident. Moreover, it will support the exchange of: (a) technical-level indicators from the baseline prevention, detection and response layer; (b) AI-generated insights, early warning alerts, contextualized CTI, models, and other shareable information from the AI-assisted situational awareness, prediction and response layer; (c) RPs from the ROAR subsystem and (d) training courses and activities from the RCR. The most important feature of this subsystem is its ability to link all the EU-level relevant parties at the strategic and political level, as well as cyber-security incident response actors (CSIRTs network, ENISA, NIS CG, CyCLONe) and relevant third parties (e.g., private entities).

Security assurance and certification

To ensure that PHOENI2X CRCs do not introduce any new vulnerabilities or issues inside the OES environments our solution includes a continuous and evidence-based Security Assurance and Certification solution. This provides a comprehensive assessment of every component inside the PHOENI2X CRCs, also covering all assets (e.g., network, compute, data, processes) and internal subsystems.

3 Resilience Orchestration, Automation, and Response

The Resilience Orchestration, Automation, and Response (ROAR) capabilities of PHOENI2X are developed by extending an incident response (IR) tool designed and provided by SPHYNX Analytics Limited [4]. This system facilitates both manual and automated execution of Collaborative Automated Course of Action Operations (CACAO [5]) security playbooks, following the latest specification. ROAR offers a graphical drag-and-drop interface for designing and editing CACAO playbooks, which can later be executed directly or exported as CACAO-compliant JSON files.

In essence, this component functions as a security orchestration, automation, and response (SOAR) solution, supporting the full lifecycle of cybersecurity operations from prevention and detection to investigation and response.

3.1 ROAR Enabler Details

The ROAR tool includes an interactive dashboard that enables users to specify, edit, and execute playbooks. It also offers real-time insights into the system's status and playbook execution, displaying both high- and low-level logs, key performance indicators (KPIs), user notifications, and other relevant information. During the project's lifecycle, ROAR's editor has been expanded to fully support the functionalities of each playbook step defined at the latest CACAO specification. Also, playbook management functionalities, such as "Save," "Save as," and "Load" were implemented into the editor. Security playbooks that follow the CACAO v2 specification can be defined within a proper GUI (graphical user interface).

3.2 Resilience Playbooks Specification, Translation, and Lifecycle Management

Resilience playbooks (RPs) are a core element of the PHOENI2X framework and its ROAR capabilities. They offer a structured, machine-readable encoding for defining sequences of actions that support an organization's business continuity, recovery, and incident response activities. Each action reflects a fundamental task (e.g., adding a rule to a firewall). Thus, through RPs, organizations will be able to specify, automate the execution (via the purpose-built execution and orchestration engine), monitor the progress, and assess the effectiveness of all their business continuity, recovery, and incident response-related processes.

RPs are designed to be adaptable and contextualized via PHOENI2X's AI-driven situational awareness and post-incident insights, RPs are also customizable to each OES's regulatory and technical environment, shareable across organizations at machine speed, capable of supporting what-if analysis for planning purposes, and translatable for use in training and cyber range simulations.

To support the above, PHOENI2X Resilience Playbooks (RPs) adopts and extends the OASIS Collaborative Automated Course of Action Operations (CACAO) standard, with version 2.0 of the specification officially approved in November 2023. This standard is a cybersecurity-specific schema and taxonomy for creating, documenting, and sharing playbooks in a structured and standardized format across organizational boundaries and technological solutions. In short, the available types of playbooks, in relation to the incident occurrence timeline, are shown in Fig. 2.

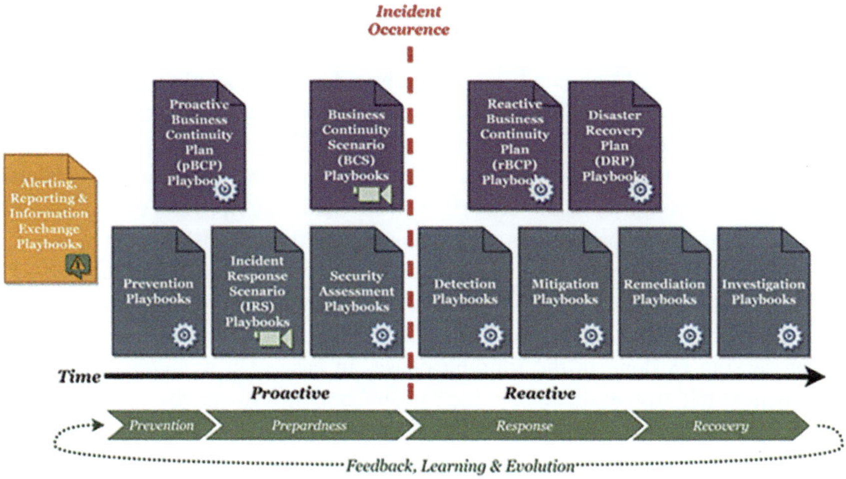

Fig. 2 Different types of resilience playbooks envisioned in PHOENI2X

In addition, it is worth mentioning that the consortium has developed and open-sourced JSON validation schemas for CACAO Version 2.0. The validation schemas are hosted at the official GitHub [6] of the OASIS CACAO technical committee and are considered the de facto validation mechanism for all implementations.

In PHOENI2X, resilience playbooks (RPs) are created, managed, and executed within the ROAR tool, which serves as the central platform for managing the entire lifecycle of RPs. This includes the initial design and specification of playbooks, their ongoing editing and refinement, as well as their automated or manual execution in response to incidents. ROAR provides a user-friendly, graphical interface that enables users to define complex workflows without requiring programming expertise.

4 AI-Assisted Situational Awareness, Prediction, and Response

4.1 User and Entity Behavior Analytics Enabler

User and Entity Behavior Analytics (UEBA) is a cybersecurity strategy designed to detect insider threats, compromised accounts, and other suspicious activities by monitoring and analyzing the behavior of users and entities within a network. It leverages machine learning algorithms and statistical models to establish a baseline of typical behavior for each user or entity and then identifies anomalies or deviations that may signal potential security threats. UEBA systems gather and analyze data from multiple sources across an organization's IT infrastructure, such as network traffic, logs, endpoints, and applications. By correlating these diverse data points,

UEBA solutions can reveal behavioral patterns that suggest malicious activity like unauthorized access, data theft, or unusual file transfers.

SphynxML UEBA uses advanced machine learning techniques to proactively identify potential security threats by analyzing historical user and entity data within the network. By developing and applying tailored UEBA use cases, SphynxML builds customized machine learning models that can detect unusual or potentially malicious behavior from both users and network devices. Using the Sphynx AutoML system, different models and parameters are evaluated, and then the user of the system can continue creating a deployment that utilizes the optimal parameters (model, etc.)

After this stage, the model is deployed and ready to receive logs for processing and prediction. Each sample (log event) that comes for prediction follows the same preprocessing as done in the training phase. In case of a malicious sample, the deployment will trigger the SPA in the backend and inform that the specific samples are possibly malicious.

4.2 CTI Discovery, Analytics, and Threat Hunting Enablers

4.2.1 CTI Discovery and Analytics

The CTI Discovery and Analytics component of PHOENI2X is designed to enhance the Cyber Threat Intelligence (CTI) process by gathering, processing, and disseminating threat information from different online sources, including dark web forums and social media. Using web scraping, machine learning, and natural language processing, it converts unstructured data into standardized formats like STIX 2.1. The processed information is then stored and sent to a TAXII server for distribution to CTI platforms such as OpenCTI, ultimately aiding in the identification of current and emerging threats.

The CTI Discovery and Analytics includes several key subcomponents: a **Crawler** that identifies new links on targeted websites; a **Scraper** for data extraction; a **PDF Decomposer** for converting PDFs to HTML; and a **Preprocessor** that organizes the extracted data. Additionally, a **Decider** module assesses the relevance of the data, while the **Internal DB Manager** structures or discards this information. The **Entity eXtraction Module (EXM)** processes and extracts STIX 2.1 entities, which are then used by a **STIX mapper** for knowledge creation. An **OpenTAXII server** facilitates sharing, supported by a custom REST API controller.

4.2.2 CTI Assessments

CTI Assessment component is a new tool that is added to the SPHYNX SPA Platform [7] and is specialized and designed to support CTI Analysts in performing dynamic threat analysis and intelligence-driven security assessments. Its primary function is to enable the creation and execution of custom queries over threat intelligence data

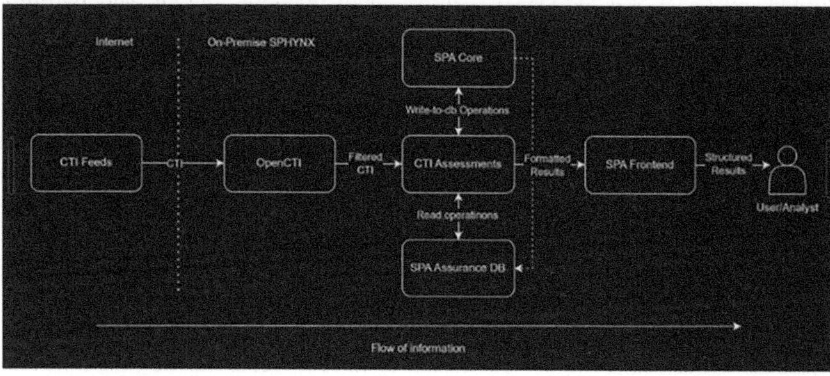

Fig. 3 Simple CTI Search Usecase

aggregated within an OpenCTI instance. By doing so, analysts can define tailored Security Assessment procedures that align with specific organizational risk profiles, threat models, or operational contexts.

Figure 3 illustrates the implementation of a simple Cyber Threat Intelligence (CTI) search use case within the SPHYNX SPA Suite. The flow of information begins with the ingestion of external CTI feeds and concludes with the delivery of structured intelligence to the end user or analyst.

External CTI feeds from the internet are first collected by the OpenCTI platform. OpenCTI serves as the central intelligence aggregator, filtering and structuring the raw data into actionable intelligence. This filtered CTI is then passed to the CTI Assessments module, which is responsible for analyzing the data and generating assessments based on predefined logic and contextual information.

During this process, the CTI Assessments component interacts with two key modules: it reads contextual data and threat assurance metrics from the SPA Assurance Database (DB), and it performs write operations to the SPA Core. The SPA Core acts as the central logic and coordination engine for the platform. Following the assessment, formatted results are generated and sent to the SPA Frontend. The frontend is responsible for presenting the information in a structured and user-friendly manner to the end user or analyst. This enables effective threat analysis, decision-making, and incident response.

This implementation supports a structured, modular approach to CTI management and ensures the integrity and relevance of the intelligence provided to users.

4.2.3 Threat Intelligence Integrator (TII)

The threat intelligence integrator (TII) is a core component of the PHOENI2X project. TII was designed to enhance cyber threat intelligence (CTI) by filtering and prioritizing information from the Malware Information Sharing Platform (MISP). TII's operational workflow begins with the event dispatcher detecting new events in MISP,

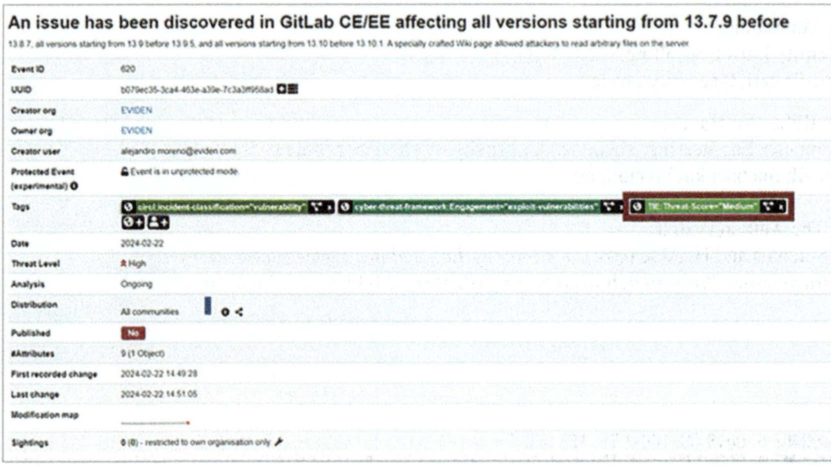

Fig. 4 Tag added by TII into a MISP event to indicate the level of threat score

subsequently relaying them to the orchestrator. The orchestrator, which contains detailed infrastructure information, determines the enrichment steps required and collects necessary data. These data are then passed to the Heuristic Engine, which uses heuristics such as timeliness, trending, completeness, and relevance to assign a threat score. The resulting score is tagged to the MISP event, as Fig. 4 shows. This scoring mechanism is vital for prioritizing threats in environments flooded with CTI data.

4.2.4 Threat Actor Context (TAC) Ontology

STIX (Structured Threat Information eXpression) [8] is a standardized, JSON-based language for representing and sharing cyber threat intelligence (CTI) in a consistent and machine-readable way. It enables better collaboration, detection, and response across cybersecurity communities. However, its flat JSON structure limits complex relationship modeling, makes schema evolution difficult, and hinders advanced analytics.

To address these challenges, the Threat Actor Context (TAC) ontology was developed by the University of Oslo and standardized through the OASIS TAC Technical Committee [9]. TAC offers a formal, semantic representation of the STIX 2.1 model using OWL (Web Ontology Language), enabling more expressive, flexible, and interoperable CTI. It supports reasoning, inferencing, and integration with heterogeneous data sources, making it better suited for advanced analytical and automation use cases.

TAC is implemented using open-source and Commercial-Off-The-Shelf (COTS) tools. Protégé [10] is used to build and visualize the ontology, while Stardog [11] enables knowledge graph construction and inferencing. The ontology is modular—each STIX Domain Object (SDO) has its own OWL file—and is publicly available on GitHub for customization and extension.

Organizations can use the TAC ontology to semantically enrich their CTI workflows. For example, it can represent threat actors using OWL classes and infer additional insights using external models like Intel's Threat Agent Library (TAL). By integrating internal asset and vulnerability ontologies, organizations can perform automated reasoning, stay threat-informed, and support risk-driven decision-making with contextual, machine-interpretable intelligence.

In the context of PHOENI2X, the TAC ontology is adopted to enable improved interoperability between CTI sources (e.g., MITRE ATT&CK, MISP Galaxies), supports context-aware analysis, and facilitates automated reasoning to derive actionable intelligence.

4.3 Attack Prediction, Response Recommendation, and Adaptation Enabler

The Attack Prediction, Response Recommendation, and Adaptation Enabler is based on the PMEM (Predictive Maintenance) component. This enabler is designed to monitor and analyze system behavior in order to minimize the risk of cyber-attacks by identifying anomalies, predicting potential threats, and recommending suitable mitigation actions. The enabler consists of three interconnected sub-modules:

- Attack Categorization: Identifies and characterizes potential attacks, estimating their scope and impact.
- Attack Detection and Prediction: Detects anomalous behaviors using machine learning algorithms, raising alerts in case of deviations from expected patterns. Recent developments include the capability to forecast infrastructure usage metrics, allowing the identification of Distributed Denial of Service (DDoS) conditions.
- Attack Response and Adaptation: Proposes mitigation and proactive actions aligned with the attack type and its estimated impact. These recommendations are based on the MITRE ATT&CK framework and aim to reduce the risk of cascading failures in critical infrastructures

Overall, PMEM is a machine learning-based tool developed to detect anomalies in systems. It is structured into two separate modules. The first focuses on detecting attacks in network traffic, while the second is focused toward predicting measurements and identifying abnormal deviations from expected values. Figure 5 illustrates the PMEM architecture, with the attack detection module on the left and the forecasting module on the right.

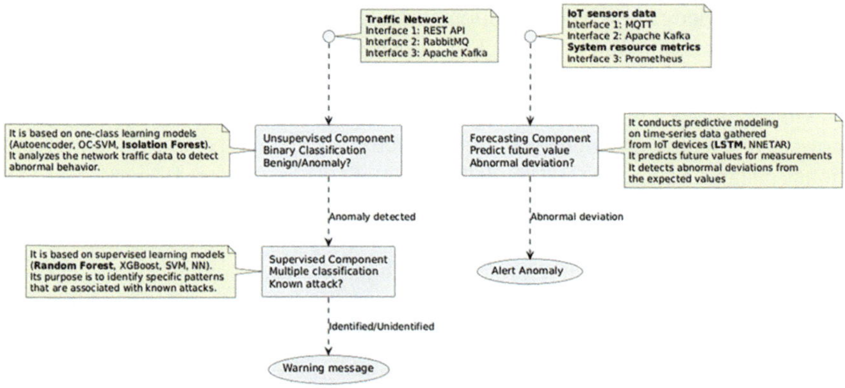

Fig. 5 PMEM architecture design

4.4 Risk Impact Assessment, Alert Triage, and Response Prioritization Enabler

The Risk Impact Assessment, Alert Triage, and Response Prioritization enablers are based on Cyber Risk Assessment Calculator (CERCA). The importance of automatic response and risk prioritization is essential for critical infrastructures, as they provide an essential service to society; partial or total disruption of their operations can cause major impacts in the economy and the livelihoods of citizens. The CERCA tool serves as a key feature in the infrastructure of the pilots, offering a holistic approach to risk assessment and management. Key features (Fig. 6) include:

1. **Dynamic assessment**: Traditional risk assessments rely on static criteria that change infrequently. This can lead to assessments that lag behind real-world changes in Cybersecurity threats and vulnerabilities. By introducing dynamic indicators with continuous monitoring of real-time data, CERCA may adapt

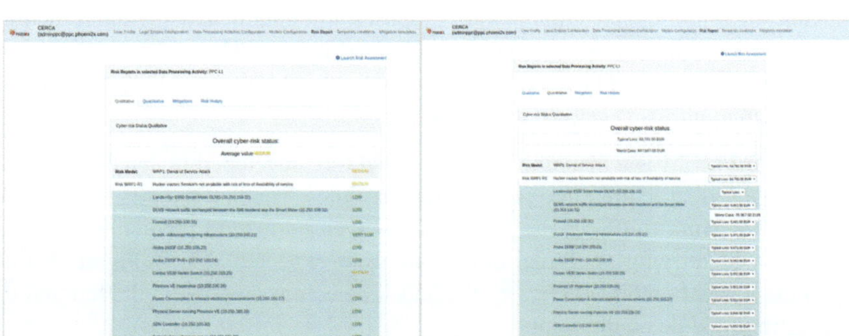

Fig. 6 Qualitative (left) and quantitative (right) risk assessment in CERCA

its results, making these evolve over time to better match up-to-date, relevant security postures.

2. **Qualitative and quantitative perspectives**: CERCA offers a comprehensive approximation to risk assessment by considering both qualitative and quantitative approximations. Qualitative assessment involves evaluating risks based on their characteristics, severity and potential impact, providing an understanding of the threat landscape. Quantitative assessment, on the other hand, involves assigning numerical values to risks, in this case monetary loss estimations, allowing for a precise risk quantification. By combining these two perspectives, CERCA enables organizations to gain a complete view of their security risks, facilitating more informed decision making.

To further support PHOENI2X, CERCA incorporates a mechanism for playbook prioritization based on simulated risk reductions: (i) When an attack is detected or vulnerability reported, CERCA dynamically prioritizes available response playbooks; (ii) Each playbook is simulated to assess its impact on current risk levels; (iii) The playbook offering the highest simulated risk reduction is prioritized for execution. This prioritization mechanism represents a shift from reactive to proactive, simulation-driven security response.

5 Preparedness

5.1 The Resilience Cyber Range (RCR)

The resilience cyber range (RCR), is a core preparedness enabler in PHOENI2X, providing realistic, hands-on training in incident response, business continuity, and cybersecurity awareness. The RCR supports Operators of Essential Services (OES) in two main ways: (1) By enabling the assessment of RPs in simulated environments that replicate real systems, helping identify and improve potential gaps. (2) By offering practical training for OES staff on business continuity, recovery, and incident response procedures defined in the RPs. The resilience cyber range (RCR) is built on the SPHYNX Cyber Range platform, which enables the deployment of realistic, virtual training environments. It supports key user roles such as trainees, who interact with emulated scenarios, and trainers, who manage sessions and monitor progress. Training programs and their scenarios are assigned by the training manager, who oversees the trainees' and trainers' training progress on the organization. The platform's content, the training programs, scenarios, and emulation sandboxes are created interactively through a user interface, by the content creator.

The platform allows for the rapid setup of custom scenarios tailored to the needs of the organization. Its modular design ensures flexibility in scenario creation and complexity adjustment. Figure 7 showcases the interface that allows users to view, filter, and select from the available training programs on the RCR. This UI not only streamlines the selection process but also offers a comprehensive overview of each

Fig. 7 A training program on the Sphynx cyber range

training program. Key information displayed includes the program title, difficulty level, duration, validity period, and training completion status. Additionally, users can view detailed descriptions of the training content, monitor the progress of individual training scenarios (e.g., "In Progress", "Completed"), and read or leave user feedback through the built-in rating and comment system. A dedicated section also outlines the program's learning objectives and assessment criteria, enabling users to make well-informed decisions based on both content and performance expectations.

5.2 *Serious Games*

The concept of serious games within the PHOENI2X framework is a module that provides an interactive, context-specific learning experience. The initiative encompasses a series of games,[8] each with a distinct focus: from the physical tabletop game HATCH, designed to foster understanding and mitigation strategies against social engineering attacks among domain experts, to digital games like PROTECT and AWARENESS QUIZ, which broaden the scope to include a wider audience. The concept is to apply that learning approach in dynamic and realistic scenarios, augmented by machine learning technologies to personalize and enhance the learning experience. This approach not only aims to increase cybersecurity awareness but also to refine the participants' ability to identify and defend against real-world security threats.

[8] https://www.social-engineering.academy/en/offers.html.

PHOENI2X project approach contains serious games that have the following goals: context-specific cybersecurity awareness training; context-specific social engineering threat elicitation and assessment; simple, engaging and target training methods; training content based on real-world security incidents; engaging graphical representation.

The gamification approach consists of one physical tabletop game that engages domain experts in an adventure to design social engineering attacks that work in their specific context. The proposed attacks are rated by the other players, who are also domain experts. The game is moderated by multiple cybersecurity experts to ensure a realistic output of threats. The tabletop game is played with a sample of stakeholders from the relevant context. By experience, playing with more players does not reveal a significant number of new threats after a certain threshold of players is reached. In addition, the game motivates the players to design realistic attacks to win the game. This triggers their motivation for learning cybersecurity.

After the threats have been elicited with the tabletop game HATCH, we use the elicited threats in a digital card game called PROTECT in which the target audience is given attack and defense cards. To win the game, they have to choose the correct defense to a social engineering attack derived from the elicited threats. Finally, the digital game AWARENESS QUIZ presents real attacks related to the elicited threats urging players to respond to this realistic scenarios.

The digital games have been augmented with ML technologies to: adapt the difficulty of the game; re-introduce attack/defense pairs that the player got wrong; choose a set of questions in the quiz related to the threats that the player got wrong before; support the player in replying to cybersecurity risk attitude test; evaluate risk attitude test and recommend training content; search for context-specific attacks on the web to create PROTECT card decks and quiz questions for AWARENESS QUIZ.

The serious game HATCH (Fig. 8) provides an entertaining, interactive group training that teaches employees to identify and successfully defend against social engineering attacks. The serious online game PROTECT enables a motivating and entertaining training that actively sensitizes employees against social engineering threats.

Training should not be boring. With the help of serious games, employees can be engaged in security activities in an entertaining and sustainable way.

6 Alerting, Reporting, and Information Exchange—The Smart Mandatory Incident Reporting Tool (SMIR)

The Smart Mandatory Incident Reporting Tool (SMIR) included in the alerting, reporting, and information exchange capabilities of PHOENI2X provides the critical infrastructures' incident reporting teams with a solution to support them during the mandatory incident reporting process for compliance with the applicable regulatory frameworks.

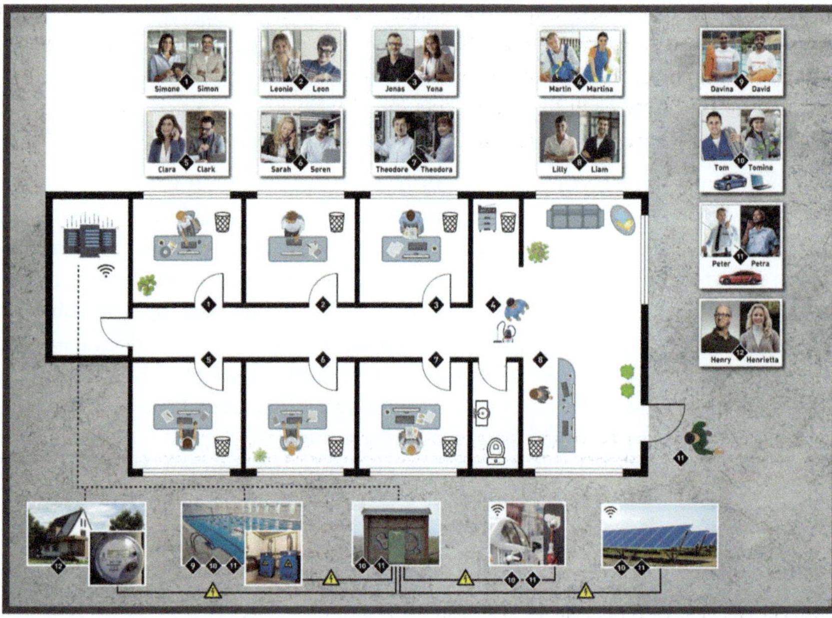

Fig. 8 Main game board of game HATCH

Key features that SMIR integrates to support PHOENI2X functionalities include:

- Automation of the notification of the incidents to the supervisory authorities, in compliance with NIS2 directive, SMIR has been extended to support the automatic sending of an early warning when an incident is classified as significant to relevant entities and the sending of the mandatory report or reports when required depending on the regulation. SMIR dashboard includes the possibility of configuring these communication channels and optionally enabling them for automatic reporting.
- Support for the reception of requests from the ROAR to automatically send an early report for cyber security teams. Further, since ROAR can register new incidents in SMIR, it also needs to receive a confirmation from SMIR when each incident is closed (e.g., to progress a playbook). As such, SMIR has been extended to support notification functionality to external endpoints.
- Inclusion of technical information in the reports received from other components within PHOENI2X to enrich the context of the incident. This includes Indicators of Compromises that are received through the CTI component or alerts generated by the monitoring components. These alerts and IoCs can be also visualized through the graphical interface provided by TheHive and integrated or merged into the incidents as Observables. SMIR has been extended to support the inclusion of these observables in the documents generated as an annex by just configuring the template to add them.

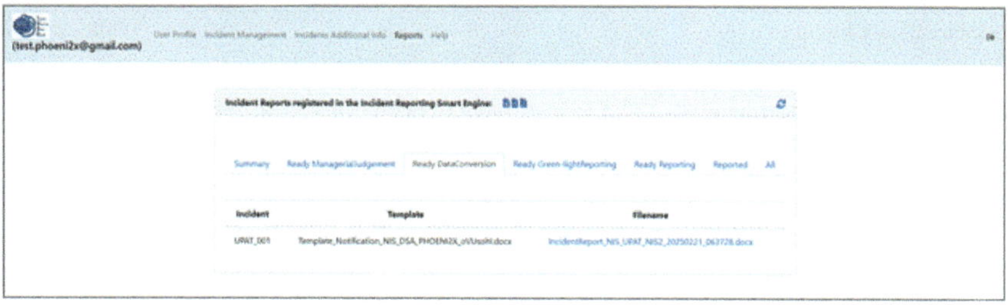

Fig. 9 SMIR—report generated for DSA according to NIS2

- Adaptation of incident reporting according to the templates provided by end users (e.g., the health use case uses a template suitable for notification of incidents to the Digital Security Authority of Cyprus—DSA—according to the NIS2 Directive; see Fig. 9).

Overall, NIS2 directive and GDPR have been considered in PHOENI2X to cover the use case needs about reporting to their respective national authorities. For NIS2, the Word template offered by the National Cyber Security Authority of Greece has been integrated into SMIR. For GDPR, an Excel template has been created based on the information required by FGC Cybersecurity procedures and regulations.

7 Conclusions

This chapter presented the overall vision of the PHOENI2X project, which developed and validated an AI-enhanced cyber resilience framework, providing orchestration, automation, and response capabilities for business continuity, incident response, and information exchange. Emphasis was given to the alignment of the solutions with the EU landscape, policy makers and National Authorities. Significant innovations were produced by the individual components comprising PHOENI2X. Additional details about them and the overall project can be found on its official website [12]. The final framework was successfully validated across three pilot cases (Energy, Transport, Health) involving two OES, a provider in the supply chain of an OES, a telecom operator and two national cyber-security authorities.

Acknowledgements The research leading to these results has received funding from the European Union's Horizon Europe program under Grant Agreement No. 101070586 (PHOENI2X project)

References

1. Markopoulou, Dimitra et al., "The new eu cybersecurity framework: The nis directive, enisa's role and the general data protection regulation," *Computer Law & Security Review*, vol. 35, no. 6, p. 105336, 2019.
2. European Union Agency for Cybersecurity (ENISA), Threat Landscape 2020 - Emerging trends
3. K. Fysarakis, et al., "A blueprint for collaborative cybersecurity operations centres with capacity for shared situational awareness, coordinated response, and joint preparedness," in IEEE Big Data, 2022.
4. https://www.sphynx.ch/products/security-privacy-assurance-suite/
5. https://www.oasis-open.org/committees/tc_home.php?wg_abbrev=cacao
6. https://github.com/oasis-open/cacao-json-schemas
7. https://www.sphynx.ch/products/security-privacy-assurance-suite/
8. https://docs.oasis-open.org/cti/stix/v2.1/stix-v2.1.html
9. https://github.com/oasis-open/tac-ontology
10. https://protege.stanford.edu/
11. https://www.stardog.com/
12. https://phoeni2x.eu/

Knowledge Sharing, Situation Awareness and Skills in Critical Infrastructure Resilience

Advancing Cyber Resilience by Timely Information Sharing and Situational Awareness, Collaborative Incident Response, and Early Identification and Management of Cybersecurity Skills Gaps

Sofia Tsekeridou, Dimitrios Skias, and Konstantinos Chisiridis

Abstract The chapter outlines the pressing needs for cyber resilience not only for critical infrastructures and essential services but also the government and private sector to provide the defense mechanisms against the growing, more sophisticated and complex cyber threats and attacks. Cyber resilience is to be achieved by a proposed holistic approach, addressing the needs of the cybersecurity regulatory landscape, and starting from ensuring secure products and services delivered to the market, after fully fledged secure software development, ensuring operational cybersecurity by putting in place cybersecurity tools and services to monitor the relevant infrastructure and services and timely sharing and reporting any types of vulnerabilities and threats to all involved stakeholders to manage the threat in the most optimal way, and reaching up to continuously training and upskilling the workforce in cybersecurity on time and to date.

Keywords Cyber resilience · Information sharing · Multi-stakeholder collaboration · Incident response · DevSecOps · Operational cybersecurity · Cyber skills · Cyber preparedness

---

S. Tsekeridou (✉) · D. Skias · K. Chisiridis
Research and Innovation Development Department, Netcompany BELUX, Athens, Greece
e-mail: sofia.tsekeridou@netcompany.com

D. Skias
e-mail: Dimitrios.Skias@netcompany.com

K. Chisiridis
e-mail: Konstantinos.CHISIRIDIS@netcompany.com

J. Soldatos and H. Abie (eds.), *Technology-Enabled Resilience: Innovations in Critical Infrastructure Protection*, https://doi.org/10.1007/978-3-032-25917-2_13

1 Introduction

Cyberattacks targeting critical infrastructures (CIs) have escalated in both frequency and sophistication in the recent years. These infrastructures—power grids, water systems, transportation, healthcare networks, smart city infrastructures, etc.—are essential for societal resilience and national security. Europe, for example, has experienced a dramatic increase in cyberattacks with incidents impacting critical infrastructures, government agencies, and major private sector organizations. Distributed Denial-of-Service (DDoS) attacks have surged by 137% in 2025, compared to the previous year, with the largest recorded attack reaching 1.4 terabits per second. These attacks are now shorter, more targeted, and technically sophisticated, often peaking within a minute and leveraging multi-vector techniques that challenge traditional defenses. The European Union Agency for Cybersecurity (ENISA) highlights that threats to availability, ransomware, and data breaches remain the most prominent risks, with attackers exploiting vulnerabilities in digital supply chains and leveraging advanced social engineering tactics [1, 2].

Several major incidents have put the spotlight on Europe's cyber vulnerabilities. In May 2025, the UK retail sector was hit by a series of attacks, including a high-profile breach at Marks & Spencer, and aggressive campaigns targeting Harrods and Co-op UK. The latter saw hackers claim access to the personal data of up to 20 million members [3]. In March 2025, the Polish Space Agency suffered unauthorized access to its IT infrastructure, prompting a rapid security response. Additionally, Europcar Mobility Group faced a significant breach when hackers stole source code and personal data from up to 200,000 customers, threatening to leak 37GB of sensitive information unless a ransom was paid [3]. These incidents underscore the significant concerns brought by threat actors.

In addition, geopolitical tensions have also driven state-sponsored cyber operations in Europe. Russian-linked groups such as Sandworm and APT28 have been associated with attacks on Ukrainian and EU infrastructure, aiming to destabilize and extract strategic information. ENISA warned in its 2024 Threat Landscape report that "*cyberattacks on critical sectors have become more persistent, with motivations ranging from financial gain to political disruption*" [2]. As these threats grow more complex, European governments are accelerating cybersecurity investments and legislation to protect their critical systems.

The financial and reputational damage from such cyberattacks is substantial. The average cost of a cyberattack on large European firms now exceeds €50,000, with losses from specific fraud schemes—such as ticket and holiday fraud—reaching tens of millions of pounds in the UK alone. Attackers are increasingly exploiting vulnerabilities in widely used software, as seen in the coordinated exploitation of SSRF flaws across Germany, Lithuania, and other countries in March 2025. Government agencies, including those in Italy and the UK, have also been targeted by state-linked actors, highlighting the geopolitical dimension of cyber threats [1, 4].

As attacks grow in frequency and sophistication, European organizations are under significant pressure to strengthen their cybersecurity strategies, invest in resilience, and comply with evolving regulatory frameworks. Hence, the chapter initially presents the advances in cyber resilience regulations in Europe the recent years and the resulting compliance requirements aiming at increasing the level of cyber preparedness and defense of critical infrastructures, governmental authorities, and the private sector. Subsequently, a holistic approach to cyber resilience is proposed that defines essential steps and processes to minimize vulnerabilities and increase cyber shields starting from developing secure software products, augmenting operational cybersecurity and cyber defenses investing on timely situational awareness and information sharing, as well as multi-stakeholder collaboration toward immediate incident response, and advancing the cyber skills of the workforce to be better prepared in managing the evolving threat behaviors of attackers. The chapter concludes by presenting the main insights and recommending follow-up actions.

2 Regulatory Landscape in Cyber Resilience

The increase of cyberattacks to critical infrastructures and essential services has led to significant advancements in the cybersecurity policy and regulatory landscape in Europe, to enable immediate preparedness actions and expedite cyber defense:

- The **NIS2 Directive** [5] is a key EU cybersecurity regulation that broadens the scope of entities subject to cybersecurity obligations, covering medium and large organizations in critical sectors such as energy, transport, health, and digital infrastructure. It mandates enhanced risk management, incident reporting, supply chain security, and accountability at the management level, aiming to strengthen the overall resilience and response capabilities of essential and important entities across the EU.
- The **Cyber Resilience Act (CRA)** [6], which entered into force on December 10, 2024, establishes uniform cybersecurity requirements for products with digital elements placed on the EU market, including hardware, software, and remote data processing solutions. It requires manufacturers, importers, and distributors to ensure cybersecurity throughout the product lifecycle, from design to maintenance, with obligations such as timely security updates and vulnerability handling. Critical products must undergo third-party assessments before market release, and compliant products will bear the **CE marking**, signaling adherence to cybersecurity standards. The CRA aims to protect consumers and businesses by making the cybersecurity level of digital products more transparent and reliable, with full compliance required by December 2027.
- The **Critical Entities Resilience Directive (CER)** [7] complements NIS2 by focusing on the resilience of critical entities, strengthening their preparedness against disruptions including cyberattacks. It imposes stricter security and incident reporting requirements to ensure continuity of essential services.

- The **Cyber Solidarity Act (CSA)** [8], which entered into force on February 4, 2025, will improve the preparedness, detection and response to cybersecurity incidents across the EU. The Act includes a European Cybersecurity Alert System, made of Security Operation Centers (SOCs) interconnected across the EU, and a comprehensive Cybersecurity Emergency Mechanism to improve the EU's cyber resilience. It promotes cooperation, mutual assistance, common situational awareness and information sharing among member states to enhance collective cybersecurity resilience and coordinated response to cyber incidents.

3 A Holistic Approach Toward Cyber Resilience

Operators of essential services, governmental authorities and the private sector have been/are called for to act upon the dictations of these regulations to ensure compliance with them, with the utmost aim to increase their cyber resilience. However, this requires specialized knowledge and coordinated effort by skilled experts in the domain, which in many cases is limited. To address this need, a **holistic approach to cyber resilience** is proposed that includes:

- **Secure software development—in line with CRA**: solid approaches and tools for security testing and hardening of software (and hardware) products in Europe before entering the market.
- **Operational Cybersecurity—in line with NIS2/CER**: methods and tools for fully-fledged vulnerability assessment and penetration testing, as well as application of cybersecurity defenses, including diverse automated threats and attack detection tools, at critical infrastructures and services in operation.
- **Information Sharing and Collaboration—in line with NIS2/CSA**: methods and tools for instant situational awareness and early warnings, incident reporting, timely information sharing and collaboration among need-to-know stakeholders.
- **Skilled workforce in cybersecurity—in line with ENISA's European Cybersecurity Skills Framework (ECSF)** [9]: tools to identify cybersecurity skills gaps, advance cyber awareness, and upskill the workforce in cybersecurity.

The methods and tools for the implementation of each step of the holistic approach to cyber resilience are subsequently described in detail.

3.1 Secure Software Development

One of the first steps in releasing secure software products is **security by design**, which is a proactive approach where security principles are integrated at the earliest stages of system or software development, rather than being added retroactively. This methodology establishes security as a foundational element—embedding controls, threat modeling, and risk assessments into the very blueprint of software and hardware products. Core principles include least privilege, defense in depth, secure defaults, fail-secure mechanisms, attack surface minimization, and robust authentication. By making security an executive-level priority and weaving secure practices throughout the lifecycle, products become resistant to exploitation and require less costly remediation down the line [10].

Secure software development capitalizes on the security by design philosophy by incorporating specific best practices at every stage of the software development lifecycle (SDLC) [11]. Engineers are expected to implement threat modeling, secure coding standards (such as the OWASP guidelines [12, 13]), and automated security testing. Practices such as the principle of least privilege, continuous updates and patching, rigorous input validation, and code reviews are essential to prevent vulnerabilities like SQL injection and unauthorized access. Adopting a security-first mindset and regularly collaborating with security experts significantly reduce the risks of data breaches, compliance failures, and financial loss [14].

Adopting a **DevSecOps approach** in SDLC integrates security even further into software development and operations pipelines, as illustrated in Fig. 1. Organizations are currently moving security "left"—embedding it from the earliest coding phases and automating it throughout the entire Continuous Integration/Continuous Deployment (CI/CD) processes. Key developments in this area include the use of AI and hyper-automation for dynamic threat detection, policy enforcement as code, and real-time vulnerability management. DevSecOps emphasizes shared responsibility, rapid risk mitigation, and compliance validation in cloud-native, containerized, and distributed environments. As emerging threats grow more complex, DevSecOps enables agile, secure development without slowing down innovation.

Both Static Application Security Testing (SAST) [16, 17] and Dynamic Application Security Testing (DAST) [18, 19] are pivotal for a comprehensive DevSecOps strategy, addressing different but overlapping areas of risk. SAST excels at catching vulnerabilities early—before code ever runs—while DAST simulates live attack

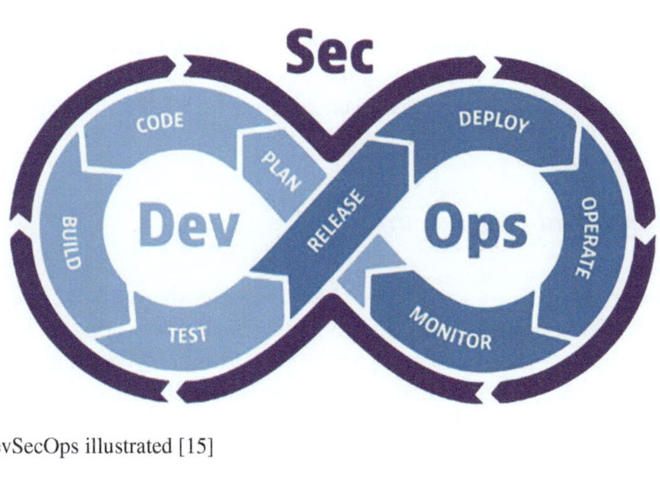

Fig. 1 DevSecOps illustrated [15]

scenarios to expose issues present in functioning applications. Using these methods together ensures that flaws missed by one approach (such as runtime misconfigurations overlooked by SAST or code-level bugs missed by DAST) are more likely to be discovered and mitigated. This complementary deployment within a DevSecOps framework drives a culture of shared responsibility, continuous feedback, and resilience throughout the SDLC, aligning with industry best practices for secure software development.

In more detail, **SAST** is an essential white-box security technique focused on scanning application source code, bytecode, or binaries to identify vulnerabilities before the application is executed. In the context of DevSecOps, SAST integrates seamlessly into early phases of the Software Development Lifecycle (SDLC), allowing developers to detect security issues like SQL injection, cross-site scripting, and authentication flaws at the code level before deployment. Integrating SAST with CI/CD pipelines enables real-time feedback within developer environments, supporting the shift-left approach and reducing remediation costs. However, SAST tools may generate false-positives and require careful tuning to the development stack to ensure actionable and relevant results. A commonly used SAST tool is SonarQube [20], which is integrated in CI/CD pipelines and provides static analysis checkers, both open-source and commercial, for multiple programming languages. It attains reporting capabilities for the overall health of a code base that can be achieved, presenting metrics on code size, code complexity, automated test coverage, outstanding static analysis findings by severity, and technical debt over time.

On the other hand, **DAST** represents a black-box security approach that evaluates running applications from the outside, just as an attacker might. Unlike SAST, DAST does not require access to the source code; instead, it interacts with deployed applications and APIs, sending crafted requests and monitoring responses to uncover runtime vulnerabilities such as authentication issues, configuration mistakes, and business logic flaws. Within DevSecOps, DAST is commonly integrated at later stages of development or during pre-production, often as part of automated CI/CD workflows. This real-world testing identifies issues that only emerge under operational conditions, complementing code-based static analysis and enhancing overall security coverage. A commonly used DAST tool is ZAP [21], an example of its use shown in Fig. 2, after scanning an initial version of the Enhanced MeliCERTes Ecosystem Platform UI [22], having resulted from the H2020 IRIS Project [23].

3.2 Operational Cybersecurity

Operational cybersecurity and resilience methods increasingly emphasize integrated, multi-layered defense frameworks that combine penetration testing, risk and vulnerability assessments, stakeholder engagement, and employee training (the latter presented in Section 3.4) as foundational elements. Regular penetration testing, risk and vulnerability assessments in operational systems identify vulnerabilities and

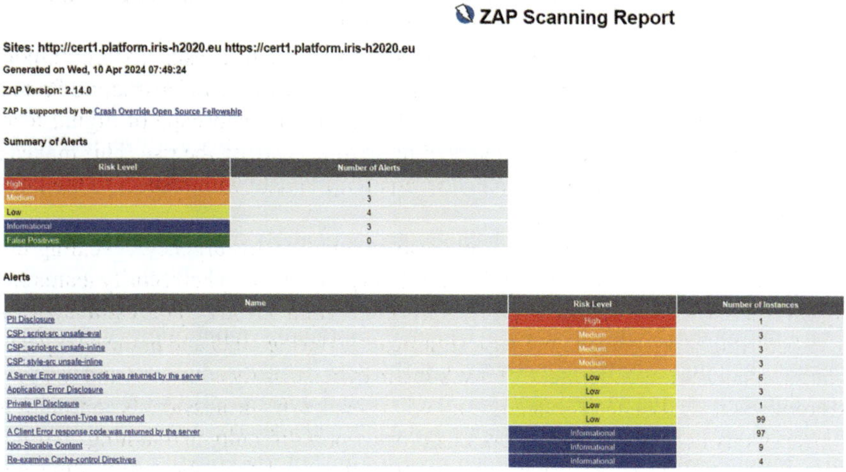

ZAP Scanning Report

Sites: http://cert1.platform.iris-h2020.eu https://cert1.platform.iris-h2020.eu

Generated on Wed, 10 Apr 2024 07:49:24

ZAP Version: 2.14.0

ZAP is supported by the Crash Override Open Source Fellowship

Summary of Alerts

Risk Level	Number of Alerts
High	1
Medium	3
Low	4
Informational	3
False Positives:	0

Alerts

Name	Risk Level	Number of Instances
PII Disclosure	High	1
CSP: script-src unsafe-eval	Medium	3
CSP: script-src unsafe-inline	Medium	3
CSP: style-src unsafe-inline	Medium	3
A Server Error response code was returned by the server	Low	6
Application Error Disclosure	Low	3
Private IP Disclosure	Low	1
Unexpected Content-Type was returned	Low	99
A Client Error response code was returned by the server	Informational	97
Non-Storable Content	Informational	9
Re-examine Cache-control Directives	Informational	4

Fig. 2 ZAP scanning report results for EME [22]

inform targeted security strategies, allowing organizations to detect early and prioritize resources effectively amid potential evolving threats. Furthermore, engaging stakeholders across departments fosters collaboration and aligns cybersecurity initiatives with organizational goals, promoting a culture of shared responsibility. Furthermore, continuous training and awareness programs are vital to mitigate human error, which remains a significant security risk, due to social engineering attacks. Collectively, these foundations underpin operational cyber resilience by ensuring organizations can prevent, detect, and adequately respond to cyber incidents in real time.

Automated vulnerability assessment in operational systems leverages sophisticated scanning tools to continuously identify security weaknesses across networks, applications, and endpoints. These tools allow for automated and scheduled scans, ensuring regular and up-to-date security evaluations with minimal manual intervention. They provide a broad coverage of attack surfaces, including outdated software, misconfigurations, and missing patches, often enhanced by integration with threat intelligence platforms to detect emerging vulnerabilities. Accuracy and low false positives are prioritized through advanced detection techniques like behavioral analysis and machine learning, enabling actionable, prioritized reporting and efficient remediation planning for security teams managing complex operational environments.

Automated penetration testing complements vulnerability assessments by simulating real-world attack scenarios to test the exploitability of identified weaknesses in operational systems. Modern solutions, aligned with international standards and best practices, such as OWASP and NIST ones, combine automated tools with manual verification to mimic attacker behavior and assess risks more rigorously. AI-powered penetration testing tools have revolutionized the field, enabling dynamic adaptation of testing strategies, predictive attack path analysis, and comprehensive exploitation that drive faster and more precise identification of critical flaws across network, cloud, and application layers. External penetration testing approaches simulate attack and infiltration efforts on an organization's network outside its network perimeter, while internal penetration testing approaches scan, investigate, analyze, and report on the level of risk associated with security vulnerabilities discovered within the organization's internal network.

Furthermore, advancements in technology, particularly the incorporation of artificial intelligence (AI) and machine learning, are reshaping operational cybersecurity by enhancing and automating threat detection and response capabilities [24]. Real-time analytics powered by AI enable continuous monitoring of vast data streams, helping security teams detect suspicious behavior, zero-day vulnerabilities, threats targeting IT, IoT, and OT systems, and discover attack patterns earlier. AI-driven tools, including autonomous assistants and virtual security advisers, optimize resource allocation and sharpen decision-making in complex environments. The integration of large language models (LLMs) into deception technologies such as honeypots adds sophistication by dynamically mimicking human interactions to mislead attackers, prolong engagement, and harvest intelligence on adversarial tactics. Such AI-powered innovations elevate the capacity for proactive defense and adaptive resilience.

According to the type of operational system or infrastructure to be protected, the optimal set of operational cybersecurity mechanisms and automated vulnerability, threat detection and early warning tools need to be put in place to cyber shield such system. In addition, such systems and tools feed results into SIEM (security incident event management), incident reporting and threat intelligence systems, described in the next subsection, to close the loop in effectively preventing or responding to emerging threats and incidents, with the utmost aim to minimize their impact and potential cascading effects.

3.3 *Incident Reporting, Information Sharing, and Collaboration*

Operational cyber resilience is further supported by technology-assisted incident reporting, information sharing and collaboration among the required stakeholders for timely awareness, knowledge sharing and collaborative coordinated response to effectively address any type of emerging threat or incident, which is one of the main aims of the Cyber Solidarity Act [8].

To this end, threat intelligence tools are used for threat information management and sharing in interoperable formats. Well known open-source threat intelligence tools, made in Europe, are the Malware Information Sharing Platform (MISP) [25] and the OpenCTI [26]. MISP facilitates the exchange and sharing of threat intelligence, Indicators of Compromise (IoCs) about targeted malware and attacks, financial fraud or any intelligence within a community of trusted members. MISP sharing is a distributed model containing technical and non-technical information which can be shared within closed, semi-private, or open communities. Exchanging such information aims to facilitate subsequent faster detection of targeted attacks within the community and improve the detection ratio, while also reducing the number of false positives. OpenCTI allows organizations to manage their cyber threat intelligence knowledge and observables, by structuring, storing, organizing and visualizing technical and non-technical information about cyber threats. Users are able to access a consolidated view of threat data from multiple sources, while information is shared along with actionable insights across teams and tools and incident response is streamlined with powerful case management capabilities.

As these tools are essentially threat intelligence tools, fully-fledged technology-assisted information sharing, collaboration, and communication among multiple involved organizations/stakeholders (including national CERT/CSIRTs and SOCs) in managing a threat or an incident are not essentially supported. However, these are essential elements in the entire operational cyber resilience lifecycle in closing the loop by responding to an incident and minimizing its impact. The EU-funded MeliCERTes v2 [27] platform has been released as open-source software to provide relevant capabilities among different CERT/CSIRT authorities. MeliCERTes has been part of the European Strategy for Cyber Security. MeliCERTes aimed to build a network for establishing confidence and trust among the national Computer Security Incident Response Teams (CSIRTs) of the Member States and for promoting a swift and effective operational cooperation within verified Trust Circles for sharing and collaborating on cyber security incidents. MeliCERTes v2 has used open-source tools developed and maintained by the participating CSIRT teams, such as MISP [25], providing incident management, threat intelligence, secure communications, and artifact analysis capabilities. The core component of MeliCERTes v2 has been Cerebrate [28]. Cerebrate is an open-source software platform meant to act as a central tool for MeliCERTes v2 and provide a central interface to manage different tools and different settings for them, thus offering tight integration with various open-source security tools. Cerebrate provides a way to build an open-source security ecosystem of tools in pursuit of enabling organizations' autonomy and independence. Cerebrate also integrates a versatile directory solution to manage constituents, members and partners in information sharing communities such as ISACs, CSIRT networks or any closed or public sharing groups.

Evidently, a distributed secure ecosystem that combines the capabilities of all aforementioned tools and is easily customizable according to the needs of a wider group of different users/stakeholders that are expected, in line with the relevant directives and regulations (e.g. NIS2, CSA), to share knowledge and collaborate for effective cyber threat and incident management, such as critical infrastructure operators, operators of essential services, CERTs/CSIRTs, SOCs, ISACs, etc., is of utmost importance. This has been the motivation for developing the Enhanced MeliCERTes Ecosystem (EME) [22] within the H2020 IRIS project [23]. It is an open source all in one integrated and distributed ecosystem with loosely coupled architecture, that enables: (i) interoperable communication and interfacing with on CI premise automated vulnerability, threat and attack detections tools, by adopting cybersecurity standards such as STIX [29], MISP standards [30], and CACAO playbooks [31], (ii) semi-automated, secure and timely cyber threat intelligence (CTI) and incidents information sharing and reporting among diverse need to know stakeholders at the time of occurrence in the case of detected threats and incidents, for better preparedness, detection and response capabilities, (iii) enhanced and timely cyber situational awareness and online collaboration among need to know stakeholders within Trust Circles to manage a threat/incident the soonest possible and in the best informed manner, (iv) semi-automated response policies customization and acknowledgment of detected vulnerabilities and threats.

EME further provides customized views of its dashboard, security incident reporting capabilities, access control and access rights to shared CTI data in accordance to the type of user/operator and the type of service/infrastructure they provide. EME follows a distributed architecture approach allowing its customized deployment at diverse stakeholders' premises. Two primary users of EME are CI operators/ operators of essential services and CERTs/CSIRTs—a demo video of the relevant EME instance for the former users is available at [32], while a demo video of the relevant EME instance users is available at [33].

As mentioned, EME capitalizes on well-known open-source software modules, such as Cerebrate [28] and MISP [25] of MeliCERTes v2 [27] but also Shuffle [34], which provides automated orchestrations of interoperable workflows and message exchanges in a standardized ways. EME thus provides a fully fledged integrated ecosystem, released as open source [22].

In terms of regulatory alignment, EME addresses core requirements of the NIS2 Directive and the Cyber Solidarity Act for information sharing, situational awareness, timely CTI and incident reporting, early warnings and alert mechanisms, collaboration among all involved stakeholders in an optimal and timely manner, as shown in Fig. 3.

The highlights of EME that strengthens operational cybersecurity are the following: (i) it helps organizations to comply with EC regulations in cybersecurity (NIS2, CSA), (ii) it is capable to interoperate with most of the existing cybersecurity solutions in the market due to the adoption of widely used standards for cybersecurity information exchange (STIX, CACAO, MISP), (iii) it leverages open-source tools to provide a cost-effective, flexible, and extensible open-source platform that can be widely adopted.

3.4 Skilled Workforce in Cybersecurity

A crucial part of guaranteeing the highest levels of Operational Cyber Resilience relates with aware and skilled human resources. However, currently, the cybersecurity industry in Europe is facing a significant cybersecurity skills shortage with

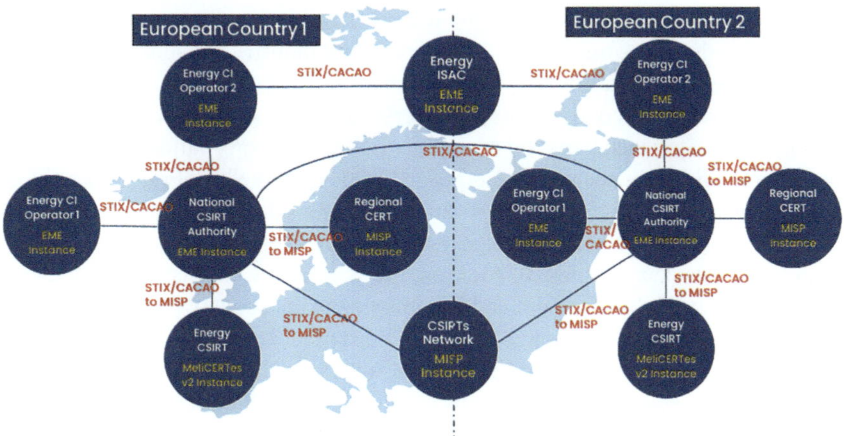

Fig. 3 Timely interoperable information sharing and collaboration across multiple stakeholders in managing a threat/incident, at regional, national, or even European level using EME deployed instances interoperating with MISP or MeliCERTes v2 instances

difficulties in hiring qualified personnel. This shortage poses a serious threat to technological advancement and digital security across the region, as organizations struggle to find qualified talent capable of addressing increasingly sophisticated cyber threats. Despite a 25% increase in cybersecurity graduates over recent years, the supply of skilled professionals remains insufficient, driven in part by rapid innovation outpacing workforce readiness and a lack of investment in reskilling and upskilling programs. The EU Agency for Cybersecurity (ENISA) stresses that human capital is critical to achieving the EU's cybersecurity goals and complying with expanding legal requirements.

Recruitment challenges are severe: 45% of companies report difficulties finding qualified candidates, while 76% of employees currently working in cybersecurity lack formal certification or training. Additionally, many cybersecurity roles are filled by professionals transitioning from non-cyber fields, further evidencing a gap in targeted skill sets. Companies also face barriers such as limited training awareness, budget constraints, and a lack of continuous professional development programs despite cybersecurity being a high priority for 71% of organizations surveyed. Diversity remains low, with more than half of firms reporting no women in cybersecurity positions despite broad recognition of its importance [35, 36].

The widening skills gap is compounded by a shortage of technical expertise across emerging technologies and operational domains, including IT operations knowledge and experience with diverse technology stacks. This shortage leaves organizations vulnerable to cyber risks, hindering compliance with regulatory frameworks and slowing digital transformation initiatives [37]. To counteract these trends, the European Commission and other stakeholders have launched initiatives such as the Cyber Skills Academy [38] and the European Pact for Skills [39], aiming to boost cybersecurity education, workforce diversification, and talent mobility. However, experts emphasize that closing the gap will require sustained multi-dimensional efforts combining education, training, policy support, and industry collaboration to address this significant skills gap.

ENISA is committed to address the cybersecurity skills gap through a comprehensive systemic approach based on education. ENISA supports national authorities by developing initiatives such as the European Cybersecurity Skills Framework (ECSF) [9]. The ECSF is a cornerstone for advancing Europe's digital future, offering a practical tool to identify and articulate the tasks, competences, and knowledge associated with key cybersecurity roles. In addition, the European Commission has identified the crucial role of skills and defined its Union of Skills strategy [40].

Further from these efforts to deliver a more skilled workforce in cybersecurity, it is of equal importance to identify gaps in the supply and demand in the cybersecurity skills sector as well as to discover gaps in available upskilling/reskilling educational programs that will need to be timely updated in accordance to the imminent market demands for advanced or specialized cybersecurity skills. To this end, the Horizon Europe SKILLAB project [41] aims to deliver an open-source skill demand/supply identification, analysis and prediction hub for citizens, enterprises, academia, and policy makers. More specifically, SKILLAB aims to: (i) develop a skills management and skills shortage identification platform, (ii) gage and monitor the European labor market, and (iii) continuously propose up-to-date strategies, required curricula development, and policies.

4 Conclusions

The increase, complexity, and severity of experienced cyberattacks the recent years in critical infrastructures, governmental services, and private sector stakeholders has intensified the need for a robust approach to cyber resilience. These is achieved only if one addresses all influencing factors which further present the weak points in the cyber resilience landscape: (i) ensuring the release into the market of fully tested secure products to be used by such entities, (ii) advancing the cyber defenses of the operational infrastructures and services of such entities by installing and using advanced cybersecurity tools for vulnerability assessment, penetration testing, automated threat and attack detection, response, and recovery, (iii) encouraging timely awareness, information sharing and collaboration at the instant a threat is detected or an incident takes place among all involved stakeholders required in managing effectively this situation, (iv) securing the level of cybersecurity skills on a continuous basis the workforce needs to have in order to secure, manage, and respond to incidents in the best manner, but also not to be the weak link in, e.g., social engineering types of attacks. Such a holistic approach to cyber resilience comes in agreement with a number of dictations of the current European regulatory landscape.

Acknowledgments The work presented in this chapter has been partially funded by the H2020 IRIS project: "Artificial Intelligence, Threat Reporting and Incident Response System," Grant Agreement

No 101021727, and the Horizon Europe SKILLAB project: "Monitoring the Demand and Supply of Skills in the European Labour Market," Grant Agreement No 101132663.

References

1. T. Ahmed (2025). *256 Cybercrime Statistics for 2025*, Bright Defense. Retrieved from: https://www.brightdefense.com/resources/cybercrime-statistics/
2. ENISA (2024). *ENISA Threat Landscape 2024*. Retrieved from: https://www.enisa.europa.eu
3. Cyber Management Alliance (2025). *May 2025: Biggest Cyber Attacks, Ransomware Attacks and Data Breaches*. Retrieved from: https://www.cm-alliance.com/cybersecurity-blog/may-2025-biggest-cyber-attacks-ransomware-attacks-and-data-breaches
4. SentinelOne (2025). *Key Cyber Security Statistics for 2025*. Retrieved from: https://www.sentinelone.com/cybersecurity-101/cybersecurity/cyber-security-statistics/
5. NIS2 Directive (2022). *Securing Network and Information Systems*. Retrieved from: https://digital-strategy.ec.europa.eu/en/policies/nis2-directive
6. Cyber Resilience Act (2024). Retrieved from: https://digital-strategy.ec.europa.eu/en/policies/cyber-resilience-act
7. Critical Entities Resilience (2022). Retrieved from: https://eur-lex.europa.eu/eli/dir/2022/2557/oj/eng
8. EU Cyber Solidarity Act (2025). Retrieved from: https://digital-strategy.ec.europa.eu/en/policies/cyber-solidarity
9. ENISA (2022). *European Cybersecurity Skills Framework (ECSF)*. Retrieved from: https://www.enisa.europa.eu/topics/skills-and-competences/skills-development/european-cybersecurity-skills-framework-ecsf
10. P. Kirvan, I. Wigmore (2025). *What is Security by Design?* Retrieved from: https://www.techtarget.com/whatis/definition/security-by-design
11. Atlassian (2025). *The complete guide to SDLC* Retrieved from: https://www.atlassian.com/agile/software-development/sdlc
12. OWASP Application Security Verification Standard. Retrieved from: https://owasp.org/www-project-application-security-verification-standard/
13. OWASP Web Security Testing Guide. Retrieved from: https://owasp.org/www-project-web-security-testing-guide/
14. M. Watson (2025). *Best practices for Secure Software Development*. Retrieved from: https://fullscale.io/blog/15-best-practices-secure-software-development/
15. J. Marsal (2025). *What is DevSecOps? And what you need to do it well*. Retrieved from: https://www.dynatrace.com/news/blog/what-is-devsecops/
16. Practical DevSecOps. *Comprehensive Guide to SAST Implementation*. Retrieved from: https://www.practical-devsecops.com/wp-content/uploads/2023/07/Comprehensive-Guide-to-SAST-Implementation.pdf
17. OWASP. *DevSecOps Guideline: Static scanning is an important part of the process*. Retrieved from: https://owasp.org/www-project-devsecops-guideline/latest/02a-Static-Application-Security-Testing
18. OWASP. *DevSecOps Guideline: Dynamic Application Security Testing (DAST)*. Retrieved from: https://owasp.org/www-project-devsecops-guideline/latest/02b-Dynamic-Application-Security-Testing
19. S. Iyer (2024). *The Top 10 DAST tools for DevSecOps in 2025*. Retrieved from: https://securityboulevard.com/2024/12/top-dynamic-application-security-testing-dast-tools-for-2025/
20. SonarQube. Available at: https://docs.sonarqube.org/latest/
21. ZAP. Available at: https://www.zaproxy.org/docs/alerts/

22. Enhanced MeliCERTes Ecosystem. Available at: https://github.com/Enhanced-MeliCERTes-Ecosystem
23. H2020 IRIS Project. Grant Agreement No. 101021727. Available at: https://cordis.europa.eu/project/id/101021727
24. World Economic Forum (2025). *Global Cybersecurity Outlook 2025*. Retrieved from: https://reports.weforum.org/docs/WEF_Global_Cybersecurity_Outlook_2025.pdf
25. Malware Information Sharing Platform (MISP). Available at: https://www.misp-project.org/
26. OpenCTI. Available at: https://filigran.io/platforms/opencti/
27. MeliCERTes, v2. Available at: https://melicertes.github.io/docs/ https://www.cerebrate-project.org/
28. Cerebrate open-source software project. Available at: https://www.cerebrate-project.org/
29. Introduction to OASIS Structure Threat Information Exchange standard (STIX). Available at: https://oasis-open.github.io/cti-documentation/stix/intro.html
30. MISP Standards. Available at: https://misp-standard.org/
31. OASIS CACAO Security Playbooks v2.0. Available at: https://docs.oasis-open.org/cacao/security-playbooks/v2.0/security-playbooks-v2.0.html
32. EME instance video demo for CI operators/operators of essential services. Available at: https://www.youtube.com/watch?v=oJFDFpszfX0
33. EME instance video demo for CERTs/CSIRTs. Available at: https://www.youtube.com/watch?v=4o5VuyfWAiE
34. Shuffle. Available at: https://shuffler.io/
35. L.G. Paule (2024). *News item at EC Digital Skills and Jobs Platform: EU faces growing cybersecurity skills gap, new Eurobarometer reveals*. Retrieved from: https://digital-skills-jobs.europa.eu/en/latest/news/eu-faces-growing-cybersecurity-skills-gap-new-eurobarometer-reveals
36. L.G. Paule (2024). *Eurobarometer on Cyberskills*. Retrieved from: https://digital-skills-jobs.europa.eu/en/inspiration/resources/eurobarometer-cyberskills
37. M. Olney (2025). *EU cyber skills gap to widen in 2025*. Retrieved from: https://digital-skills-jobs.europa.eu/en/inspiration/resources/eurobarometer-cyberskills
38. Cyber Skills Academy at EC Digital Skills & Jobs Platform. Available at: https://digital-skills-jobs.europa.eu/en/cyber-skills-academy-get-involved
39. European Pact for Skills. Available at: https://pact-for-skills.ec.europa.eu/index_en
40. European Commission Union of Skills Strategy. Available at: https://commission.europa.eu/topics/eu-competitiveness/union-skills_en
41. Horizon Europe SKILLAB Project. *Monitoring the Demand and Supply of Skills in the European Labour Market*. Available at: https://skillab-project.eu/

Enhancing Critical Infrastructure Resilience Through a Decision Intelligence Distributed Situational Picture and Incidents Mitigation

Nicola Gregorio Durante, Lorenzo Eccher, Gabriele Giunta, Ionut Vladut Dinu, and Cristian Raul Vintila

Abstract Critical Infrastructure (CI) operators increasingly face systemic challenges due to interdependencies between physical and digital domains, heterogeneous data, delayed threat detection, and limited predictive capabilities. The ATLANTIS project addresses these barriers by delivering an integrated, modular framework for enhanced situational awareness, cross-sector data fusion, proactive response, and secure data processing. It leverages a combination of Digital Twin models, semantic reasoning, and cognitive decision intelligence to support real-time, cross-CI risk management. This chapter presents the ATLANTIS architecture, focusing on its key components: the Digital Twin paradigm for CI modelling; SAFER for event correlation and situational picture generation; semantic integration strategies; cognitive decision support tools; and AI-agent systems for dynamic mitigation. The solution prioritizes interoperability among diverse systems and intelligent automation to effectively address complex cyber-physical threats. The chapter concludes by outlining ATLANTIS's contributions to CI resilience, including its decision-intelligence approach, semantic interoperability through STIX/TAXII and ontologies, and LLM-driven agentic AI for coordinated risk response. Policy and regulatory alignment are also discussed, with attention to GDPR-compliant AI, data

N. G. Durante · L. Eccher · G. Giunta (✉)
Engineering Ingegneria Informatica SpA, Rome, Italy
e-mail: gabriele.giunta@eng.it

N. G. Durante
e-mail: nicola.durante@eng.it

L. Eccher
e-mail: lorenzo.eccher@eng.it

I. V. Dinu · C. R. Vintila
Siemens S.R.L., Bucharest, Romania
e-mail: vladut.dinu@siemens.com

C. R. Vintila
e-mail: cristian.vintila@siemens.com

J. Soldatos and H. Abie (eds.), *Technology-Enabled Resilience: Innovations in Critical Infrastructure Protection*, https://doi.org/10.1007/978-3-032-25917-2_14

sharing, and standardization. Future work will enhance scalability, trust, and explainability, and pursue operational validation in real-world settings to support broader deployment across CI domains.

Keywords Situational awareness · Digital twin · Cyber-physical systems · Semantic interoperability · Cognitive decision support · AI-agent systems · Risk mitigation · Data fusion · Resilience framework

1 Introduction

Critical Infrastructure (CI) operators face growing complexity in enhancing resilience due to tightly coupled physical and digital systems, fragmented and heterogeneous data sources, delayed threat detection, and limited predictive simulation capabilities. These challenges hinder timely decision-making and effective incident response, particularly in interconnected environments where localized events may trigger systemic impacts.

The ATLANTIS project[1] proposes an integrated, modular framework to support decision-making and risk mitigation in such complex CI ecosystems. It enables real-time situational awareness, cross-sector data fusion, and proactive threat response through secure, interoperable, and privacy-preserving data processing. By combining Digital Twin models, semantic technologies, and AI-driven decision support, ATLANTIS advances the management of cyber-physical risks in critical domains.

This chapter outlines the architectural and functional components of the ATLANTIS solution:

- Section 3 introduces the system architecture, detailing the Digital Twin paradigm for CI modelling, the SAFER component for event correlation, and the semantic data model supporting interoperability.
- Section 4 describes the shift from traditional situational awareness to proactive mitigation, emphasizing simulation, knowledge graph integration, and strategy adaptation.
- Section 5 presents the decision intelligence framework, covering cognitive models, event-risk association, predictive analytics, and decision support techniques.
- Section 6 focuses on agentic AI systems built on large language models (LLMs), highlighting their architecture, integration mechanisms, and role in orchestrating mitigation.
- Section 7 concludes the chapter and discusses future directions for intelligent CI protection.

By embedding intelligence and interoperability into the CI management lifecycle, ATLANTIS provides a foundation for resilient, secure, and adaptive infrastructure operations.

2 Background and Related Work

In recent years, the growing interconnection between IT systems and physical infrastructure has highlighted the need for integrated risk management, overcoming the traditional separation between cybersecurity and physical security. This convergence is particularly relevant in critical areas such as smart cities, industrial plants (ICS/SCADA), digital healthcare, and energy infrastructure, where a cyber threat can have tangible physical impacts and vice versa.

Projects such as CSIM (Converged Security and Information Management) offer architectures capable of collecting data from physical sensors (CCTV, access, environmental) and IT systems (SIEM, endpoints, logs) to generate a holistic view of urban and corporate security.[2] Similarly, initiatives such as the VISOR project in the Netherlands have demonstrated the possibility of integrating IoT and AI technologies for real-time monitoring of large public events, combining physical and cyber surveillance.[3]

However, current systems such as CSIM and VISOR typically lack advanced AI-driven correlation mechanisms, real-time decision support, or large-scale cross-border operability. Their architectures are often domain-specific, limiting extensibility and multi-stakeholder collaboration.

Guidelines published by the Cybersecurity and Infrastructure Security Agency (CISA) in the United States encourage the adoption of a unified approach to cyber-physical threat management through tools such as the Cybersecurity and Physical Security Convergence Action Guide.[4] In the industrial sector, the integration of OT and IT is now a priority, with the adoption of zero-trust architectures, network segmentation, and AI-based solutions for real-time anomaly detection.[5]

Today's challenge is twofold: on the one hand, a cultural change is needed within organizations, with collaboration between traditionally separate teams; on the other hand, a review of technological architectures is necessary to support the correlation of cyber and physical events in a predictive and resilient manner. The convergence of the two domains is not only desirable but essential to address modern threats in a systemic manner.

This convergence still faces key barriers, such as fragmented incident handling across domains, limited predictive capabilities, and the absence of interoperable tools for systemic risk analysis and coordinated mitigation.

In brief:

- In the smart building/city domain, converged systems like CSIM and VISOR provide integrated visibility across physical and digital security layers.
- In critical infrastructure and OT (ICS/SCADA): digital twins and threat modelling combine cyber and physical risk analysis.
- In the manufacturing sector: OT–IT integration is achieved through rugged endpoint security and anomaly detection.
- At the governmental and regulatory level: organizations like CISA promote frameworks and scenarios for cyber-physical convergence.

- Emerging technologies like Machine Learning and AI are enabling advanced detection systems that operate across both cyber and physical domains.

ATLANTIS addresses these challenges by introducing an integrated framework that combines digital twins, machine reasoning, and LLM-based decision support to correlate and mitigate complex hybrid threats in real time.

Within this context, the European ATLANTIS project represents a strategic initiative that aims to strengthen the resilience and protection of critical European infrastructure against increasingly complex and interconnected threats, both physical and digital.

The project aims to achieve four key objectives: improving situational awareness of systemic risks; enhancing prevention, response and mitigation capabilities; strengthening cooperation between operators and authorities; and developing an open-source technology platform based on innovative approaches to artificial intelligence, predictive modelling and intelligent decision-making tools.

From an architectural point of view, ATLANTIS is based on an integrated framework that combines data from physical sensors, computer networks and monitoring systems to provide a unified view of threats. The solutions implemented include digital twins for impact simulation, vulnerability assessment tools and AI modules for anomaly detection.

Key innovations include:

- The use of large language models (LLMs) to assist in incident narrative synthesis and operator support;
- AI modules that fuse cyber and physical indicators to detect hybrid attack patterns;
- A simulation engine powered by digital twins to assess cascading effects across critical systems.

The project is being developed through three major pilot trials involving different sectors (energy, transport, finance, healthcare and logistics), with the aim of validating the proposed solutions in real-world contexts, including cross-border ones.

Expected impacts include greater capacity for operators to respond in a coordinated manner to critical events, improved efficiency in hybrid risk management, and the adoption of interoperable tools capable of supporting cyber-physical security in a systemic and sustainable manner.

At the core of ATLANTIS's architecture is a Digital Twin model of critical infrastructures, represented as graphs of assets (nodes) and their interdependencies.

This structure enables contextual understanding during cyber-physical incidents.

These graphs, combined with various cybersecurity and physical protection data sources, are embedded into a knowledge graph. This provides contextual information to a large language model (LLM), which can reason over it to identify actionable items for critical infrastructure (CI) operators. These actions can be applied over time to increase the resilience of the affected infrastructure in the event of an attack.

Agentic systems, which are autonomous AI agents, offer emerging capabilities to create reactive environments for critical infrastructure protection.

Given the broad capabilities of LLMs, including image, video, and audio analysis, as well as inference over complex information, fine-tuning is often the preferred approach when addressing the security needs of a CI. This is because sensitive data cannot be published online or provided to publicly available LLM services, as doing so could compromise the stability of the CI or, in the worst case, the integrity of a country.

The ATLANTIS framework combines state-of-the-art safety mechanisms and novel AI-based approaches, including LLM-powered agents and human-in-the-loop decision support, to effectively manage complex cyber-physical events.

3 Architectural Framework

Enhancing resilience in Critical Infrastructure requires addressing fragmented monitoring, limited interoperability, and insufficient automation in threat detection and response. Existing systems struggle with integrating heterogeneous data sources, forecasting cascading failures, and supporting timely, informed decisions across domains.

The ATLANTIS architectural framework is designed to overcome these limitations through a modular, secure, and interoperable solution. It supports real-time situational awareness, predictive simulation, and cognitive decision-making by combining Digital Twin models, semantic integration, and AI-driven components. The following sections detail the key elements of this architecture, starting with the Digital Twin concept for CI modelling.

3.1 Digital Twin Concept for Critical Infrastructure

This section outlines the foundational principles of a Digital Twin (DT), key interoperability paradigms, and the overall architectural strategy. It includes the instantiation view and the component library used to develop domain-specific DTs.

A DT is a computational model that persistently represents a physical structure, system, or process. Within the CI context, a DT reflects real-time and historical data flows across interconnected assets. DTs are instrumental in conducting "*what-if*" simulations, anticipating failures, supporting early anomaly detection, and enabling predictive analytics.

A DT also serves as a multi-physics, multi-scale, probabilistic simulation of a real-world system. It integrates diverse data sources—sensor feeds, models, historical logs—across different lifecycle stages to produce a dynamic representation of physical assets.

Figure 1 provides a representative overview intended to exemplify the typical structure and functional components of a DT.

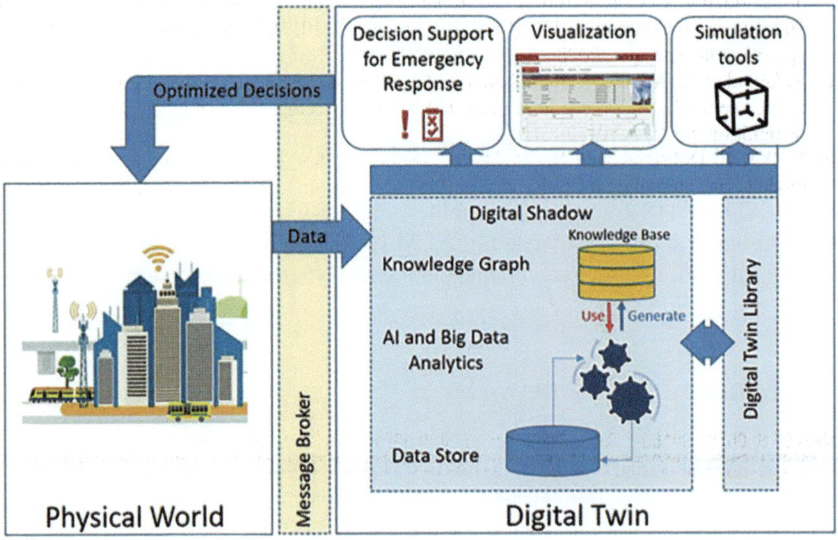

Fig. 1 Digital twin overview

DTs are virtual counterparts of physical systems that maintain a dynamic, real-time connection with their real-world counterparts. Unlike earlier static models, DTs support bidirectional data exchange via APIs and messaging protocols. This allows not only continuous monitoring but also simulation, control, and adaptive optimization based on live system states and contextual data.

To fully exploit DT capabilities, three core tool categories are provided:

- **Simulation Tools**—Run scenario-based simulations for risk assessment and model pretraining.
- **Visualization Tools**—GUI interfaces for data display, threat prediction, serious gaming, and emergency plan visualization.
- **Decision Support Tools**—Advanced reasoning and analytics modules for operational response planning and risk mitigation.

The following sections describe real-world implementations that operationalize the Digital Twin concept. We illustrate how the ATLANTIS framework integrates the core tool categories—simulation, visualization, and decision support—into dynamic event-driven contexts for CI protection.

Effective DT deployment requires interoperability among heterogeneous systems. The strategy adopted in the project involves standardized protocols and data formats, enabling consistent communication and integration across distributed components.

Event-Driven Architecture (EDA)[6,7] is a design paradigm centered on the production, detection, and reaction to events—significant changes in system state or user actions. It decouples components by allowing them to interact through events, rather than direct calls, enabling asynchronous processing and enhancing system responsiveness. EDA is particularly suited for complex, distributed systems where scalability, modularity, and real-time responsiveness are essential.

The adopted paradigm is based on EDA, offering:

- Loose coupling between producers and consumers
- Asynchronous data exchange
- Scalable and modular integration.

EDA ensures efficient, resilient communication between microservices and subsystems. Each event in an EDA represents a state change of a specific entity—such as a device, application component, or user session—transitioning from one condition to another (e.g., from idle to active, offline to online, or threshold not reached to threshold exceeded). Events may be triggered by internal processes (like service responses or internal metrics) or external stimuli (such as sensor inputs, user interactions, or third-party system alerts). In this model, producers are components that detect and emit these state changes by publishing event messages to designated Event Channels, while consumers are decoupled services or modules that subscribe to specific event types and react accordingly. Events typically include discrete, time-stamped facts representing relevant occurrences, such as telemetry updates, configuration changes, system faults, or user commands, which the architecture is designed to detect, propagate, and act upon. From a technological view, the Kafka Message Broker[8] is employed as the event streaming platform, offering:

- High throughput and fault tolerance
- Persistent storage via distributed, append-only logs
- Topic partitioning and offset-based message consumption
- Support for Transport Layer Security (TLS) encryption and mutual authentication.

Kafka's partitioning model ensures horizontal scalability and message ordering for keyed data. This design supports large-scale processing and real-time analytics across distributed infrastructures. Each topic is divided into partitions, which are distributed across brokers, allowing parallel processing and ensuring that messages with the same key are routed to the same partition to preserve order.

Event sourcing extends EDA by treating the event log as the source of truth, enabling components to reconstruct state or replay history. Producers and consumers must adhere to agreed schemas to maintain semantic compatibility and manage evolution over time.

This architecture enables resilient system behaviour, where services can fail and recover independently without disrupting the overall flow. The message broker stores undelivered events, ensuring guaranteed delivery when services resume.

3.2 ATLANTIS Framework and SAFER Component

The ATLANTIS Digital Twin implements a systemic, continuous risk analysis process, essential for enabling dynamic situational awareness across critical infrastructures. It integrates physical and cyber data sources to maintain an up-to-date security posture and inform decision-making.

ATLANTIS Digital Twin integrates multiple data sources, including:

- Physical sensors (Industrial Control Systems—ICS, Supervisory Control and Data Acquisition—SCADA)
- Cybersecurity tools (Intrusion Detection Systems—IDS, Endpoint Detection and Response—EDR)
- Intelligence feeds (Open-Source Intelligence—OSINT, Cyber Threat Intelligence—CTI)
- Digital Twin models and simulations.

This multi-source fusion ensures a comprehensive and contextualized understanding of the operational environment.

At the core of the ATLANTIS Digital Twin lies the SAFER component, a middleware component responsible for data aggregation, normalization, and dissemination (Fig. 2).

SAFER collects, standardizes, and processes data from all architectural layers using the IDMEFv2 (Intrusion Detection Message Exchange Format v2),[9] a structured IETF-defined format for representing security alerts and events. Its

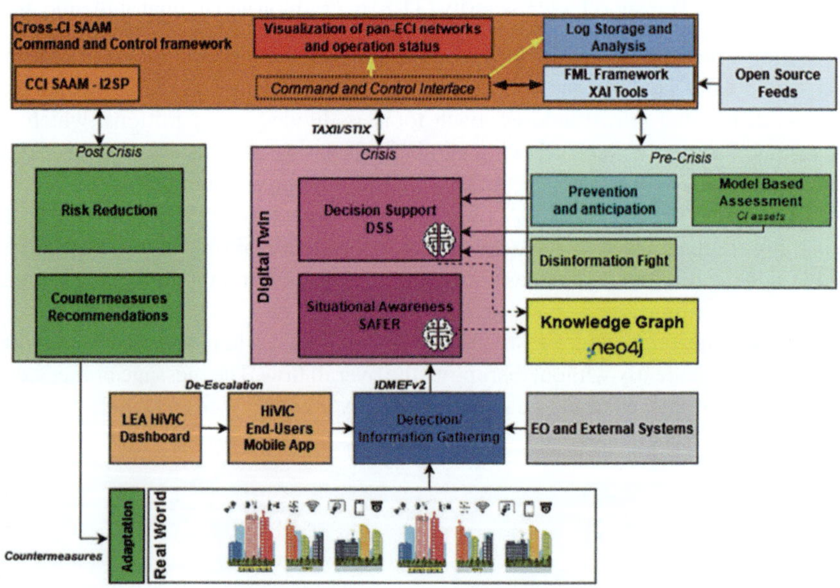

Fig. 2 ATLANTIS security framework: integrated component interaction

design ensures interoperability across components and aligns with cyber situational awareness standards.

The module leverages Apache Kafka for:

- High-throughput message exchange
- Scalable topic-based data ingestion
- Fault-tolerant streaming architecture.

SAFER continuously updates the Situational Picture (SP)—a dynamic data structure capturing:

- Real-time risk assessments
- Vulnerability metrics
- Asset state and alerts
- Incident metadata.

This unified risk landscape supports proactive threat mitigation, automated decision support, and resilience planning.

ATLANTIS Digital Twin embeds a Complex Event Processor (CEP) to perform real-time event correlation. It ingests alerts from cyber-physical sensors and threat correlators, executing multi-layer anomaly detection, pattern-based predictive analysis, and early identification of cascading effects and systemic risks.

The CEP utilizes temporal rules, threshold analysis, and state transition modelling to detect interrelated anomalies.

The framework maintains a graph-based interdependency model, enabling analysis of asset interconnections, detection of high-risk propagation pathways, and enhanced understanding of systemic vulnerabilities.

Events are associated with CI assets through geospatial and network correlation. SAFER applies heuristic evaluation to estimate risk levels based on:

- Asset criticality
- Threat type (cyber, physical, hybrid)
- Propagation vectors
- Impact likelihood and severity.

Structured risk tables define event categories, impact chains, and priority rules. These tables are used by SAFER to compute warning levels and trigger notifications, leading to real-time updates of the SP.

As a result, the ATLANTIS Digital Twin enables automated risk prioritization, continuous monitoring, and adaptive response—ensuring that operators maintain full awareness and control over complex CI ecosystems.

3.3 Situational Picture Data Model Overview

The SP is the central security state representation within the ATLANTIS framework, produced as the output of the SAFER pipeline. SAFER leverages an EDA to collect, standardize, and process heterogeneous, multi-domain data streams in real time. Through EDA, SAFER enables asynchronous, decoupled ingestion of cyber-physical events, which are then correlated and fused into the SP. This continuous event-driven flow ensures the SP remains a dynamic, machine-processable snapshot of the infrastructure's evolving risk profile, supporting timely decision-making and operational response.

The SP data model is designed to:

- Aggregate heterogeneous inputs from cyber-physical monitoring systems
- Preserve semantic consistency across layers
- Enable machine-processable correlation and inference.

Key structural elements include:

- Asset descriptors: unique identifiers, type, location, interdependencies
- Event descriptors: timestamps, source, type (cyber, physical, hybrid), severity
- Risk metrics: likelihood, impact, threat actor, propagation vectors.

Each SP instance represents a snapshot of the infrastructure's risk profile, continuously updated by the SAFER pipeline.

4 From Situational Awareness to Proactive Mitigation

Traditional incident management in Critical Infrastructure relies heavily on reactive strategies, often triggered after disruptions have occurred. While situational awareness tools have improved the ability to monitor and detect anomalies, they remain limited in anticipating threat evolution and enabling coordinated, preventive action.

This section presents the conceptual and technical transition from static awareness to proactive mitigation. It examines the evolution of incident management approaches, introduces the implementation of the distributed situational picture, and explores how simulation, semantic reasoning, and risk modelling enable anticipatory response. The goal is to support early intervention and minimize cascading effects across interconnected systems.

4.1 Evolution of Incident Management Approaches

Historically, incident management in CIs has been reactive, triggered by detected anomalies or ongoing events. ATLANTIS reshapes this model by embedding simulation, prediction, and learning within a hybrid digital environment, aligning with the EU's strategic vision for resilience through intelligence-driven mitigation. The framework enhances this evolution through:

- **Risk Anticipation**: Using synthetic data, complex event processing, and interdependency graph analysis to simulate and identify latent risks.
- **Continuous Learning**: Incorporating post-incident feedback loops to update models and refine mitigation strategies.
- **Digital Twin Enablement**: Providing a virtual replica of CI systems to test and validate responses in advance.

This shift supports a full lifecycle management—from pre-planning and real-time support to post-event analysis—positioning ATLANTIS beyond traditional reactive paradigms.

4.2 The Situational Picture Implementation

SP within ATLANTIS represents a dynamic, multi-source fusion platform that integrates alerts and data across different CI sectors. Its core function is to combine real-time cross-CI alerts and intelligence to form a coherent and actionable situational picture.

Key features include:

- **Cross-CI Alert Integration**: The SP ingests and correlates alerts from diverse CI domains (e.g., energy, transport, communications), synthesizing them into a unified view for operators. This integration allows for a holistic understanding of threats impacting multiple sectors, aiding in faster decision-making and response coordination.
- **Virtual Sensor and Synthetic Data Streams**: In addition to real-time alerts, synthetic data streams generated by virtual sensors simulate CI behaviour under normal and threat conditions. These synthetic inputs supplement real-world data, particularly where direct access is limited due to operational or security constraints.
- **Hybrid Digital Twin Environment**: The SP leverages a Hybrid Digital Twin—a virtual model combining both physical system data and simulated behaviours—to continuously monitor and simulate CI systems in real-time. This mapping provides contextual awareness, supporting proactive scenario testing and risk assessment.
- **Environmental and Contextual Data Integration**: External sources such as Earth observation and climate model data enhance the SP with insights into environmental risks (e.g., flooding, extreme weather), contributing to a comprehensive threat landscape.
- **STIX Standardization for Interoperability**: Alerts, intelligence, and simulation outputs are encoded in STIX format to facilitate seamless sharing and understanding across systems and stakeholders.

The implementation of the SP is driven by the SAFER component, which processes data from the Detection Layer (e.g., IoT sensors) and Risk Analysis Layer, delivered using IDMEFV2 messages—a standardized format commonly used in Critical Infrastructure for conveying security alerts and event data.[10] SAFER employs machine learning algorithms to tackle tasks including classification and regression, alongside correlation analysis, using large volumes of contextual data. These models incorporate differential privacy techniques to preserve data confidentiality while enabling reliable predictions and classifications.

The SP is continuously updated by correlating and aggregating raw or processed events. It is enriched with threat intelligence, disinformation mitigation, and infrastructure context models. The Decision Support System (DSS) further augments the SP with indicators such as risk severity, supporting prioritization during risk management.

The resulting situational understanding is delivered to the Risk Reduction Layer and shared via the CCI-SAAM platform—a cross-critical infrastructure sharing engine—using TAXII[11] and STIX,[12] which are standardized formats and transport protocols for cyber threat intelligence exchange.

4.2.1 Enabling Proactive Mitigation Through Simulation

The ATLANTIS framework enhances its proactive mitigation capabilities by incorporating advanced simulations that inform both operator training and real-time decision-making. This functionality is underpinned by:

- **Simulation-Based Training**: Virtual sensors and the Digital Twin enable realistic, scenario-based training for operators. These simulations replicate real-world and threat scenarios, helping operators to test responses and refine procedures in a safe environment.
- **Real-Time Decision Support**: The SP, combined with DSS, provides operators with actionable insights and predictions, improving situational awareness. Predictive models and cross-sector threat assessments support decision-making by offering recommendations on risk mitigation and response prioritization.
- **Proactive Risk Analysis**: ATLANTIS uses simulation to test various risk scenarios, helping identify weaknesses in infrastructure and response mechanisms. The results of these tests feed into the framework's risk analysis pipeline, informing future mitigation priorities and refining response strategies.

4.3 Knowledge Graph and Semantic Integration

The ATLANTIS framework leverages a Knowledge Graph[13] (KG) to improve DSS results and enable proactive risk analysis with high accuracy. The decision-making can be substantially improved using the KG that offers a structured semantic representation of the CI, where interdependencies and relationships of assets, systems, and process data are explicitly modelled. The integration of KG data in the entire decision-making process enhances situational awareness and targets a superior CI resilience when semantic technologies are applied to reason over the complex CI relationships and their dependencies, by formalizing knowledge, enabling inference and anomaly detection. The semantic technologies layer transforms data into actionable intelligence, creating a robust foundation for proactive mitigation and the creation of realistic training scenarios for complex operational scenarios, revealing hidden patterns, potential vulnerability chains and cascading effects with respect to cyber-physical attacks.

On the other hand, the KG is used within the DSS to:

- Understand and query data about the CIs: data has been integrated and harmonized to map CI entities and their relationship by a common schema, also known as ontology.
- Retrieve accurate insights by understanding context. The DSS uses the semantic web of connected data to understand the meaning and connections within the data.
- Reason by inferring new knowledge and relationships from existing data, leading to meaningful proactive risk assessment with actionable mitigation measures as a result of the AI-driven recommendations.

To support seamless decision-making with the support of KG knowledge, an exchange mechanism has been innovated by implementing LLM[14] models to provide the enhanced situational context necessary for the DSS. The model specialized for knowledge retrieval and optimized for performing in extracting the relevant information has been selected to be QWEN 2.5.[15] Despite its good performance, the Llama 3.1[16] model has been outstanding in generating explanatory content in natural language. Specifically for the multi-model visual interpretation and reasoning tasks, a LLava[17] model has been selected to operate along with more models needed for complex reasoning tasks and decision support, like the Llama 3.2[18] who all have been integrated as agents within a multi-agent framework that will be detailed in Sect. 1.

The integration of LLMs has ensured efficient access to structured knowledge that can be utilized immediately to generate relevant and accurate semantic descriptions of the situational picture, enhancing the information for threat assessment and mitigation action recommendations.

A bidirectional information exchange with the KG works seamlessly, LLMs being able not only to retrieve but also to add new relevant information, continuously updating and improving the situational awareness of the system.

The specialized models are able to use Natural Language Processing and Multi-Modal data processing to support the understanding of unstructured data, from various sources, including real-world and virtual sensors. Using this semantic representation of the cross-domain and cross-border CI environments, the framework achieves a unified understanding of the complex interdependencies and the cascading effects that may occur during cyber-physical attacks.

4.4 Transition from Reactive to Proactive Strategies

ATLANTIS shifts from reactive approaches to proactive strategies by using historical and streamed data coming from the real world together with data from virtual sensors to forecast potential threats and identify emerging risk vectors, enabling the human operators to apply mitigation actions before these threats can materialize into incidents. The human operators are provided with precise recommendations and transparent explanations by the Risk Reduction and Incident Mitigation (RRIM) component due to the advanced reasoning of the LLM agents and deployed predictive AI tools.

The recommendations generated by RRIM enhance awareness for the humans-in-the-loop who can better understand not only the recommended actions by also the context in which the recommendations can be applied as mitigations for emerging threats and therefore can become actionable measures to enhance the resilience and security of the CIs.

ATLANTIS introduces proactive mitigation of risk using tools and methodologies that implement interdependency mapping together with synthetic data attributed to virtual sensors used for the:

- Anticipation of potential attacks by simulating threat scenarios in an IF–THEN paradigm and concluding about the vulnerabilities of real assets to be used for predicting disruptions.
- Assessment of complex systems vulnerabilities for a comprehensive understanding of CI interdependencies and cascading effects to support informed decision-making.
- Support of informed mitigation actions by the operators who can prioritize mitigation strategies effectively by using the results of the recommendation pipeline, which implies predictive analytics, risk assessment, recovery decision support, and recommendation engines.

The data streams attributed to virtual sensors are a realistic mirror of the CI and can be used for:

- Robust verification and validation of mitigation strategies in a safe and controlled environment
- Training of human operators through realistic simulations
- Refining Emergency Response Plans with meaningful insights derived from a deep understanding of complex threat vectors in the context of the simulated scenarios.

Proactive strategies in comparison to reactive strategies can reduce downtime because of ensuring that early detection knowledge is accessible, allowing human operators to initiate efficient mitigation measures that prevent prolonged operational interruptions. The detail of information in the Cyber-Physical Security Space improves preparedness for the operators and enhances the resilience of the CIs, being compliant with regulatory requirements at the same time.

5 Decision Intelligence Framework

Proactive risk mitigation in complex CI environments requires more than data aggregation. It demands the ability to interpret evolving situations, anticipate threats, and support timely, informed decisions. This is where decision intelligence becomes a critical layer, bridging situational data with cognitive reasoning and predictive analytics.

This section introduces the decision intelligence framework developed within ATLANTIS. It presents the cognitive models used to represent and interpret CI scenarios, mechanisms for associating events with impacted assets, and techniques for assessing and forecasting risk. Furthermore, it details the decision support algorithms and visualization tools that enable operators to evaluate response options under uncertainty and time constraints.

5.1 Cognitive Models for Situational Understanding

The ATLANTIS framework makes use of artificial intelligence (AI) and machine learning cognitive models to efficiently process the heterogenous data describing the complex operational landscape within the CIs. By integrating LLMs, the situational picture of the current CI state can be enhanced, managing information from various sources (Fig. 3).

LLMs will process heterogenous data that:

- Describes the Cyber-physical Event (sensor data, alerts, detected anomalies, indicators of compromise, etc.)
- Defines the Critical Infrastructure Asset; the information is queried from the KG
- Updates the context with new information from public sources
- Describes the up-to-date Guidelines and Emergency Response Plans
- Describes the nearby geographical proximity and dependency information.

The tremendous advantage of LLM models is that mitigation recommendations are formulated in natural language and crafted on demand with transparent explanations it enhances explainability and trust towards stakeholders. That for utilizing Natural Language Processing (NLP) techniques is an advantage not only for extracting meaningful information from documents but also for enhancing the deep understanding of the generated content for the humans-in-the-loop. The advantage of multi-modal models, like particularly LLava, is that images and time-series can also be interpreted for a holistic understanding of diverse data, going beyond only text-communicated information.

The Risk Reduction and Incident Mitigation component implements, for this reason, the cognitive models to interpret the operation context when assessing the threats and generate recommendations considering heterogeneous data spaces for detecting interdependencies and potential cascading effects.

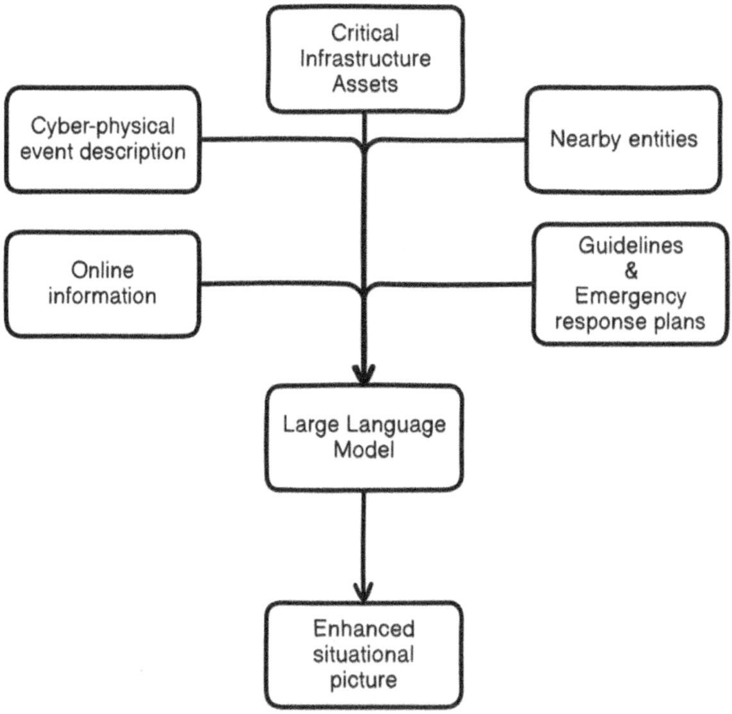

Fig. 3 Creating the enhanced situational picture using various data sources

Additionally, to support informed decision-making for the humans-in-the-loop by providing transparent explanations for the recommended actions in natural language, the ATLANTIS Framework also deploys Explainable AI (XAI) tools that enhance operator confidence towards the generative process of recommendation making.

5.2 *Event-Asset Association and Risk Assessment*

The faithful context assessment is key to provide comprehensive understanding of the events by accurately mapping the event to the affected asset, understanding the dependencies, and the different implications of the asset type. Implementing the correct contextual interpretation of criticality enables effective risk management with prioritized response strategies with respect to enhanced situational awareness (Fig. 4).

Associating events with CI assets in early stages of data retrieval, when cyber-physical sensor data is collected, is important for the generation of threat reports with:

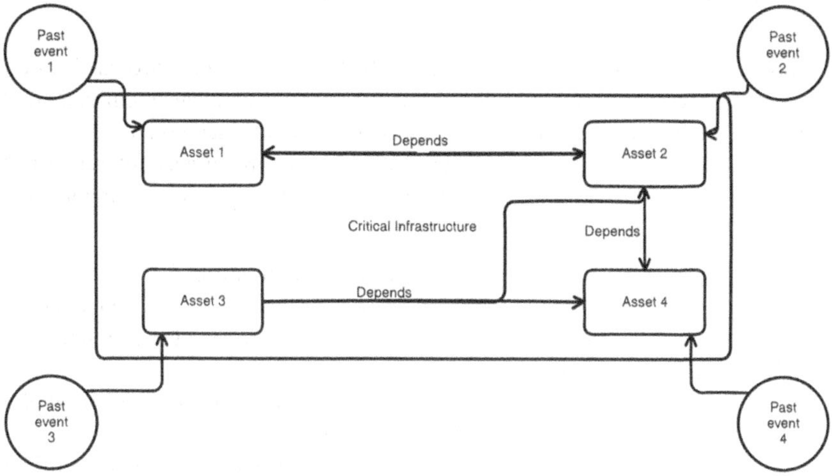

Fig. 4 Past events associated with assets within a critical infrastructure

- **Knowledge Graph Utilization** when the Knowledge Graph serves as a central semantic repository to retrieve the information of relationships between assets, associated events, and their interdependencies.
- **LLM Support** because LLMs such as Llama 3.2 and Qwen 2.5 interact with the KG to query information that enables the system to process context-rich analysis for generating faithful reports about the complex CI structures and assess the cascading effects of events.

The LLM support for the **Risk Assessment Process** is important for:

- **Threat Analysis** when continuously evaluating the nature and severity of the event with respect to the interconnected assets and known patterns.
- **Vulnerability Identification** in assets that could offer attack surfaces or amplify the impact of cyber-physical attacks on specific CIs.
- **Impact Evaluation** on the intended operational environment with respect to functionality, safety, and compliance requirements. To efficiently predict the severity of such an impact, the use of Fine-tuning LLMs with data labelled with respect to protection goals is recommended.
- **Mitigation Strategy Development** that reflects policy intent by using risk assessment, policies and strategies to maintain currency, verify and validate the generated recommendations.

The advantages of the seamless integration are:

- **The results are Actionable Recommendations**: Providing operators with faster, more informed and more reliable prioritized mitigation actions tailored to the specific CI.

- **Transparent Explanations of the results**: when utilizing LLMs to also analyse and generate understandable, transparent reasoning for the resulting recommendation and apply XAI in the process to maintain trust and human acceptance during decision-making processes.

5.3 Predictive Analytics for Threat Detection

Predictive Analytics tools to analyse threats and forecast potential incidents have been developed and implemented to be used in a collaborative architecture with the named LLMs. The predictive analytics tools based on machine learning techniques, which include decision trees and neural networks, are used by the LLMs to get forecasts of incidents and use them to analyse the situational picture. The interplay between the LLMs and these predictive analysis tools is orchestrated in an agentic architecture that will be presented later in the chapter.

By nature, a cyber-physical attack is a very rare event that is hard to predict because of its infrequent appearance. To overcome this, synthetic data and simulations have been employed to create capabilities of evaluation and forecast based on virtual recreation of threat scenarios. Using virtual sensors or synthetic data, we addressed to overcome telemetry gaps and support "what-if" analysis for threat forecasting.

The predictive analytics for threat detection is part of the proactive risk mitigation paradigm, improving response times due to threat prediction, detection and continuous threat monitoring that should be part of any optimized and continuously improving defence strategy.

5.4 Decision Support Algorithms and Visualization

The complex cyber-physical CIs need to be protected by the actions of human operators who need to take decisions under time pressure, incomplete information and the uncertainty of a continuously changing security landscape. To assist human operators in decision-making, ATLANTIS has integrated advanced decision support algorithms and visualization tools.

5.4.1 Advanced Decision Support Algorithms

- **Multi-Criteria Decision Analysis** (MCDA[19]): Are used by the DSS to evaluate multiple competing factors and quantify response options with respect to prioritizing them. This methodology supports making balanced decisions that consider trade-offs between options to gain operational effectiveness, allocate resources and target.

- **Retrieval-Augmented Generation** (RAG[20]): The framework implements a retrieval-generation pattern using vector store and KG alongside LLMs to ensure that generated outputs are aligned with the knowledge for ATLANTIS, the regulatory considerations and expert knowledge. This ensures the quality and reliability of recommendations by synthesizing information from multiple knowledge sources.
- **Explainable AI** (XAI[21]): By integrating XAI techniques, the recommendations generated are explained, showing how the attention mechanism contributed to recommendation generation. Knowing the focus of LLMs human operators can oversee the complex analytical process, and that for build trust, and accept suggestions.

The ATLANTIS Framework enhances human operators' support through advanced visualization tools, integrating multiple interactive dashboards—such as the one shown in Fig. 5—that display real-time information on key events projected on assets and their interdependencies, also the ongoing incidents, communication and operations of first responders and stakeholders. The visualization combines text, images, and time-series data to provide a comprehensive and contextualized understanding of the operational environment.

The advantage of enabling operators with visualization of the semantic structure known from the KG is that it not only delivers context-rich insights, but also that human operators can interactively explore relationships among assets, events and vulnerabilities, looking into an intuitive graph visualization.

Interfacing the human with decision support algorithms to apply visual analytics is the basis for:

- Improved decision-making by intuitively analyzing and understanding complex scenarios and selecting optimal responses.
- Rapid comprehension of the situation, which accelerates response times and reduces errors.
- Operators prioritize critical tasks without experiencing information overload.

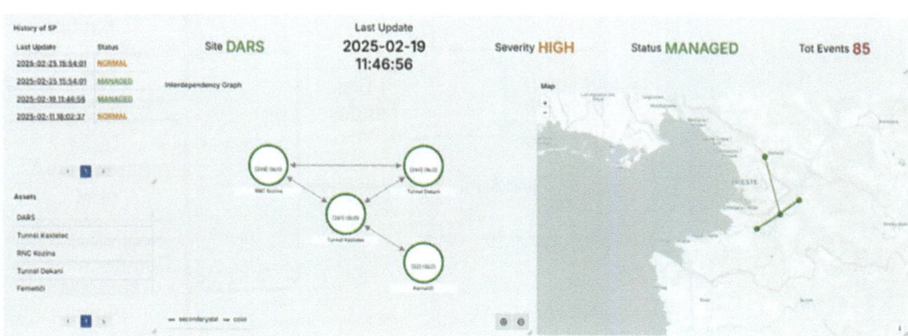

Fig. 5 SAFER situational picture dashboard

6 Agentic LLM Systems for Incident Mitigation

As Critical Infrastructure scenarios grow in complexity, traditional automation struggles to adapt to dynamic threat conditions and heterogeneous information sources. To address this, ATLANTIS integrates agentic systems based on Large Language Models (LLMs) to support flexible, context-aware decision-making.

This section introduces the architecture and implementation of these agentic systems. It describes the multi-model agent design, inter-component communication strategies, and the orchestration mechanisms that govern coordination and task delegation. Emphasis is placed on how these agents enhance mitigation workflows through reasoning, adaptability, and real-time interaction with other framework components.

6.1 Agentic Architecture for Flexible Communication

The ATLANTIS framework implements agentic architecture, an orchestrated approach where autonomous agents communicate flexibly and operate coordinated to protect the Critical infrastructure. The interaction of AI models, databases, and operator interfaces enhances threat detection, analysis, and response capabilities.

6.1.1 Multi-AI-Model Agentic Architecture

Figure 6 represents a multi-agent orchestration system designed for analyzing the situational picture in a given context to generate countermeasures.

The architecture implements the five primary components:

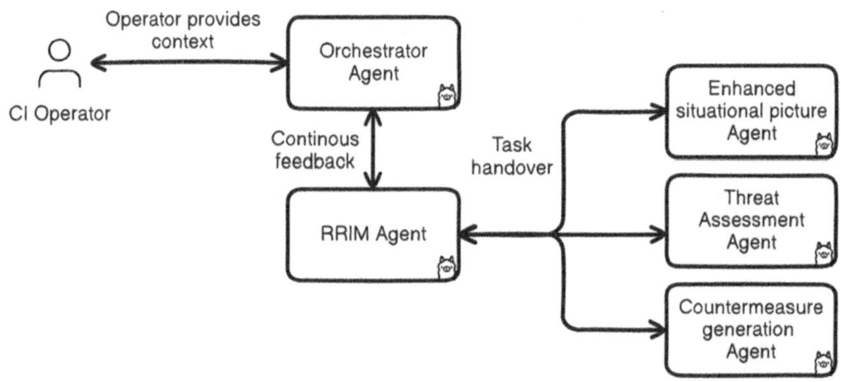

Fig. 6 Multi-agent event analyzer and countermeasure generation using feedback loops

- **Interconnected through Bidirectional Information Flow**: the information flow is management among the tasks executed by the specialized agents.
- **Coordinated to Inter-Agent Collaboration**: The Orchestrator Agent manages the central coordination and breaks down complex tasks into overseeable topics. It makes use of the specialized agents for a specific task:
  - Risk Reduction and Incident Mitigation (RRIM) Agent;
  - Enhanced Situational Picture Agent;
  - Threat Assessment Agent;
  - Countermeasure generation agent.

  The benefits of this architecture are:

- Centralized control that ensures a logical progress of the workflow.
- Modularity by breaking down complex risk and threat analysis tasks into manageable topics.
- Parallel task execution to speed up overall processing time.
- Scalability when deploying multiple instances of specific agents to handle increased workloads
- Complexity management: Encapsulates specific functionalities within agents, making the system easier to manage and understand.
- Encapsulation: encapsulates the logic required to interact with the knowledgebase.

The Orchestrator Agent assesses the outputs and behaviour of the RRIM agent, checking performance such as factual accuracy, adherence to instructions and overall quality. This agent is used to automatically ensure the quality of the recommendations, detect risks with respect to disruptions to RRIM confidentiality, integrity and availability.

In a wider sense, it supervises the RRIM agent to act as the master agent, interacting with the following components:

- The *Threat Assessment Agent* acts as a specialized tool that the central RRIM Agent can call upon to understand the nature of a security event, as it has been implemented to identify and classify security attacks according to the standardized MITRE ATT&CK[22] Framework. It integrates the Natural Language Processing Model to process incoming event information. It transforms textural event data into numerical representations named embeddings, which are fed into a RandomForestClassifier machine learning model. Based on the patterns learned from the embeddings the RandomForestClassifier makes the final classification of the attack type.
- The *Enhanced Situational Picture Agent* acts as a tool for the RRIM Agent to access and retrieve structured information from STIX format. The agent interacts with the KG using generated queries used to retrieve specific information about Critical Infrastructure Assets, and the countermeasures and attack types known from the past. With this information, it enhances the Situational Picture available to the RRIM agent.

- The *Countermeasure Generation Agent* serves as a tool for the RRIM agent to generate specific, actionable countermeasure recommendations in response to an analyzed security event. It uses the classified attack type, contextual information, historical information and policies to formulate recommendations. The historical information describes past events, established procedures of the Emergency Response Plans, conformity rules of Policy and Regulations Collection and other context information collected in the past with respect to Cyber-Physical Human protection. The data has persisted in a vector store that has been optimized for searching. The output is structured in a specific JSON format. The generated JSON recommendations are ready to be sent to a Kafka Broker to be distributed.
- The *Orchestrator Agent* is specialized also in overseeing and improving the quality of the countermeasure generation by the Recommender Engine Agent. It acts on the meta-level, assessing the performance and output of the analysis and recommendation flow. It can reconfigure parameters as for example, the time window of the analysis when it's necessary to gather more attack-related information, because initial data can sometimes be insufficient.

6.1.2 Inter-component Integration

Successful integration tests have been verifying the reliability of the interactions between components of SAFER, DSS and RRIM following the diagram in Fig. 7. The observed advantages of this implementation have been the following:

- **Seamless Operation**: The interconnected agents may communicate directly having a high flexibility in exchanging messages between LLMs who are more independent and makes the process open-ended. This flexibility allows for dynamic problem-solving, which is advantageous at analysis in the cold-start phase of the project. The interconnected agents must quickly respond to novel and unexpected situations, adaptability being a key advantage in critical CPH protection. Threats are constantly evolving and LLM agents must be able to dynamically

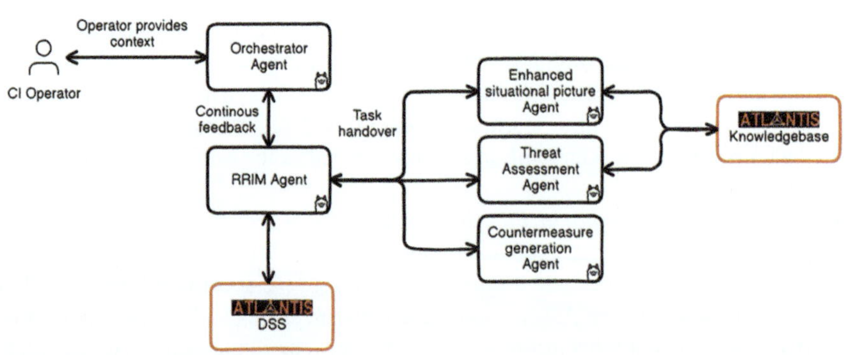

Fig. 7 Inter-component integration diagram

formulate response strategies that can be a significant advantage in comparison to rigid, rule-based systems.

- **Data Integrity**: To ensure integrity and availability, it's important to maintain data consistency across components and ensure accurate and timely information flow. The ability of LLMs to contend and communicate complex situations in an acceptable time frame has been carefully and intensively evaluated during testing and benchmarking. The mixed results have highlighted that highlight a careful selection and integration of the LLMs with respect to the task-specifics is needed and can be done by practicing benchmark-informed model selection.

The key benefits of the proposed agentic architecture are:

- Adaptability because this modular design allows to update models and agents with the newest developments as technology evolves.
- Scalability of the system means that increased data flow and functionalities can be accommodated without significant refactoring.
- Enhanced Collaboration of the agents will result in consistent responses.

6.2 Design Principles and Implementation Advantages

The ATLANTIS framework design principles implement adaptability, security, and interoperability to achieve effectiveness and resilience.

Adaptability has been achieved through a modular design that allows components to be interchanged or updated, and through API-driven interfaces that enable flexible communication between components or tools.

The most significant innovation is the introduction of a Tool Inventory powered by LLMs, as a central registry of tools and plug-ins that can be dynamically utilized to execute discrete tasks. Within the Risk Reduction and Incident Mitigation (RRIM), the agentic flow functions sequentially, triggered by the incoming messages. The RRIM will accumulate the data attributed to a time window to be processed for enhancing the Situational Picture, execute the Threat Assessment and generate countermeasures recommendations. The sliding time window will define how long RRIM waits for additional event data to reiterate the process. The results are published on the communication bus with the other RRIM components connected to the Kafka Broker.

Challenges towards full decision-making autonomy of the LLMs can manifest in uncontrolled processing loops that have been addressed using the tool inventory and process-driven methodologies, as shown in Fig. 8. The RRIM Agent orchestrates a defined operational sequence using the tool inventory to execute the specialized tasks. The tools included in the inventory can be extended in a flexible way, adding features additional to the already existing attack classification, query of knowledge graph, query of Emergency Response Plans and Countermeasure Generation tools. For example, the inventory tool has been used to seamlessly integrate semantic and

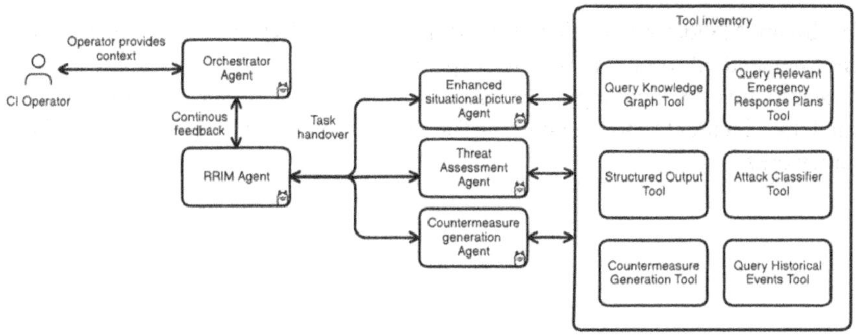

Fig. 8 Tool inventory diagram showing the tools available for the AI agents

advanced analytics tools, visualization tools, and data adaptation and standardization tools.

6.3 Validation Mechanisms and the Orchestrator Agent Role

The ATLANTIS framework integrates a robust validation mechanism to ensure the reliability, security, and effectiveness of its components, such as DSS and RRIM. These validation mechanisms include:

- Functional testing and integration validation, to assess architectural decisions and the component interoperability.
- STIX payload validity to address threat data validation.
- Empirical performance evaluation, through assessments and benchmarks that confirm which LLM components meet ATLANTIS requirements.

The life cycle of the system, operations are monitored by the Orchestrator Agent, who continuously oversees the agent performance and inter-agent interaction, validating the outputs and persisting the monitoring results into traceability information.

The Orchestrator Agent enforces predefined rules like temporal rules, corroboration rules, data quality controls, relevance criteria and a set of RAGAS[23]-based criteria to assess the quality of countermeasures, such as relevance, completeness, factual consistency, and contextual alignment. The central coordination is done by the Orchestrator Agent, who validates the outputs and manages the submission of results. A workflow sequence is exemplified in Fig. 9, where the Orchestrator Agent:

- Initiates communication with the RRIM Agent in the input phase
- Completes the first validation cycle when RRIM Agent submitted generated countermeasures

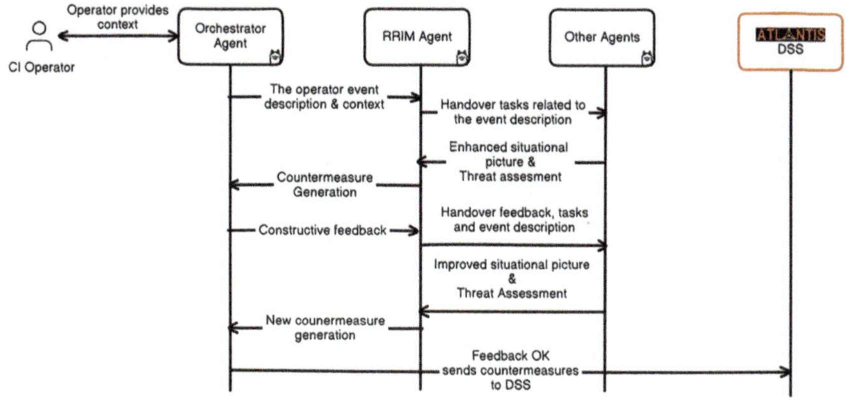

Fig. 9 Orchestrator agent feedback loop diagram

- Continues the feedback loop by providing constructive feedback and triggering result refinements
- New countermeasures are finally validated by the Orchestrator Agent who will give "Feedback OK" to validate countermeasures that will be used by the DSS.

The benefits of these mechanisms are trustworthiness, traceability and output guardrails, as validation processes and oversight improve all these aspects with respect to recommendations.

7 Conclusion and Future Directions

This chapter concludes the description of the ATLANTIS framework by emphasizing its contributions to CI resilience and identifying directions for future development. ATLANTIS adopts a decision-intelligence-centric approach to situational awareness and incident mitigation, offering key capabilities such as an integrated situational picture that fuses real-time alerts, synthetic data, and Digital Twin simulations; semantic interoperability through standardized formats like STIX/TAXII and ontological models; an AI-powered DSS for risk assessment and mitigation prioritization; and agentic AI systems leveraging LLM-based agents for coordinated planning and execution of countermeasures. Together, these components enhance preparedness, responsiveness, and coordination in complex cyber-physical environments.

From a policy perspective, deploying frameworks like ATLANTIS necessitates regulatory alignment, particularly in areas such as cross-sector data sharing agreements, privacy-preserving AI compliant with GDPR[24] and national laws, and standardization to enable interoperable data exchange. ATLANTIS supports policy goals related to digital sovereignty, coordinated risk response, and CI resilience in accordance with EU and international directives.

Future research and development will focus on improving scalability and self-adaptation in edge and cloud environments, integrating trust metrics and AI explainability, advancing predictive capabilities through hybrid digital twins and cognitive models, and conducting operational validation in real-world, multi-stakeholder scenarios. Further efforts will aim to expand coverage to additional CI domains and develop certification-ready deployments aligned with EU and national resilience frameworks.

Definitions and Acronyms

AI	Artificial Intelligence
API	Application Programming Interface
CCTV	Closed-Circuit Television
CI	Critical Infrastructure
CEP	Complex Event Processing
CISA	Cybersecurity and Infrastructure Security Agency
CSIM	Converged Security and Information Management
DSS	Decision Support Systems
DT	Digital Twin
EDA	Event-Driven Architecture
EDR	Endpoint Detection and Response
GDPR	General Data Protection Regulation
GUI	Graphical User Interface
ICS	Industrial Control Systems
IDMEFv2	Intrusion Detection Message Exchange Format version 2
IDS	Intrusion Detection Systems
IoT	Internet of Things
IT	Information Technology
JSON	JavaScript Object Notation
KG	Knowledge Graph
LLaVA	Large Language and Vision Assistant
LLM	Large Language Model
MCDA	Multi-Criteria Decision Analysis
NLP	Natural Language Processing
OT	Operational Technology
RAG	Retrieval-Augmented Generation
RRIM	Risk Reduction and Incident Mitigation
SCADA	Supervisory Control and Data Acquisition.
SIEM	Security Information and Event Management
SP	Situational Picture
STIX	Structured Threat Information eXpression
TAXII	Trusted Automated eXchange of Intelligence Information
TLS	Transport Layer Security

XAI eXplainable AI

Notes

1. https://www.atlantis-horizon.eu/.
2. Zhao et al. (2022)—"CSIM: Converged Security and Information Management System as a Tool for Smart City Infrastructure Resilience Assessment,", Smart Cities, MDPI.
3. De Pascale et al. (2023)—"Internet-of-Things Architectures for Secure Cyber-Physical Spaces: the VISOR Experience Report", Journal of Software: Evolution and Proces, July 2023.
4. CISA (2021). Cybersecurity and Physical Security Convergence Action Guide. U.S. Department of Homeland Security.
5. Gary Barlet. (2024). "How to harness AI and Zero Trust segmentation to boost cyber defenses." *C4ISRNET*.
6. https://link.springer.com/rwe/10.1007/978-0-387-39940-9_570.
7. https://www.oreilly.com/library/view/fundamentals-of-software/978149204 3447/.
8. https://kafka.apache.org/.
9. https://www.rfc-editor.org/rfc/rfc4765.html.
10. https://www.rfc-editor.org/rfc/rfc4765.html.
11. https://oasis-open.github.io/cti-documentation/taxii/intro.html.
12. https://oasis-open.github.io/cti-documentation/stix/intro.html.
13. https://neo4j.com/product/neo4j-graph-database/.
14. Brown, T. B., Mann, B., Ryder, N., et al. (2020). Language Models are Few-Shot Learners. arXiv preprint arXiv:2005.14165.
15. https://huggingface.co/Qwen/Qwen2.5-1.5B-Instruct.
16. https://huggingface.co/meta-llama/Llama-3.1-8B.
17. https://llava-vl.github.io/.
18. https://huggingface.co/meta-llama/Llama-3.2-1B.
19. Taherdoost, H., & Madanchian, M. (2023). Multi-Criteria Decision Making (MCDM) Methods and Concepts. *Encyclopedia, 3*(1), 77–87. https://doi.org/10.3390/encyclopedia3010006.
20. Lewis, P., Perez, E., Piktus, A., Petroni, F., Karpukhin, V., Goyal, N., … Joulin, A. (2020). Retrieval-augmented generation for knowledge-intensive NLP tasks. In H. Larochelle, M. Ranzato, R. Hadsell, M. Balcan, & H. Lin (Eds.), Advances in Neural Information Processing Systems 33 (pp. 9459–9474). Curran Associates, Inc.
21. Arrieta, A. B., Díaz-Rodríguez, N., Del Ser, J., Bennetot, A., Tabik, S., Barbado, A., García, S., Gil-López, S., Molina, D., Benjamins, R., Chatila, R., & Herrera, F. (2019). Explainable artificial intelligence (XAI): Concepts, taxonomies, opportunities and challenges toward responsible AI. arXiv. https://arxiv.org/abs/1910.10045.
22. https://www.ibm.com/think/topics/mitre-attack.

23. https://github.com/explodinggradients/ragas.
24. Voigt, P., & von dem Bussche, A. (2017). The EU General Data Protection Regulation (GDPR): A practical guide. Springer International Publishing. https://doi.org/10.1007/978-3-319-57959-7.

Acknowledgements This project has received funding from the European Union's Horizon Europe research and innovation programme under grant agreement No. 101073909. The authors would like to thank the participating CIs partners enabling the implementation and testing of the ATLANTIS methods and tools. This article reflects only the authors' views, and the Research Executive Agency and the European Commission are not responsible for any use that may be made of the information it contains.

Upskilling and Reskilling Critical Infrastructure Protection Professionals: Learning Paths for 20 Important Skills Profiles

John Soldatos, Marianna Charalambous, and Rauno Pirinen

Abstract This chapter presents structured learning paths tailored to twenty critical skills profiles in the domain of Critical Infrastructure Protection and Resilience (CIP/CIR). It aims to support organizations in addressing the growing cybersecurity and resilience skills gap by outlining curated educational pathways aligned with real-world capability needs. The specification of the presented profiles has been driven by an extensive analysis and stakeholder input from the Horizon Europe EU-CIP project. Each profile integrates technical, non-technical, and soft skills, which are essential for CIP/CIR professionals. Overall, the chapter serves as a practical guide for human resources departments, training providers, and security organizations that wish to implement effective upskilling and reskilling strategies for CIP/CIR professionals.

Keywords Critical Infrastructure · Skills gap · Cybersecurity training · Learning pathways · Workforce resilience · Reskilling · Upskilling · CIP · CIR · EU-CIP

1 Introduction

The European Commission (EC) emphasizes the growing need for cybersecurity professionals in Europe. Specifically, an analysis published in 2023 highlighted a shortage of skilled workers, which is estimated at one million in Europe and 3.4 million globally [1]. In 2022, the shortage of cybersecurity professionals in the EU ranged between 260,000 and 500,000. The EU's cybersecurity workforce needs were estimated at 883,000 professionals [2].

J. Soldatos (✉) · M. Charalambous
INNOV-ACTS Limited, Nicosia, Cyprus
e-mail: jsoldat@innov-acts.com

M. Charalambous
e-mail: mariannacharal@innov-acts.com

R. Pirinen
Laurea University of Applied Sciences, Vantaa, Finland
e-mail: Rauno.Pirinen@laurea.fi

J. Soldatos and H. Abie (eds.), *Technology-Enabled Resilience: Innovations in Critical Infrastructure Protection*, https://doi.org/10.1007/978-3-032-25917-2_15

According to the analysis in [1], the demand for cybersecurity skills is rising, especially following the COVID-19 pandemic outbreak which led to a proliferation of digital infrastructures and digital activities within modern industrial organizations. Moreover, the same analysis report highlights a significant gender disparity, with women comprising less than 25% of the cybersecurity workforce. Furthermore, it underscores the importance of educational initiatives and frameworks to bridge the skills gap, while at the same time enhancing cybersecurity training and workforce diversity. This requires the establishment of novel training programs and inclusive policies, which will equip more individuals with the required cybersecurity competencies. This approach is destined to fill the proclaimed deficit of skilled professionals, but also to foster a more diverse workforce. The EC is already implementing measures in this direction. For instance, it has established a Cyber Skills Academy that coordinates and promotes upskilling and reskilling initiatives for cybersecurity professionals, with emphasis on professionals working in critical infrastructure. The academy responds to the increasing demand for qualified staff due to new EU cybersecurity regulations and the rise in cyberattacks [2].

The World Economic Forum has also identified upskilling and reskilling as critical drivers for the future of jobs, emphasizing that the 2020s will be pivotal for workforce development. Employers are encouraged to invest in these areas as a significant number of workers' skills will be disrupted within the next few years. This focus is not only essential for individual career resilience but also for maintaining competitive economies in a technologically advancing world [3].

The rapid evolution of threats to critical infrastructure asks for continuous upskilling and reskilling of professionals in the field of Critical Infrastructure Protection and Resilience (CIP/CIR). In today's fast-changing landscape of digital technologies, the complexity and frequency of these threats are on the rise. This makes it imperative for CIP/CIR professionals to remain up to date with respect to the latest advancements and methodologies.

Other recent market analysis reports also underscore this urgency, revealing that nearly 60% of organizations recognize the skills gap in their cybersecurity teams as a significant risk factor [4]. This is particularly concerning given that the estimated cost of cybercrime is projected to increase significantly during 2025 and in the next years. This emphasizes the importance of investing in cybersecurity education and training. Moreover, the challenge of retaining skilled cybersecurity professionals continues to appear intensified. About 60% of organizations report difficulty in retaining qualified cybersecurity professionals, which underlines the need for robust upskilling and reskilling initiatives to maintain a proper cybersecurity workforce [5].

In this landscape, the introduction of new technologies such as artificial intelligence (AI), machine learning (ML), quantum computing (QC), and the Internet of Things (IoT) is continually broadening the scope of potential vulnerabilities. These developments require CIP/CIR professionals to master new tools and technologies based on a proactive approach to threat detection and response. Consequently, upskilling and reskilling becomes a strategic imperative for organizations aiming to safeguard their critical infrastructure.

This goal of the present chapter is to present a series of tailored learning paths that can guide learners toward acquiring specific CIP/CIR skill profiles, while facilitating training organizations and human resources departments of critical infrastructure operators to structure relevant training and career development activities. The presented learning paths are designed to bridge the current skills gap and to foster innovation in CIP/CIR practices. Specifically, the chapter highlights skills profiles and learning paths that support novel security roles in the contemporary CIP/CIR landscape. Moreover, it highlights the need for a holistic educational approach that integrates technical, non-technical, and soft skills toward growing a cohort of skilled professionals that can cope with modern CIP/CIR challenges.

2 Skills Profiles and Learning Paths Construction Methodology

2.1 Skills Profiles Identification

Our approach to developing learning paths for CIP/CIR professionals was driven by an analysis of capability gaps for critical infrastructure (CI) operators and entities, which has been carried out in the scope of the EU-CIP project. This analysis revealed emerging areas and themes that are important for protecting assets within CI entities, while at the same time identifying the importance of training CIP/CIR professionals in understanding and fully leveraging capabilities in these areas. Hence, the first step of our methodology involved analyzing these capability gaps toward identifying the professionals' capabilities and skills profiles required to fill them. Each of the identified skills profiles highlights the skillsets that are required to cope with the identified gaps such as gaps in the areas of cyber-physical threat intelligence and emergency response. Given that the capability gaps have been identified by CI practitioners and experts, the presented approach provided a sound basis for aligning educational and training plans with industry requirements. This alignment was further audited by soliciting feedback from relevant stakeholders (e.g., CI experts, technology experts, human resources professionals, cybersecurity education experts) following an earlier publication on skills profiles [6]. Specifically, we solicited feedback from 15 CI experts about the relevance of the identified skills profiles and their alignment to CIP/CIR stakeholders' training needs. Following this feedback, we fine-tuned some of the initial profiles and included a list of five additional skills profiles that were deemed identified as important by the experts. The outcome of this process was an expansion of the initial list of skills profiles [6] by five additional profiles, i.e., to a total of twenty distinct skills profiles for CIP/CIR professionals. These profiles are listed in Table 1.

Table 1 Important skills profiles for CIP/CIR

Cybersecurity Specialist	The learning path for a cybersecurity specialist starts with basic cybersecurity courses and progresses through specialized training in encryption methods, vulnerability assessments, and incident response
Risk management analyst	A risk management analyst's path focuses on courses in risk assessment methodologies, developing risk management strategies, and enhancing communication skills to effectively convey risk information
Emergency response coordinator	The learning path for an emergency response coordinator includes training in emergency management protocols, effective communication during crises, and familiarity with incident command systems
Physical security specialist	A physical security specialist's path covers courses in designing and implementing physical security measures, understanding access control and surveillance technologies, and developing security policies and procedures
Network infrastructure engineer	The learning path for a network infrastructure engineer entails courses in network design and maintenance, network protocols and technologies, and network security principles
Critical infrastructure resilience planner	A critical infrastructure resilience planner's path focuses on developing resilience plans, understanding resilience frameworks, conducting risk assessments, and applying strong analytical skills
Incident response manager	The learning path for an incident response manager includes managing incident response teams, developing incident response plans, forensic analysis, and coordination with law enforcement
Compliance auditor	A compliance auditor's path would focus on courses in auditing, understanding regulatory frameworks, compliance assessments, and developing strong analytical skills
Industrial control systems (ICS) engineer	The learning path for an ICS engineer covers designing and maintaining industrial control systems, understanding SCADA systems, and assessing cyber threats targeting ICS
Crisis communication specialist	A crisis communication specialist's path involves developing crisis communication plans, managing communication channels during emergencies, and crafting clear messages for various stakeholders
Business continuity planner	The learning path for a business continuity planner includes developing business continuity plans, understanding continuity frameworks, identifying critical business functions, and conducting continuity exercises.
Data protection officer (DPO)	A data protection officer's path would focus on ensuring compliance with data privacy regulations, understanding data protection laws, assessing data privacy risks, and managing data breach incidents.
Physical infrastructure engineer	The learning path for a physical infrastructure engineer entails designing and maintaining critical physical infrastructure, understanding building codes and standards, and assessing structural vulnerabilities

(continued)

Table 1 (continued)

Supply chain security manager	A supply chain security manager's path includes securing supply chains, understanding supply chain risk management principles, assessing supplier security practices, and developing contingency plans
Cyber threat intelligence analyst	The learning path for a cyber threat intelligence analyst starts with courses in cyber threat intelligence collection and analysis, understanding threat actors and their methods, and proactive threat mitigation strategies
Climate resilience adaptation specialist	Assesses climate-related risks to infrastructure (e.g., floods, wildfires), integrates adaptation strategies into design, applies sustainable and green infrastructure approaches, and uses tools such as GIS-based risk mapping
Digital twin systems engineer	Designs and manages digital twins for physical infrastructure, integrates IoT sensors, models cyber-physical systems, simulates threat/resilience scenarios, and applies AI for predictive maintenance
Cross-sector interdependency analyst	Maps interdependencies across sectors (e.g., energy–water–food nexus), models cascading failures, uses network analysis tools, and supports crisis coordination across infrastructure systems
Behavioral security culture architect	Develops security awareness and training programs rooted in behavioral science, applies gamification, measures security culture, and promotes human-centric approaches to risk mitigation
Quantum-resilient cryptography specialist	Specializes in post-quantum cryptographic techniques, evaluates systems for quantum vulnerabilities, understands quantum key distribution, and manages cryptographic migration strategies

2.2 *Linking of Skills Profiles with Specific Skill Sets*

The second step of our methodology was focused on breaking down the identified skills profiles into individual skills. The latter profiles were accordingly used to create clear and structured pathways for career advancement and professional development toward these skills. The analysis covered the needs of various stakeholder groups such as first responders, governments, analysts, and engineers. These insights indicated a clear path forward for stakeholders involved in CIP/CIR.

2.3 *Construction of Learning Paths*

The individual skills assigned to each profile led to construction of specific learning paths for each profile. To make each learning path more practical, we also mapped directly each path to one or more courses of the EU-CIP training catalogue, which is available within EU-CIP's knowledge hub.[1] At the time of writing of this version

[1] https://knowledgehub.eucip.eu/training-courses/

of the chapter, the EU-CIP training catalogue comprises over 160 CIP/CIR related courses, including courses developed by the EU-CIP project and courses produced by third-party providers. This made it generally possible to identify relevant courses that could match the skills of the various paths. However, it is also possible to accomplish the indicated learning paths based on other relevant courses, i.e., courses outside the EU-CIP training catalogue. Overall, the presented learning paths can be seen as a guide for CI operators and security organizations to structure and deliver upskilling and reskilling activities for their employees. The delivery of these activities may or may not benefit from the EU-CIP courses and the EU-CIP training catalogue.

3 Learning Paths Description

The tables in this section provide detailed learning paths for the above-listed twenty CIP/CIR skills profiles. Each learning path description is structured in three sections:

- **Individual Skills of the Profile:** This section lists the essential skills required for professionals to effectively fulfill roles associated with each CIP/CIR skills profile. The skills listed are indicative and can be expanded based on emerging trends and the evolution of the professional roles that are linked to the profile.
- **Mandatory Courses of the Learning Path:** This part lists a curated list of courses that are essential for acquiring the skills specified in the profile. The courses are sourced from the EU-CIP Training Courses Catalogue and provide foundational pathway for skill acquisition. Given the dynamic nature of CIP/CIR, other similar courses from platforms like Udemy, Coursera, and edX may also be suitable to support the acquisition of the key skills of the learning path.
- **Other Optional Courses:** In addition to the mandatory courses, this section suggests courses that could enhance the learning path. These courses are optional but recommended as they provide deeper insights or broader knowledge that complements the core mandatory skills (Tables 2, 3, 4, 5, 6, 7, 8, 9, 10, 11, 12, 13, 14, 15, 16, 17, 18, 19, 20 and 21).

4 Outlook: How to Use the Skills Profiles and how to Engage

Security organizations, training organizations, CI operators, and other CIP/CIR stakeholders can use the above-listed learning paths to plan training, upskilling, and reskilling activities. Moreover, human resources departments can consider the presented learning paths in the specification of career development roadmaps. Apart from fostering the development of special skillsets for critical infrastructure protection, most of the listed learning paths also serve the following objectives:

Table 2 Skills and learning path for the "cybersecurity specialist" skills profile

Cybersecurity specialist
Individual skills of the profile: • Proficiency in identifying and mitigating cyber threats • Knowledge of encryption methods and security protocols • Ability to conduct vulnerability assessments and penetration testing • Familiarity with incident response procedures and forensic analysis
Learning path (mandatory courses): The Foundations of Cybersecurity, Coursera, University of Maryland, College Park Introduction to Cybersecurity & Risk Management Specialization, Coursera/UCI Critical Infrastructure Protection, TEEX—Texas A&M University Build Security Incident Response for GDPR data protection, Udemy Cybersecurity Compliance Framework & System Administration, Coursera/IBM Cybersecurity Risk and Strategy, edX/University of Connecticut Cybersecurity and Privacy in the IoT, edX/Curtin University
Learning path (optional courses): Cybersecurity Threat Hunting for SOC Analyst, Udemy Understanding Indicators of Compromise, cybersecurity and infrastructure security agency (CISA) Science Communication: Communicating Trustworthy Information in the Digital World, Coursera/Erasmus University Rotterdam AWS Certified Security Specialty 2023, Udemy

Table 3 Skills and learning path for the "risk management analyst" skills profile

Risk management analyst
Individual skills of the profile: • Expertise in assessing risks to critical infrastructure • Ability to develop risk management strategies and plans • Proficiency in analyzing data to identify potential vulnerabilities • Strong communication skills to convey risk information effectively
Learning path (mandatory courses): Introduction to Risk Management, edX/ColumbiX ISO 31000—Enterprise Risk Management for the Professional, Udemy Risk Management Tools and Practices, edX/NYIF Cybersecurity Foundations for Risk Management, Coursera/Georgia University Detect and Mitigate Ethical Risks, Coursera/CertNexus Cybersecurity Compliance Framework & System Administration, Coursera/IBM Introduction to Business Continuity E-Learning Course, the business continuity institute (BCI)
Learning path (optional courses): Data: Law, Policy and Regulation, edX/London School of Economics Dominant Risk Management Standards and Frameworks, Coursera/university system of Georgia Understanding Indicators of Compromise, cybersecurity and infrastructure security agency (CISA)

Table 4 Skills and learning path for the "emergency response coordinator" skills profile

Emergency response coordinator
Individual skills of the profile: • Experience in coordinating response efforts during emergencies • Knowledge of emergency management protocols and procedures • Ability to communicate effectively with internal and external stakeholders during crisis situations • Familiarity with incident command systems and emergency communication systems
Learning path (mandatory courses): Introduction to Cybersecurity & Risk Management Specialization, Coursera/UCI Security Analyst Fundamentals Specialization, Coursera/IBM ISO 27001:2022 Lead Implementer, Udemy IBM Cybersecurity Analyst Professional Certificate, Coursera/IBM Leadership in Emergency Management, Udemy Cybersecurity Compliance Framework & System Administration, Coursera/IBM MGT551: Building and Leading Security Operations Centers, SANS institute Crisis communications, Coursera/IESE business school
Learning path (optional courses): Critical Infrastructure Protection, TEEX—Texas A&M University Misinformation and Disinformation Training, management and strategy institute Science Communication: Communicating Trustworthy Information in the Digital World, Coursera/Erasmus University Rotterdam Build Security Incident Response for GDPR data protection, Udemy

Table 5 Skills and learning path for the "physical security specialist" skills profile

Physical security specialist
Individual skills of the profile: • Proficiency in designing and implementing physical security measures • Knowledge of access control systems and surveillance technologies • Experience in conducting security assessments and audits • Ability to develop security policies and procedures
Learning path (mandatory courses): Intro to the Physical Security Industry, Udemy The Foundations of Cybersecurity, Coursera, University of Maryland, College Park Physical and Advanced Side-Channel Attacks, edX/ TUGrazX The Complete Cyber Security Course: Hackers exposed!, Udemy Dominant Risk Management Standards and Frameworks, Coursera/ university system of Georgia CCTV Cameras From Scratch: Security Camera System, Udemy MGT551: Building and Leading Security Operations Centers, SANS institute
Learning path (optional courses): Build Security Incident Response for GDPR data protection, Udemy Applied Control Systems 3: UAV drone (3D Dynamics & control), Udemy SEC511: Continuous Monitoring and Security Operations, SANS institute Endpoints and Systems, Coursera/Cisco

Table 6 Skills and learning path for the "network infrastructure engineer" skills profile

Network infrastructure engineer
Individual skills of the profile: • Expertise in designing and maintaining network infrastructure • Knowledge of network protocols and technologies • Ability to troubleshoot network issues and optimize performance • Familiarity with network security principles and best practices
Learning path (mandatory courses): The Foundations of Cybersecurity, Coursera, University of Maryland, College Park Networking Fundamentals, Coursera/Akamai Introduction to Network Protocols, Coursera/university system of Georgia VPC networking: Cloud HA-VPN, Coursera/Google cloud Real-time cyber threat detection and mitigation, Coursera/NYU Introduction to Software Defined Networking, edx/CurtinX Google Cybersecurity Professional Certificate, Coursera/Google
Learning path (optional courses): SEC511: Continuous Monitoring and Security Operations, SANS institute Build Security Incident Response for GDPR data protection, Udemy Applied Control Systems 3: UAV drone (3D Dynamics & control), Udemy Endpoints and Systems, Coursera/Cisco

Table 7 Skills and learning path for the "critical infrastructure resilience planner" skills profile

Critical infrastructure resilience planner
Individual skills of the profile: • Experience in developing resilience plans for critical infrastructure • Knowledge of resilience frameworks and methodologies • Ability to conduct risk assessments and prioritize mitigation efforts • Strong analytical and problem-solving skills
Learning path (mandatory courses): System Administration and IT Infrastructure Services, Coursera/Google Protecting Critical National Infrastructure, Cranfield University Introduction to Cybersecurity & Risk Management Specialization, Coursera/UCI Cybersecurity Risk and Strategy, edX/University of Connecticut Cybersecurity Capstone: Breach Response Case Studies, Coursera/IBM Crisis communications, Coursera/IESE business school Cybersecurity Compliance Framework & System Administration, Coursera/IBM
Learning path (optional courses): Build Security Incident Response for GDPR data protection, Udemy Introduction to Business Continuity E-Learning Course, the business continuity institute (BCI) Organizational Crisis Leadership E-Learning Course, the business continuity institute (BCI) Dominant Risk Management Standards and Frameworks, Coursera/ university system of Georgia

- They empower CIP professionals to learning to use and facilitate new tools and technologies.
- They boost resilient learning such as learning by following unexceptional event flows, anomalies, and early warning signs, which is important in roles that have to proactively and intelligently cope with security incidents.

Table 8 Skills and learning path for the "incident response manager" skills profile

Incident response manager
Individual skills of the profile: • Proficiency in managing incident response teams • Experience in developing and implementing incident response plans • Knowledge of forensic analysis techniques and evidence preservation • Ability to coordinate with law enforcement and regulatory agencies during investigations
Learning path (mandatory courses): Introduction to Cybersecurity & Risk Management Specialization, Coursera/UCI IBM Cybersecurity Analyst Professional Certificate, Coursera/IBM Leadership in Emergency Management, Udemy Cybersecurity Compliance Framework & System Administration, Coursera/IBM MGT551: Building and Leading Security Operations Centers, SANS institute Crisis communications, Coursera/IESE business school Cybersecurity Capstone: Breach Response Case Studies, Coursera/IBM
Learning path (optional courses): Critical Infrastructure Protection, TEEX—Texas A&M University Misinformation and Disinformation Training, management and strategy institute Science Communication: Communicating Trustworthy Information in the Digital World, Coursera/Erasmus University Rotterdam Build Security Incident Response for GDPR data protection, Udemy

Table 9 Skills and learning path for the "compliance auditor" skills profile

Compliance auditor
Individual skills of the profile: • Expertise in auditing critical infrastructure for compliance with regulations and standards • Knowledge of relevant regulatory frameworks and industry guidelines • Ability to conduct compliance assessments and identify gaps • Strong attention to detail and analytical skills
Learning path (mandatory courses): Cybersecurity Compliance Framework & System Administration, Coursera/IBM IBM: Cybersecurity Compliance and System Administration, edX/IBM Enterprise and Infrastructure Security, Coursera/NYU Data Analysis and Visualization Foundations Specialization, Coursera/IBM Modern Internal Audit Leadership, Udemy Fundamentals of Compliance Auditing, the Institute of Internal Auditors Science Communication: Communicating Trustworthy Information in the Digital World, Coursera/Erasmus University Rotterdam
Learning path (optional courses): Data Ethics, AI and Responsible Innovation, edX/ University of Edinburgh Data: Law, Policy and Regulation, edX/ London School of Economics Crisis communications, Coursera/IESE business school CertNexus Certified Ethical Emerging Technologist Professional Certificate, Coursera/ CertNexus

Table 10 Skills and learning path for the "industrial control systems (ICS) engineer" skills profile

Industrial control systems (ICS) engineer
Individual skills of the profile: • Experience in designing and maintaining industrial control systems • Knowledge of SCADA (supervisory control and data acquisition) systems and protocols • Ability to assess and mitigate cyber threats targeting ICS • Familiarity with industry-specific regulations and best practices
Learning path (mandatory courses): Applied Control Systems 3: UAV drone (3D Dynamics & control), Udemy AlaskaX: Unmanned Aerial Systems (UAS): Fundamentals, edX/AlaskaX Learn SCADA from Scratch - Design, Program and Interface, Udemy PLC Fundamentals (Level I), Udemy HMI Interfacing with PLC, Udemy Cybersecurity Compliance Framework & System Administration, Coursera/IBM Advanced Process Control & Safety Instrumented Systems SIS, Udemy
Learning path (optional courses): Science Communication: Communicating Trustworthy Information in the Digital World, Coursera/Erasmus University Rotterdam Data: Law, Policy and Regulation, edX/ London School of Economics Crisis communications, Coursera/IESE business school Networking Fundamentals, edX/CurtinX

Table 11 Skills and learning path for the "crisis communication specialist" skills profile

Crisis communication specialist
Individual skills of the profile: • Communicating messages about crisis • Clear and concise messaging about security issues • Rapid response and adaptability • Customizing communication for diverse audiences • Emotional intelligence and empathy • Trusted communications
Learning path (mandatory courses): Crisis communications, Coursera/IESE business school Science Communication: Communicating Trustworthy Information in the Digital World, Coursera/Erasmus University Rotterdam Organizational Crisis Leadership E-Learning Course, the business continuity institute (BCI) Build Security Incident Response for GDPR data protection, Udemy Reputation Management in a Digital World, edX/Curtin Crisis Communication Management Certification Course, Institute for Crisis Management Fighting Misinformation: Digital Media Literacy, International Research & Exchanges Board
Learning path (optional courses): Data Ethics, AI and Responsible Innovation, edX/ University of Edinburgh Misinformation and Disinformation Training, management and strategy institute Data: Law, Policy and Regulation, edX/ London School of Economics Fake News, Facts, and Alternative Facts, edX/University of Michigan

Table 12 Skills and learning path for the "business continuity planner" skills profile

Business continuity planner
Individual skills of the profile: • Expertise in developing business continuity plans for critical infrastructure • Knowledge of continuity planning frameworks and methodologies • Ability to identify and prioritize critical business functions • Experience in conducting continuity exercises and simulations
Learning path (mandatory courses): Introduction to Business Continuity E-Learning Course, the business continuity institute (BCI) Organizational Crisis Leadership E-Learning Course, the business continuity institute (BCI) Cybersecurity: Developing a Program for Your Business Specialization, Coursera/Georgia University Dominant Risk Management Standards and Frameworks, Coursera/ university system of Georgia Cybersecurity Foundations for Risk Management, Coursera/Georgia University Cybersecurity Compliance Framework & System Administration, Coursera/IBM Data: Law, Policy and Regulation, edX/ London School of Economics Crisis communications, Coursera/IESE business school
Learning path (optional courses): Data Ethics, AI and Responsible Innovation, edX/ University of Edinburgh Data Analysis and Visualization Foundations Specialization, Coursera/IBM Science Communication: Communicating Trustworthy Information in the Digital World, Coursera/Erasmus University Rotterdam Business Continuity Basics, the business continuity institute (BCI)

Table 13 Skills and learning path for the "data protection officer (DPO)" skills profile

Data protection officer (DPO)
Individual skills of the profile: • Proficiency in ensuring compliance with data privacy regulations • Knowledge of data protection laws and standards • Ability to assess data privacy risks and develop mitigation strategies • Experience in managing data breach incidents and conducting investigations
Learning path (mandatory courses): Cybersecurity Compliance Framework & System Administration, Coursera/IBM Build Security Incident Response for GDPR data protection, Udemy Cybersecurity Capstone: Breach Response Case Studies, Coursera/IBM Data Ethics, AI and Responsible Innovation, edX/ University of Edinburgh IBM: Cybersecurity Compliance and System Administration, edX/IBM Data: Law, Policy and Regulation, edX/ London School of Economics How to succeed in a Data Protection Officer Role (GDPR DPO), Udemy
Learning path (optional courses): Gen AI for Data Privacy & Protection, Coursera/Edureka Dell Technologies Data Protection Design, Udemy IBM Cybersecurity Analyst Professional Certificate, Coursera/IBM AI and Disaster Management, Coursera/DeepLearning.ai

Table 14 Skills and learning path for the "physical infrastructure engineer" skills profile

Physical infrastructure engineer
Individual skills of the profile: • Experience in designing and maintaining physical infrastructure for critical facilities • Knowledge of building codes and construction standards • Ability to assess structural vulnerabilities and recommend improvements • Familiarity with disaster-resistant building techniques and technologies
Learning path (mandatory courses): Introduction to Engineering and Design, edX/BrownX Transportation, sustainable buildings, green construction, Coursera/John Hopkins Management of urban infrastructures, Coursera/EPFL Intro to the Physical Security Industry, Udemy Energy Within Environmental Constraints, edX/HarvardX EPFLx: A Resilient Future: Science and Technology for Disaster Risk Reduction, edX/EPFL Understanding Indicators of Compromise, cybersecurity and infrastructure security agency (CISA)
Learning path (optional courses): Sustainable Construction Management, edX/University of Maryland Build Security Incident Response for GDPR data protection, Udemy Dominant Risk Management Standards and Frameworks, Coursera/ university system of Georgia Physical and Advanced Side-Channel Attacks, edX/ TUGrazX

Table 15 Skills and learning path for the "supply chain security manager" skills profile

Supply chain security manager
Individual skills of the profile: • Expertise in securing supply chains for critical infrastructure • Knowledge of supply chain risk management principles • Ability to assess supplier security practices and enforce compliance • Experience in developing contingency plans for supply chain disruptions
Learning path (mandatory courses): Supply Chain Fundamentals, edX/MIT Cybersecurity Compliance Framework & System Administration, Coursera/IBM Cybersecurity Foundations for Risk Management, Coursera/Georgia University Supply Chain Risk Management(SCRM) ISO/IEC27036 / ISO28000, Udemy Gen AI for Data Privacy & Protection, Coursera/Edureka IT Security: Defense against the digital dark arts, Coursera/Google Introduction to Business Continuity E-Learning Course, the business continuity institute (BCI)
Learning path (optional courses): Introduction to Physical Security in Critical Infrastructure Protection, EU-CIP Strategies for Supply Chain Digitalization, edX/ IMD Data Ethics, AI and Responsible Innovation, edX/ University of Edinburgh Business Continuity Basics, the business continuity institute (BCI)

Table 16 Skills and learning path for the "cyber threat intelligence analyst" skills profile

Cyber threat intelligence analyst
Individual skills of the profile: • Proficiency in collecting and analyzing cyber threat intelligence • Knowledge of threat actors and their tactics, techniques, and procedures (TTPs) • Ability to produce actionable intelligence reports for decision-makers • Experience in threat hunting and proactive threat mitigation strategies
Learning path (mandatory courses): The Foundations of Cybersecurity, Coursera, University of Maryland, College Park Endpoints and Systems, Coursera/Cisco The Complete Cyber Security Course: Hackers exposed!, Udemy MGT551: Building and Leading Security Operations Centers, SANS institute Cyber Threat Intelligence, Udemy IBM Cybersecurity Analyst Professional Certificate, Coursera/IBM Google Cybersecurity Professional Certificate, Coursera/Google Cybersecurity Capture the Flag (CTF) Competition Training, TONEX
Learning path (optional courses): Digital forensics concepts, Coursera/Infosec Data Analysis and Visualization Foundations Specialization, Coursera/IBM Cybersecurity Threat Hunting for SOC Analyst, Udemy Understanding Indicators of Compromise, cybersecurity and infrastructure security agency (CISA)

Table 17 Skills and learning path for the "climate resilience adaptation specialist" skills profile

Climate resilience adaptation specialist
Individual skills of the profile • Expertise in assessing climate-related risks to infrastructure (e.g., floods, wildfires) • Ability to integrate climate adaptation strategies into infrastructure design • Knowledge of sustainable materials and green infrastructure solutions • Proficiency in climate modeling tools (e.g., GIS-based risk mapping)
Learning path (mandatory courses): *Natural Disaster and Climate Change Risk Assessment* (Coursera/Yonsei University) *From Climate Science to Action* (Coursera/the World Bank Group) Green Building Design for LEED and other rating system (Udemy) Synergizing Disaster Risk Reduction and Climate Change Adaptation (United Nations system staff college (UNSSC)
Learning path (optional courses): Renewable Energy Specialization (Coursera/University of Colorado) Sustainable Design Practices in Building Design Specialization (Coursera/L&T EduTech)

- Most of the courses that are associated with the presented learning paths are available through the EU-CIP knowledge hub, which provides a single point of access to the educational resources referenced within this chapter. This can substantially help training development stakeholders to locate relevant courses for the presented learning paths.

Table 18 Skills and learning path for the "digital twin systems engineer" skills profile

Digital twin systems engineer
Individual skills of the profile: • Proficiency in creating and managing digital twins of physical infrastructure with emphasis on their security and resilience aspects • Ability to simulate cyber-physical threats and resilience scenarios • Knowledge of IoT sensor integration and real-time data analytics • Familiarity with AI-driven predictive maintenance systems
Learning path (mandatory courses): Digital Twins (Coursera/University of Michigan) Cybersecurity for IoT Internet of Things (Udemy) Digitalisation in Predictive Maintenance (Helixconnect Europe/EIT deep tech talent initiative) Cyber-Physical Systems: Modeling and Simulation (Coursera/UC Santa Cruz)
Learning path (optional courses): Technical Training on EBSI (European Blockchain services infrastructure) Natural Disaster and Climate Change Risk Assessment Specialization (Coursera/BID)

Table 19 Skills and learning path for the "cross-sector interdependency analyst" skills profile

Cross-sector interdependency analyst
Individual skills of the profile: • Expertise in mapping infrastructure interdependencies (e.g., energy-water-food nexus) • Ability to model cascading failure scenarios • Proficiency in network analysis tools (e.g., graph theory applications) • Experience in multi-stakeholder crisis coordination
Learning path (mandatory courses): Sustainable Infrastructure Specialization (Coursera/ Inter-American Development Bank) Complex Problem Structuring: a Socio-Technical Perspective (edX/DelftX) Resilience Engineering and Leadership in Crisis Specialization (Coursera/edX) Social and Economic Networks: Models and Analysis (Coursera/Stanford)
Learning path (optional courses): Supply Chain Risk Management(SCRM) ISO/IEC27036 / ISO28000 (Udemy) Investing in Quality Infrastructure for a Green, Inclusive & Resilient Recovery (World Bank/ edX)

EU-CIP encourages stakeholders to engage with the proposed learning paths in order to contribute to their evolution, while proposing other important learning paths that are missing from the initial list. To this end, a dedicated section on the EU-CIP Knowledge Hub can be leveraged to provide feedback and suggestions. The section comprises a feedback form, which enables stakeholders to propose new skills profiles or suggest modifications to existing ones.

Table 20 Skills and learning path for the "behavioral security culture architect" skills profile

Behavioral security culture architect
Individual skills of the profile: • Ability to design human-centric security training programs • Expertise in measuring and improving organizational security culture • Knowledge of behavioral psychology in risk perception • Proficiency in gamified learning platforms
Learning path (mandatory courses): Human Factors & Usability Engineering: Designing for Humans Specialization (Coursera/Arizona State University) Enterprise and Infrastructure Security (Coursera/NYU) Gamification for Learning, Performance & Change (Udemy) CIP Communication & Collaboration (EU-CIP)
Learning path (optional courses): The Neuropsychology of Decision Making (edX/University of Cambridge) Leadership in Multinational, Cross-Cultural Teams (Coursera/UCDAVIS)

Table 21 Skills and learning path for the "quantum-resilient cryptography specialist" skills profile

Quantum-resilient cryptography specialist
Individual skills of the profile: • Mastery of post-quantum cryptographic algorithms • Ability to audit existing infrastructure for quantum vulnerabilities • Knowledge of quantum key distribution (QKD) systems • Experience in migration strategies for legacy systems
Learning path (mandatory courses): *Introduction to Post-Quantum Cryptography* (edX/university system of Maryland) Quantum Communication and the Quantum Network Explorer (edX/TUDelft) Cryptography I (Coursera/Stanford) Applied Cryptography & Information Security (Coursera)
Learning path (optional courses): Introduction to Quantum Computing (MIT professional education/xPRO ICS Cybersecurity In-Depth(SANS ICS612) Video Surveillance for CI Security (EU-CIP)

References

1. European Commission. (2023) Mind the Cyber Skills Gap: a deep dive into Cybersecurity. Digital Skills & Jobs Platform. Available at: https://digital-skills-jobs.europa.eu/en/latest/briefs/mind-cyber-skills-gap-deep-dive (Accessed: 16 May 2024).
2. European Commission. (2023) Cybersecurity Skills Academy. Digital Skills & Jobs Platform. Available at: https://digital-skills-jobs.europa.eu/en/cybersecurity-skills-academy (Accessed: 16 May 2024).
3. World Economic Forum (2024) 'The 2020s will be a decade of upskilling: Employers should take notice', World Economic Forum, 5 January. Available at: https://www.weforum.org/agenda/2024/01/the-2020s-will-be-a-decade-of-upskilling-employers-should-take-notice/ (Accessed: 12 May 2024).

4. ISACA (2023) 'New ISACA research: 59 percent of cybersecurity teams are understaffed', ISACA, 25 April. Available at: https://www.isaca.org/about-us/newsroom/press-releases/2023/new-isaca-research-59-percent-of-cybersecurity-teams-are-understaffed (Accessed: 12 May 2024).
5. Alouffi, B., Hasnain, M., Alharbi, A., Alosaimi, W., Alyami, H., Ayaz, M. (2021). A systematic literature review on cloud computing security: threats and mitigation strategies. *IEEE Access 9*, 57792-57807. https://doi.org/10.1109/ACCESS.2021.3073203. Available at: https://ieeexplore.ieee.org/document/9404177
6. Soldatos, I., Charalambous, M., & Pirinen, R. (2024). Upskilling and Reskilling Critical Infrastructure Protection Professionals: Learning Paths for 15 Important Skills Profiles. Zenodo. doi:https://doi.org/10.5281/zenodo.13150759

Securing Cyber Resilience Through Situational Awareness: The Resilmesh Approach

Brian Lee, Martin Husák, Ekam Puri Nieto, Jorge Bernal, Maxime Vinh Hoa La, Branka Stojanovic, and Tengfei Yang

Abstract The threats toward digital services and infrastructures continue to mushroom and advanced persistent attacks continue to proliferate. Increasingly service providers are realizing attack prevention is no longer sufficient to prevent infiltration and they have to adopt a resilience-based approach that will allow them to recover from the almost inevitable service breach. In this chapter, we therefore outline a contributing approach to achieving cyber resilience based on the use cyber situational awareness and present how this approach is implemented in the Resilmesh system. Given the fact that cyber infrastructures are complex, heterogeneous, and dispersed, and exposed to advanced persistent threats, there is a need to map and understand them well to effectively protect them. Therefore, Resilmesh addresses these challenges by developing a resilient and collaborative event aggregation and processing mesh across dispersed digital infrastructures, giving CSIRTs better awareness of the service and asset dependencies in their network, developing AI-based algorithms and

---

B. Lee (✉) · T. Yang
Software Research Institute, Technological University of the Shannon, Athlone, Ireland
e-mail: brian.lee@tus.com

T. Yang
e-mail: tengfei.yang@tus.ie

M. Husák
Institute of Computer Science, Masaryk University, Brno, Czech Republic
e-mail: husakm@ics.muni.cz

E. P. Nieto · J. Bernal
University of Murcia, Murcia, Spain
e-mail: ekam.purin@um.es

J. Bernal
e-mail: jorgebernal@um.es

M. V. H. La
Montimage sarl., Paris, France
e-mail: vinhhoa.la@montimage.eu

B. Stojanovic
Joanneum Research Digital, Graz, Austria
e-mail: branka.stojanovic@joanneum.at

J. Soldatos and H. Abie (eds.), *Technology-Enabled Resilience: Innovations in Critical Infrastructure Protection*, https://doi.org/10.1007/978-3-032-25917-2_16

tools for early and ongoing attack detection and prediction, and developing a situation assessment system for system risk comprehension, projection, and response.

Keywords Cyber situational awareness · Resilience · Intrusion detection · Attack mitigation · Cybersecurity mesh architecture

1 Introduction

Resilmesh was developed to meet the expected outcomes of the Horizon Europe call HORIZON-CL3–2022-CS-01-01-"Improved monitoring of threats, intrusion detection and response in complex and heterogeneous digital systems and infrastructures". The need to increase the resilience capability of digital infrastructures and the resilience capacity of the organizations that run them was a key thread in the call.

Resilmesh thus aims to address this demand to improve resilience in infrastructures and organizations in a systematic and structured manner. This chapter describes our approach to the former issue of improving resilience of digital services and infrastructure by improving the capabilities of improving Security Operation Center (SOC) operations. Resilmesh adopts two key principles to achieve this:

- Placing Cyber Situational Awareness (CySA) at the heart of the Resilmesh approach.
- Grounding our approach to resilience on a sound, state-of-the-art resilience engineering framework, devised by NIST [1].

Cyber situational awareness is an approach to cyber defense that entails perceiving the cyber environment, understanding the current security situation, and being able to anticipate how the situation may evolve [2]. It provides a cyber-defense team with powerful capabilities to anticipate and withstand adversarial attacks and lays the basis for recovery of services and critical infrastructure, which is the core of the resilience lifecycle. NIST defines a systematic approach to engineering resilience capabilities into the network and enshrines several types of engineering constructs to develop cyber resilient solutions. Some of these constructs relate to improving CySA and Resilmesh has adopted these practices and developed functions to implement them.

Resilmesh includes a number of novel contributions:

- The system has is specifically designed to address cyber situational awareness aided resilience and this is reflected in the architecture where the functional layers map to clear CySA functions. Moreover, novel CySA functions such as network situation awareness and mission awareness are provided to give operator powerful insight into infrastructure a service situational awareness. Resilmesh also integrates various CySA functions into a consolidated dash board framework to enable network engineers to opt to only see network status but to drill down to specific items to investigate underlying root causes.

- The system also has baked in several industry best practices to improve both interoperability following the Open Cybersecurity Alliance Open XDR architecture [3] and cyber resilience engineering best practices from NIST [1].
- A comprehensive set of integration reference points (IRP's) are provided to enable easy extensibility (at all layers) to adapt Resilmesh to a wide array of digital service domains. For example this new asset types or customized Artificial Intelligence algorithms and functions can be added and as well as new attack mitigations to address new threats.

This chapter is structured into eight sections, including this introduction. Section 2 outlines the background on cyber situational awareness and Sect. 3 introduces the concept of cyber resilience, which are the two building blocks of Resilmesh. Section 4 outlines the general approach taken by Resilmesh, while its key components are discussed in follow-up sections. Section 5 describes the contextual awareness components, Section 6 describes the analytic monitoring, and Sect. 7 describes the adaptive response. Section 8 concludes the chapter.

2 Cyber Situational Awareness

Cyber situational awareness (CySA) is an application of situational awareness into the cyber domain and a frequent research topic. In essence, it is a holistic approach to cyber defense and corresponds to perceiving the cyber environment, understanding the current security situation, and being able to project how the situation will evolve [2]. While the notion of situational awareness is very old, the theoretic foundations were laid during the last decade of the twentieth century. One of the fundamental works and definitions of SA was published by Mica Endsley [4], who proposed the three-level model definition. The definition consists of three ascending components referred to as levels. The definition can be formulated as follows: Situational Awareness is the perception of the elements in the environment within a volume of time and space, the comprehension of their meaning and the projection of their status in the near future [4]. Figure 1 illustrates how projection builds on comprehension, which builds on perception.

At the first level, the level of perception of the elements in the environment, status attributes and dynamics of relevant attributes of an environment are perceived [4]. The analysis and interpretation of the data is left to the next level. Nevertheless, if the data are biased or faulty, the operator gains a misleading notion of the situation and actual state. Without the correct basic perception of relevant information, the chances of making wrong conclusions increase dramatically [5]. A common problem, especially in CySA, is the data overload [6]. An operator is overwhelmed with large volumes of data from sensors, and omitting or not even encountering relevant pieces of information.

The second level, comprehension, goes beyond the perception of the data and attempts to understand the situation correctly. The comprehension of a situation is

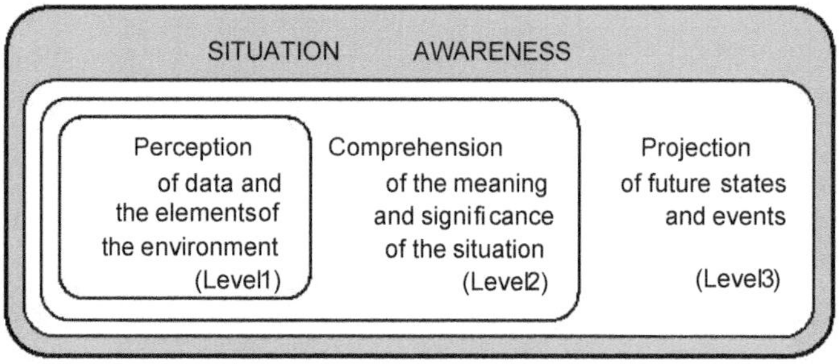

Fig. 1 Three-level model. Adapted from [4]

based on a synthesis of disjoint elements perceived at Level 1 [4]. The elements are combined, interpreted, assigned with significance concerning given goals, and combined with knowledge of an operator to form a holistic picture of an environment. The difference between the levels is analogous to reading individual words (Level 1) and understanding the sentences (Level 2) [2].

The projection of future status is the third and highest level of situational awareness, but also the hardest to reach. In the majority of fields, where situational awareness is of importance, including cyber, the experienced operators rely heavily on future projection. The future projection based on current events and dynamics enables them to anticipate future events and their implications, which allows them to make decision in time [5].

Although CySA shares a lot with general situational awareness, there are several specific features of the cyber environment that need to be considered. First, the cyber environment has no borders. This dynamic world is highly malleable and potentially scale-free, not limited by the law of physics as in other domains. If one wishes to set any boundaries and avoid noting that everything can be everywhere, the physical location of the network or system is usually used as a spatial boundary for CySA [7]. However, even that can have its limitations due to wide use of, for example, cloud computing or network tunnels.

Second, the perception in cyber environment is more complicated. We can no longer rely on physical sensors or human perception, but use specialized hardware or software sensors and rely solely on them, since it is not possible to observe the information directly. Each sensor is a complex system that is prone to errors, measurement artifacts, or false positives, in case of various detection systems. It is also complicated to observe the adversary directly—we may only observe the activity or traces of the adversary, who typically stays hidden in both cyber and physical environment. Related factors to consider concerning the performance is the speed of the events and resources required to process them. The speed of the events is orders of magnitude quicker than in case of physical conflicts. The resources needed for the processing of such large volumes of information are significantly higher in the case of CySA [2].

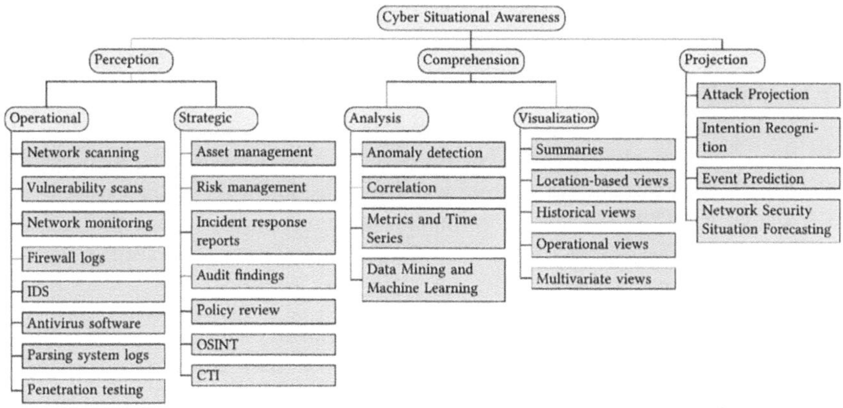

Fig. 2 Taxonomy of cyber situational awareness tools and components [2]

Finally, contrary to traditional military doctrine, in which defender gains numerous advantages [6], a cyber attacker overtakes all advantages in the case of cyber conflicts. The advantages of a cyber attacker include, but are not limited to, anonymity (it can hide across national sovereignty boundaries), global reach to probe weaknesses, social engineering to exploit human weaknesses, and a possibility to pick a time, place and tools for an attack. A related issue is the disproportion of resources required for launching the attack compared to resources required to successfully detect it [2].

Figure 2 illustrates which cybersecurity tools and techniques belong to are used to reach which level of CySA [2]. Readers interested in further details on the topic of CySA are kindly referred to fundamental book by Kott et al. [6], an older, yet exhaustive literature survey by Franke and Brinielsson [8], or a more recent review by Husák et al. [2]. Nevertheless, CySA is no longer a purely academic term, but is gaining attention and recognition in the industry and commercial cybersecurity products.

3 Engineering Cyber Resilience

Cyber resilience is commonly understood as the ability of a system, or component, to recover its capability or performance after this capability has been diminished as a result of a cyber attack [6]. Resilience is related to but should not be confused with other cyber concepts such as robustness, risk and security. Traditionally, risk management is used to identify vulnerable components of a system and then apply security controls to harden these parts. As attacks have become more sophisticated and underlying technologies have become more complex and diverse it has become increasingly difficult and expensive to harden every entry point. This has led to an acceptance that prevention is almost impossible and that organizations must adapt to attacks to bounce back from the effects of these attacks. Viewed in this this way cyber resilience can be seen as the capacity of a system to:

1. *Anticipate*: Maintain a state of informed preparedness for adversity.
2. *Withstand*: Continue essential mission or business functions despite adversity.
3. *Recover*: Restore mission or business functions during and after adversity.
4. *Adapt*: Modify mission or business functions and/or supporting capabilities in response to predicted changes in the technical, operational, or threat environments.

NIST has created a systems engineering framework to develop cyber resilient solutions for cyber systems [1]. Figure 3 shows how an organization's risk management strategy defines the priorities and requirements for its cyber resiliency goals and objectives, and how these, in turn, guide the selection and application of desired cyber *resilience techniques* to complete the required cyber resilient solution. Goals and objectives define WHAT capabilities are needed while techniques inform HOW the systems engineering framework can be used to develop cyber resilient systems. A cyber resiliency technique is thus ".. a set or class of practices and technologies intended to achieve one or more goals or objectives by providing capabilities". The cyber resilience framework defines 14 such techniques, a sample of which are described in Table 1 to illustrate the concept. Techniques are stated at a high level and have one or more implementation approaches, i.e., "a subset of the technologies and processes included in a technique that are defined by how the capabilities are implemented," e.g., see the privilege restriction implementation approaches in Table 1.

The standard develops the techniques and implementation approaches in greater depth and the reader is referred to the standard for a more complete description.

A subset of these techniques is used as the basis for achieving resilience in Resilmesh.

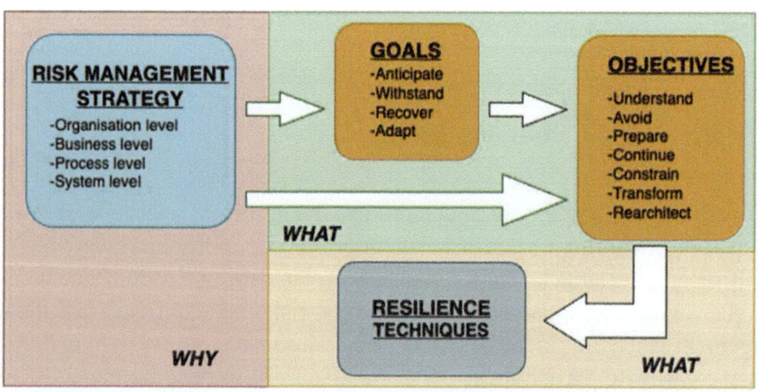

Fig. 3 Cyber resilience solution. Adapted from [1]

Table 1 Cyber resilience techniques and implementation approaches in Resilmesh

Technique	Description	Implementation approach
Coordinated protection	Ensure that protection mechanisms operate in a coordinated and effective manner.	– Trust based privilege management – Attribute based usage restriction – Dynamic privileges
Privilege restriction	Restrict privileges based on attributes on environmental factors.	Restriction of users and system elements, as well as on environmental factors.
Unpredictability	Make changes randomly or unpredictably.	– Temporal unpredictability
Segmentation	Define and separate system elements based on criticality and trustworthiness etc.	–

4 Resilmesh Approach

4.1 Challenges and Objectives

Resilmesh was developed to meet the expected outcomes of the Horizon Europe call HORIZON-CL3-2022-CS-01-01- "Improved monitoring of threats, intrusion detection and response in complex and heterogeneous digital systems and infrastructures" for which we identified the following challenges:

1. *Complex, heterogeneous and dispersed*: with multiple infrastructure layers with many types of technologies spread over wide geographical areas (cloud/edge/endpoint), creating dependencies between services and assets and thus leading to multiple attack vectors.
2. *Exposed to advanced persistent threats (APTs)*: that seek to exfiltrate information or impede critical aspects of a mission or organization based on multiple attack vector approaches.
3. *Slow adoption of best practices*: such as Zero Trust and Secure System Edge enabling easier access to attackers.

Resilmesh aims to address these challenges by improving cyber resilience through the following:

- Developing a resilient and collaborative event aggregation and processing mesh across dispersed digital infrastructures.
- Giving CSIRTs better awareness of the service and asset dependencies in their network.
- Developing AI-based algorithms and tools for early and ongoing attack detection and prediction (counter APT's).

- Developing a situation assessment system for system risk comprehension, projection and response.

In order to achieve these goals, Resilmesh adopts the NIST resilience engineering framework [1] as described in the next section.

4.2 Resilience in Resilmesh

In this section, we describe how a subset of NIST cyber resiliency techniques—and their related implementation approaches—form the basis of the Resilmesh resilience approach.

The relevant techniques are found in Table 2. While the techniques define the high-level approach the implementation approaches actually develop the Resilmesh functional components that implement the resilience solution. These are described in detail in subsequent sections, after we first describe the Resilmesh architecture design.

4.3 Situational Awareness in Resilmesh

Resilmesh extends the CySA process described in Fig. 1 and 'unfolds' it into an end-to-end processing pipeline with steps from when the event is emitted by the endpoint to its final consumption by situation assessment—see Fig. 4.

This CySA 'logical architecture' is crystallized by the Resilmesh reference architecture, as shown in Fig. 5. This is realized as a number of distinct management planes:

- *Infrastructure Plane*: the network and system being monitored.
- *Aggregation Plane*: collects, aggregates, normalizes, and streams data and events from multiple heterogeneous sources including logs, Intrusion Detection Systems, and other network sensors.
- *Collaboration Mesh*: integration and interoperation to enable the operation of the system across the dispersed digital infrastructure.
- *Threat Awareness Plane*: suite of analytics functions to manage event correlation and alarming, attack detection and prediction, CTI sharing, and threat hunting.
- *Situation Assessment Plane*: captures dependencies between services onto the IT/OT resources that realize them; visualizes the current network risk status; forecasts the near-term situation evolution.
- *Security Operations Plane*: automates and orchestrates response and mitigation actions.

Table 2 Cyber resilience techniques and implementation approaches

Technique	Implementation approach	Description
Contextual awareness	Resource awareness	Maintain information about systems assets, resources, services and their connectivity.
	Mission awareness	Maintain information about enterprise business functions and their dependencies on IS resources and services as well as the threat status of resources.
	Threat awareness	Ensure awareness of threat actors and adversaries. Maintain ongoing observation and analysis of threat event and indicators and enable better prediction of cyber security threats.
	Risk awareness	Ensure awareness of the risk of the end point and networking components. Enable a hierarchical risk estimation for the network by aggregating the risk upwards.
Adaptive response	Dynamic reconfiguration	Make changes to systems, system elements and resources to change functionality or behavior without interrupting service.
	Adaptive management	Change how mechanisms are used based on changes in the operational environment as well as changes in the threat environment.
Analytic monitoring	Monitoring and damage assessment	Monitor and analyze behavior and characteristics of components and resources to look for indicators of adversary activity or precursor conditions or indications of other threat events and to detect and assess damage from adversity.
	Sensor and fusion analysis	Fuse and analyze monitoring data and analysis results from different information sources or at different times together with externally provided threat intelligence.
	Forensics and behavioral analysis	Analyze indicators and adversary TTPs, including observed behavior, malware, and other artifacts left behind by adverse events.

The Aggregation plane and Collaboration mesh represent the Resilmesh 'platform' and provide a base for the application-oriented upper layers. The threat awareness layer provides the essential basic detection capability and provide hooks for the situation assessment and security operations layer which consequently depend on the threat awareness layer. Resilmesh functions to improve resilience are primarily focused in the upper planes of the architecture (Threat Awareness (Analytic Monitoring), Situation Assessment (Contextual Awareness) and Security Operations (Adaptive Management) as these address improving resilience of enterprise services and infrastructure. However, the project also seeks to improve the resilience of the Resilmesh platform itself through the Collaboration Mesh (Adaptive Management).

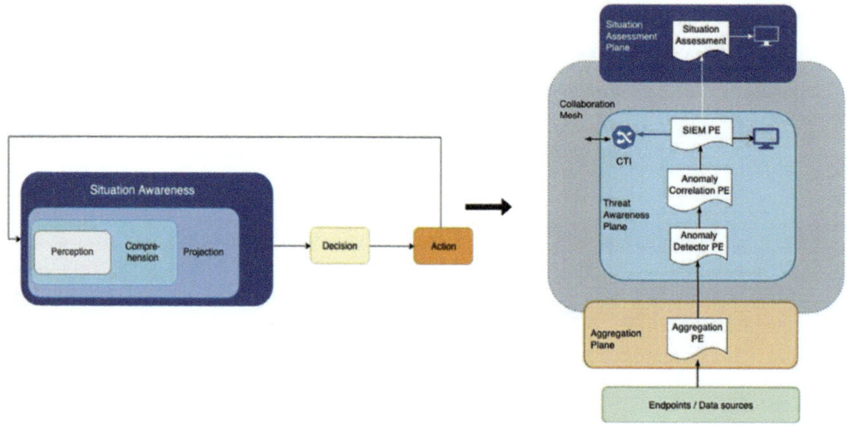

Fig. 4 CySA in Resilmesh

5 Contextual Awareness

Contextual awareness corresponds to the awareness of the cyber environments, including the cyber assets, actors, and threats. Resilmesh divides the contextual awareness into three parts: resource awareness focusing on the cyber assets, such as hosts and services, in purely technical context; mission awareness that links cyber assets to business objectives and missions, including the notion of their criticality and dependencies; and risk awareness, that approaches the risks based on cyber threats and malicious behavior.

5.1 Resource Awareness

Resource awareness is paramount to the perception level of CySA as its main goal is to map the cyber environment and enumerate the assets. Resilmesh implements two closely connected components, CASM and ISIM, to conduct periodical network scans and store the results in a comprehensive way to enable resource awareness.

The **Infrastructure and Service Information Model (ISIM)** is the foundational component, which materializes the data model and provides the central point for the other components. It consists of the database, where all the data collected by CASM are stored. The database conforms to an elaborated ISIM data model, which is enforced by data quality checks during every insert operation or on demand.

We aimed to represent all assets from the cyber defense matrix (CDM)[1]—devices, applications, networks, data, and users. Networks are represented as subnets with Classless Inter-Domain Routing (CIDR) notations. Moreover, their relationships in

[1] https://cyberdefensematrix.com/

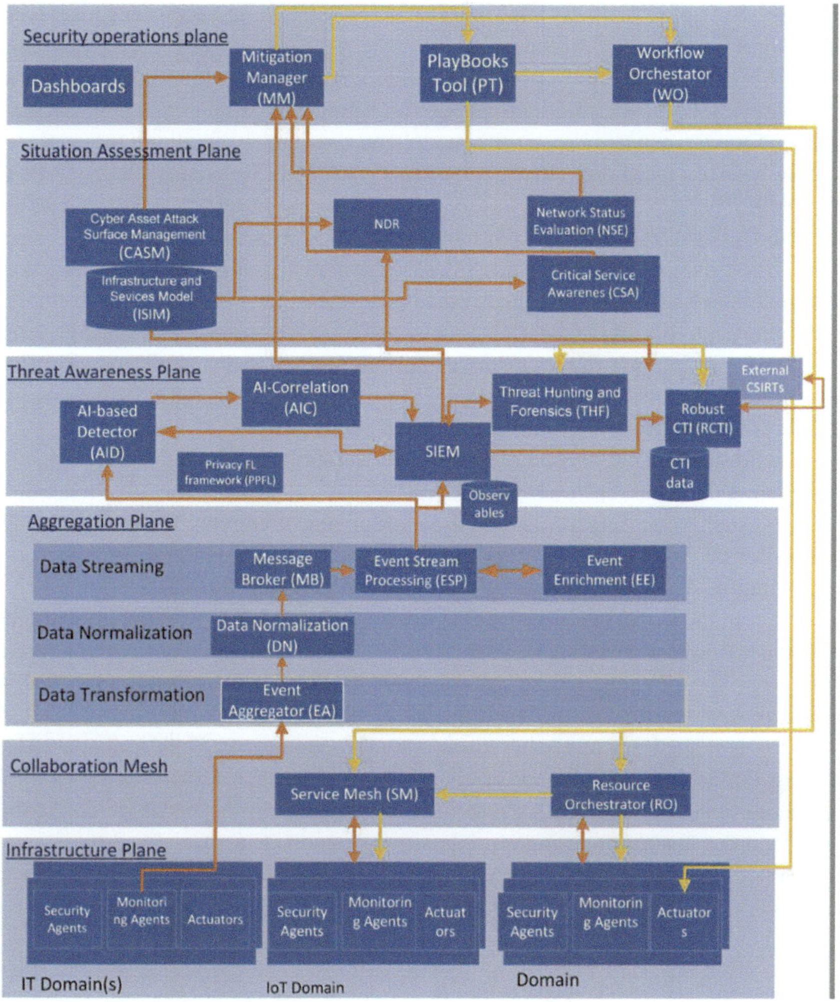

Fig. 5 Resilmesh reference architecture

the data model express the domain names of hosts in the network, their IP addresses, and connections. All types are connected with components they are processed by or stored at. Such a representation can express captured indicators and sensitive data, which helps model cyber threats and helps with situation assessment. Subsequently, we added two additional layers to the data model to support a robotics use case. The layers represent hardware (e.g., hardware versions and device groups) and physical locations. The current draft was based on data models and ontologies for industrial robotics [9]. The model aims to support storing information about cloud devices. We added a Uniform Resource Identifier (URI) to the data model. Furthermore we added a labeling attribute to distinguish types of network nodes based on MITRE

D3FEND's Digital Artifact Ontology (DAO).[2] We store timestamps for some vertices and edges for situation awareness.

ISIM implementation uses Neo4j, one of the most widely known and used graph databases, since the data model is modeled as a graph and because the graph structure allows for easy extensibility of the model and the data. To facilitate the communication between ISIM and other components, two APIs are created, REST API and GraphQL.

The **Cyber Attack Surface Management (CASM)** is the data collection component. The principal role of the CASM is to monitor the organization's internal and external attack surface and security posture. It carries out domain (and sub-domain) enumeration to discover so-called 'shadow-IT', and to identify and manage threats discovered in internet-facing assets using independent scans of the organization's attack surface. CASM can detect changes of asset status and trigger required actions or alerts to the Mitigation Manager (MM) as required.

CASM component is orchestrated with Temporal.io tool,[3] which allows for scheduling the scans, ensuring all tasks end successfully, and automating reactions to failures, such as waiting a specified time period and attempting the task again. This allows for continuous, reliable, and almost zero-touch scanning of the infrastructure, providing fresh data for the use by the other components.

We built the CASM around the EasyEASM.[4] EasyEASM scans the infrastructure using a set of various tools. CASM enumerates the subdomains of a given domain and scans the IP range of the infrastructure to enumerate the active devices and network services. For scanning internal networks and non-Internet-facing assets, we also implemented a scanner based on common Nmap tool. Both tools parse CPE strings describing the software on the discovered hosts. CASM then looks up the vulnerabilities of the given CPE strings. Depending on the needs of deployment environment, for example in the realms of IoT and OT, a specialized tools can be integrated and orchestrated to collect additional data (e.g., specialized IoT or OT scanners).

5.2 Mission Awareness

Mission awareness is an innovative approach that effectively combines cyber situational awareness with business objectives of the organization operating the infrastructures protected by Resilmesh. A mission is decomposed into mission-supportive processes that are mapped to IT services that are consequently mapped to cyber components. The functional requirements are posed on IT services by the mission-supportive processes and on cyber components by the IT services. A satisfying mission configuration is a subset of mission-supportive processes, IT services, and

---

[2] https://d3fend.mitre.org/dao/

[3] https://temporal.io/

[4] https://github.com/g0ldencybersec/EasyEASM

cyber components that align with the mission's functional requirements. Finally, mission resilience is the mission's ability to continue its operations while maintaining all the required functional and security requirements.

The mission decomposition model is a graph structure with the mission-supportive processes, IT services, and cyber components as nodes [10]. The functional requirements are represented as edges between mission-supportive processes and IT services or between IT services and cyber components. The alternative mission configurations are expressed via the AND/OR nodes [11]. An AND node in a mission configuration poses a requirement for the presence of all of its child nodes. An OR node expresses a mission configuration option since it poses a requirement for the presence of at least one of its child nodes. A mission-supportive process may be satisfied by more mission configurations, i.e., combinations of IT services and cyber components. For example, a mission-supportive process dependent on one of the two IT services produces three satisfying mission configurations, one with the first IT service, one with the second IT service, and one with both. See Fig. 6 for an illustrative example of a mission decomposition model.

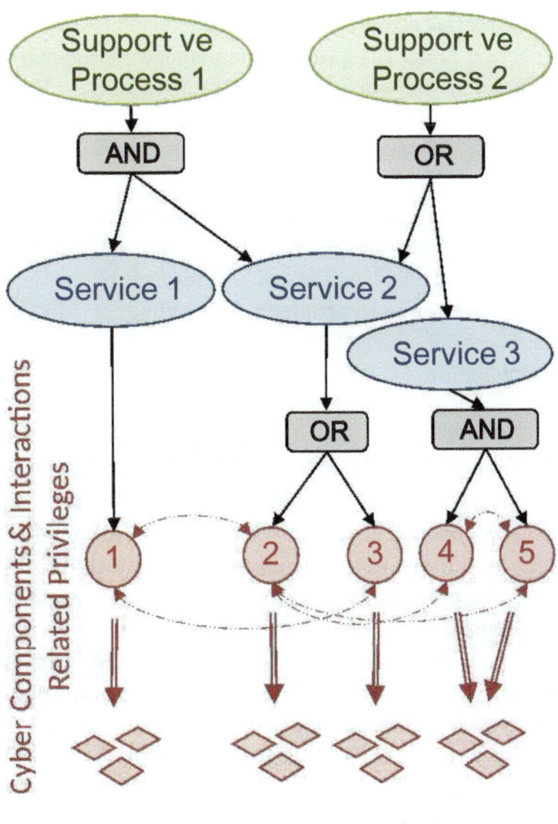

Fig. 6 Mission decomposition model [10]: supportive processes of business mission are mapped to abstract services (e.g., information systems), which are then mapped to specific cyber components (e.g., devices and services)

In Resilmesh, the mission awareness is implemented in the **Critical Service Awareness (CSA)** component. This component provides hierarchical risk assessment to aggregate infrastructure risk into a risk for the critical service or business mission. It also uses the data stored in ISIM and forward the outputs to the Network Situation Evaluation (NSE). The critical services or business missions can be represented as a tree structure, in which nodes representing cyber assets (devices or services) or children of nodes representing the missions or its logical component. In addition, AND/OR nodes can be placed between missions and assets to model situations, in which the mission relies on multiple assets simultaneously (AND) or one of interchangeable assets (OR). Tree structure can be provided as JSON files, stored in ISIM, and visualized in the dashboard.

5.3 Risk Awareness

This subsection introduces the Risk Situation Awareness function within the Resilmesh framework, designed to provide a comprehensive understanding of network risk posture and enable timely detection and mitigation of cyber threats. The function integrates two complementary components: **Network Status Evaluation (NSE)** and **Network Detection and Response (NDR)**.

The NSE component is responsible for assessing the overall security status of the network. It interacts with the SIEM to receive threat information and with the ISIM to retrieve systems information, e.g., node information including vulnerability status. It computes risk scores based on asset criticality, threat activity, and vulnerability status, providing a high-level view of risk across the infrastructure. NSE also supports hierarchical analysis, allowing operators to drill down into specific network segments or entities for more detailed risk insights.

The NSE application server, forming the core of the logic layer, manages the computation and visualization of network risk. It constructs a virtual representation of the monitored infrastructure by aggregating individual nodes based on user-defined criteria. For IPv4 networks, aggregation is performed using Classless Inter-Domain Routing (CIDR), which groups IP addresses into supernets for hierarchical analysis. It calculates risk scores for each aggregated node using a defined model that takes into account asset criticality, threat status, and vulnerability information. These scores are then passed to the front end for visualization. While IPv4-based aggregation is fully supported, IPv6 aggregation methods remain under consideration for future development.

The NDR component, built on top of the Montimage AI Platform (MAIP) delivers advanced threat detection and mitigation capabilities by continuously monitoring, analyzing, and responding to network traffic anomalies.

At the core of the NDR system is its AI-based behavioral anomaly detection capability. MAIP provides a suite of pre-trained machine learning models, including supervised, unsupervised, and deep learning approaches that are optimized for identifying irregularities in network traffic. These models detect deviations from baseline behaviors and known threat signatures, thus allowing for a hybrid detection approach. To ensure transparency and interpretability of detection outcomes, MAIP integrates Explainable AI (XAI) techniques such as SHAP and LIME. By bridging the gap between high-performance automated detection and human decision-making, XAI fosters trust and accelerates incident response. A fundamental strength of the NDR system lies in its Root-Cause Analysis (RCA) module. Rather than stopping at identifying anomalous activity, the system performs in-depth analysis to trace the origin of the issue. It leverages technologies such as graph databases (e.g., Neo4j) to map dependencies between assets, services, and network segments. This allows for pinpointing the underlying causes of alerts, which is crucial for understanding how threats propagate and where mitigation efforts should be focused.

6 Analytic Monitoring

Analytic Monitoring involves adjusting the use of mechanisms in response to changes in both the operational and threat environments. Resilmesh divides analytic monitoring into three parts: *monitoring and damage assessment*, focusing on monitoring and analysis of components and resources to identify threat indicators, *sensor and fusion analysis*, focusing on fusion and analyzes of monitoring data with threat intelligence across sources and timeframes; and *forensics and behavioral analysis*, including indicators and adversary TTPs. These functions are encapsulated within the Threat Awareness Layer of Resilmesh—a key subsystem responsible for anomaly detection, event correlation, alerting, and predictive analysis.

At the core of this layer is the **Security Incident and Event Manager (SIEM)**, a central component built on Wazuh.[5] The SIEM plays a vital role in collecting, logging, visualizing, and correlating events from across the infrastructure. It supports both rule-based and AI-assisted correlation to highlight anomalies and generate alerts. The SIEM acts as the main conduit for integrating the output of various detection models, forwarding correlated events to security operators and feeding additional analytics pipelines.

The SIEM is in turn supported by a number of other layer functional components as described in the following sessions. Together, these integrated analytic functions—anchored by the Wazuh-based SIEM—enable Resilmesh to deliver responsive, intelligent, and resilient cybersecurity monitoring across heterogeneous, distributed infrastructures.

[5] https://wazuh.com/

6.1 *Monitoring and Damage Assessment*

Monitoring involves observing and analyzing the behavior and characteristics of system components and resources to identify signs of adversary activity, early threat indicators, and to detect and assess any resulting damage. Resilmesh addresses two main components: artificial intelligence (AI) machine learning (ML)-based anomaly detection and correlation.

6.1.1 Anomaly Detection

The Anomaly Detection Component (AID) in the Resilmesh framework is designed to detect suspicious events and potential cyberattacks in both IT and OT environments. Utilizing advanced ML and AI techniques, AID delivers robust anomaly detection across diverse applications and network protocols.

This component supports deployment on endpoints, edge devices, or in the cloud, offering flexibility based on operational needs and resource constraints. Resilmesh emphasizes the use of deep learning techniques—such as hierarchical feature extraction using autoencoders and decision-level ensemble models—to ensure accurate and efficient detection at the edge. These methods enable localized processing with minimal latency and reduced energy consumption during both training and inference.

Traditional anomaly detection methods, particularly signature-based systems, are often ineffective in identifying zero-day attacks or novel adversary behaviors. In contrast, ML and deep learning models can identify previously unseen anomalies by learning normal behavior patterns and flagging deviations from them. However, deploying these models at scale poses significant challenges due to the need for centralized datasets, substantial compute power, and compliance with privacy and data governance constraints.

To address these limitations, Resilmesh explores the use of federated learning, a distributed AI paradigm that allows multiple decentralized nodes to collaboratively train a shared model without transferring raw data. This method enhances privacy, as local data remains confined to its origin, while enabling collaborative learning across diverse systems and environments. It also improves detection latency, since the trained model is directly deployed where it is needed—at the edge.

Federated learning offers additional benefits, such as improved robustness and generalizability. By training across a wide variety of local conditions and usage patterns, the resulting model is better equipped to detect sophisticated or previously unseen threats. Moreover, distributing training workloads enhances scalability and reduces the risk of bottlenecks or single points of failure associated with centralized systems.

Nevertheless, federated learning introduces new security concerns, including model poisoning and inference attacks. Resilmesh addresses these by evaluating privacy-preserving techniques such as Differential Privacy, which ensures that the contribution of individual data points remains untraceable within the trained model.

The approach also supports massive scalability—federated learning frameworks can span thousands of nodes, enabling high-throughput, real-time monitoring of critical systems.

By combining localized deep learning, privacy-aware distributed training, and efficient edge deployment, the AID component provides a high-performance, scalable, and secure solution for anomaly detection in modern cyber-physical infrastructures.

6.1.2 Anomaly Correlation

The integration of AI in cybersecurity anomaly detection serves to enhance the correlation of events, thus reducing the number of false positives, and, moreover, facilitating context-aware anomaly detection. AIC is based on the DeepCASE framework [12], which combines contextual learning, deep learning models, and Large Language Models (LLMs) to identify and interpret complex event sequences.

This system begins with the **Context Builder** module, whose function is to extract and encode contextual information from raw event logs. The model incorporates a transformer-based architecture to produce an attention vector for the event context, which has been developed to more effectively identify sequential dependencies and enhance the model's sensitivity to minor alterations in event behavior. This enriched representation of event context forms the foundation for the subsequent analysis stages.

Then, the **Sequence Detector** module leverages transformer networks to detect and classify event sequences. The use of attention vectors ensures that the detector considers both the semantic and temporal relationships between events, thereby enhancing detection accuracy.

Using the attention vectors generated by the Context Builder module and the detection result from the Sequence Detector module, the **Interpreter** module groups similar event trajectories into distinct clusters. These clusters function as reference sets, representing known behavior patterns. The Interpreter enables the system to contextualize incoming event sequences against a repository of known patterns. This, in turn, supports the rapid identification of novel or anomalous activities.

The final module is an **LLM-based explanation** module. Initially, an LLM was employed to extract relevant information from a large corpus using a hybrid rule-based approach. In subsequent phases of development, integration between the LLM and the event clusters has been implemented to generate targeted, context-aware explanations for detected anomalies. This not only assists analysts in their understanding and decision-making processes, but also addresses the requirement for Explainable AI in security operations.

6.2 Sensor and Fusion Analysis

Sensor and fusion analysis involves combining and analyzing data from diverse monitoring sources and timeframes, integrating it with external threat intelligence to generate a more comprehensive understanding of potential threats. In complex IT and OT environments, where systems operate across different technologies and protocols, effective fusion of sensor data is critical to uncovering hidden or coordinated attack patterns.

Resilmesh extends the state of the art by incorporating advanced multi-view learning techniques specifically designed to address the challenges of analyzing heterogeneous data sources. These techniques enable the system to integrate and correlate information from various types of monitoring components—such as operational technology (OT) devices, industrial control protocols (e.g., Modbus), and IT intrusion detection systems (IDS)—to identify anomalies that may not be evident when sources are analyzed in isolation.

A core focus of this component is the use of deep learning methods for both feature-level and decision-level fusion. Hierarchical model architectures, including deep stacked networks (DSNs) and autoencoders, are employed to learn and merge features from different views in a structured way. In parallel, ensemble learning methods support decision fusion, enhancing model robustness and improving detection accuracy. This layered approach allows the system to account for the natural hierarchies and distribution patterns commonly found in edge and fog computing environments.

By supporting distributed deployment, fusion analysis can be performed locally at the edge or aggregated in the cloud, depending on the needs of the deployment. This ensures scalable performance and efficient resource usage, while preserving the richness of data-driven insights across domains.

6.3 Forensics and Behavioral Analysis

Forensics and behavioral analysis in Resilmesh focuses on Threat Hunting and Digital Forensics for Robotics.

6.3.1 Threat Hunting

Threat hunting is a cybersecurity practice that involves the systematic search for hidden threats, anomalous behaviors, and advanced attack techniques within the digital environment of an organization. Although it greatly enhances an organization's capacity to detect and mitigate threats at an early stage, the current approach to threat hunting is largely manual and dependent on SIEM systems and analyst expertise. A promising shift in this domain is presented by recent advances in LLMs. The role of LLMs in structured threat hunting processes has not yet to be fully explored. To address this issue, this study focuses on integrating LLMs into the 3C (Construct, Conduct, Conclude) methodology developed by the HORIZON Resilmesh project.

Construct the Hunt. The initial phase of the 3C threat hunting process involved the establishment of the objectives of the hunt, which were typically initiated by recent incidents, CTI reports, or shifts in the threat landscape. **Conduct the Hunt**. In the second phase, analysts collect and analyze security data to test the hunt hypothesis. In the event of confirmation, the findings are documented; if inconclusive, the hypothesis may be refined and the search repeated to enhance detection accuracy. **Conclude the Hunt**. The final phase involves the documentation and reporting of findings, and the preservation of the hunt for future reference.

Each phase—Collect, Construct, and Conclude—can benefit from LLMs, especially when deployed as part of agent-based systems that support task automation and decision-making. In the **Construct the Hunt**, LLMs are capable of employing a range of techniques to gather and interpret CTI from both internal and external sources. Depending on the nature of the hunt, LLMs can operate within a conversational, single-agent framework or interact with multiple tools or agents. More research is required to identify the optimal agent architectures for different hunting scenarios and determine how LLM capabilities can be best integrated into existing security workflows. In the **Conduct the Hunt**, LLMs can dynamically configure data collection agents (e.g., Filebeat), establish SIEM ingestion pipelines, and generate detection rules tailored to the hypothesis. By reasoning across this multi-source data, LLMs provide insight into whether the hypothesis is supported. If it is not supported, the LLM has the capacity to revise the hypothesis or refine the search parameters. Given the largely autonomous, tool-rich nature of this phase of development, a multi-agent system is particularly suited to support the diverse tasks involved. **Conclude the Hunt**, it is possible to utilize the generative capabilities of the LLM to a considerable extent. An LLM is able to summarize and document the results and outputs of the search and communicate these findings to all relevant parties, as well as archive the results. It appears that a single-agent approach is adequate for this stage.

When integrated within an agent-based architecture, LLMs can operate autonomously across the 3C phases. This orchestration not only streamlines the threat hunting process, but also enables adaptive, continuous analysis at scale, enhancing the analyst's ability to detect and respond to emerging threats with greater speed and precision.

7 Adaptive Response

Adaptive Response refers to the capability of the system to take action to adjust to changes in its environment though *dynamic reconfiguration* at the collaboration mesh layer and *adaptive management* at the security and operations layer to apply attack mitigations.

7.1 Dynamic Reconfiguration

Resilmesh builds resiliency into its aggregation functions though the use of cloud native reconfiguration capabilities based on Service Mesh[6] capabilities. A service mesh is a specialized infrastructure layer that manages and secures communication between microservices within a virtualised application. It provides features like traffic routing, load balancing, security, and observability without requiring changes to the application code itself. When used with appropriate resource provisioning the service mesh allows a platform service to be effectively reconfigured seamlessly across dispersed digital infrastructures with minimal effort.

The Resilmesh service mesh is implemented using the NATS message broker.[7]

7.2 Adaptive Management

The **Mitigation Manager (MM)** functional component—see Fig. 7 is responsible for deciding which mitigation actions, if any, should be taken in response to a detected incident. To achieve this, it follows the Reactive/Mitigation Flow:

1. As the SIEM component collects and analyzes system behavior, such traffic might trigger one or more predefined SIEM rules when it is considered to be anomalous. When this happens, the SIEM generates and fires an alert containing related information. This alert is directly consumed by the Mitigation Manager, and is the initial step in attack mitigation.
2. In order to determine a suitable mitigation for the received alert, the Mitigation Manager will utilize Cyber Situational Awareness information, the purpose of which is to provide critical and up-to-date information about the current status of the network including risk scores. This information is provided by the Infrastructure and Services Model (ISIM) component, which keeps an updated representation of the complete system explained above.
3. With the information gathered in the previous steps, the Mitigation Manager evaluates the collection of actions available for mitigation—the Course of Action (CoA) Playbooks—and selects the one most appropriate for the current situation. The Mitigation Manager (MM) optimizes its decision based on the cost of executing playbooks in order to ensure that the selected mitigation is the least disruptive.
4. Once the decision process has landed on a CoA playbook, the Mitigation Manager requests the execution of the strategy. This process requires the sourcing of playbook parameters used to determine the target/s of a mitigation strategy, the degree

[6] https://www.cncf.io/blog/2017/04/26/service-mesh-critical-component-cloud-native-stack/

[7] https://dale-bingham-soteriasoftware.medium.com/using-nats-to-implement-service-mesh-functionality-part-1-service-discovery-5f2e6088843b/

to which a dynamic action is performed, a list of additional targets for the mitigation, and so on. These parameters are extracted, inferred, or computed from both the threat alert that triggered the attack mitigation process and the information retrieved from the ISIM.

5. After determining the course of action, the Mitigation Manager requests the execution of a playbook. This mitigation can be applied by two components: Playbook Tools or Workflow Orchestrator. The kinds of mitigation actions to be enforced as part of the playbooks might include: network filtering (diverting/drop-blocking traffic), network traffic mirroring, channel protection setup, Ansible script execution, host quarantining, remote software upgrading, service restoration, SIEM indicator/rule updating (YARA rules, Sigma rules for detection), and CTI Sharing.

For its main decision-making algorithm, the Mitigation Manager uses a **graph-based approach** where an attack is divided into steps, with each edge representing an attacker's action as progress toward a goal such as exfiltrating data, or causing damage to critical infrastructure. This graph structure has been modeled after the MITRE Attack Flow data model, which defines attack graphs as directed acyclic graphs (DAGs).

Combining attack graphs and workflows, the Mitigation Manager is able to evaluate threat alerts and select optimal mitigations. The complete **mitigation process** consists of the following steps:

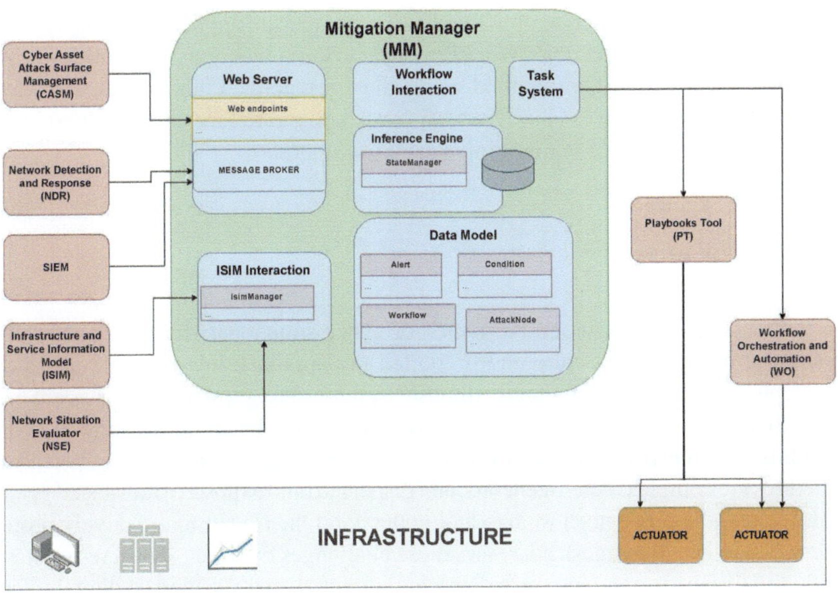

Fig. 7 Mitigation Manager architecture and overview

1. **Alert parsing**: When an alert is received, the Mitigation Manager dispatches a task to handle the event. SIEM alerts contain a large amount of fields, and different alert types contain different fields.
2. **Introduction of new attack graphs:** The Mitigation Manager can track multiple concurrent attacks taking place, thanks to the attack graph structure. When evaluating an alert, the attack graph database is first checked to locate any potential new attacks triggered by the alert and added to a data structure known as the "attack front", a collection of Attack nodes considered to be the next steps taken by adversaries.
3. **Attack front advancement:** After any new attack graphs are included, the attack front is evaluated for ATT&CK Technique matches. When a technique matches a node in the attack front, the node is treated as a "compromised node"—meaning that it might have been exploited by an attacker.
4. **Updating probabilities:** Alongside attack front advancement, all nodes in attack graphs belonging to the attack front receive an update to their risk value. Risk is dynamic and changes based on multiple factors including total attack graph progression, number of conditions to fulfill, and proportion of conditions fulfilled. **Node filtering:** When selecting mitigations to apply, the Mitigation Manager distinguishes between three types of nodes: those that are directly impacted by the threat alert, those that are highly likely to be progressed to, and those that have been "historically risky".
5. **Workflow filtering**: For each node in the previously mentioned collection, a workflow is selected. Workflows are filtered using the previously described procedure: filtering based on ATT&CK Technique and applicability, and reduced by lowest associated cost. **Workflow execution**: Once a list of workflows has been selected, they are applied using their respective components: PT workflows are forwarded to Shuffle, and WO workflows are forwarded to the selected orchestrating service in question, such as Temporal.

8 Conclusion

In this chapter, we discussed the issues of cyber situational awareness and cyber resilience in IT infrastructure and presented the approach toward achieving those concepts implemented in the Resilmesh system. We reprised the fundamental background cyber situational awareness and cyber resilience to illustrate how the Resilmesh system implements these concepts. Given the fact that current cyber infrastructures are complex, heterogeneous, and dispersed, and exposed to advanced persistent threats, there is a need to map and understand them well to effectively protect them. Therefore, Resilmesh addresses these challenges by developing a resilient and collaborative event aggregation and processing mesh across dispersed digital infrastructures, giving CSIRTs better awareness of the service and asset dependencies in their network, developing AI-based algorithms and tools for early and ongoing attack detection and prediction, and developing a situation assessment system for system risk comprehension, projection, and response.

Declarations.

This project has been funded by the European Union as part of the Horizon Europe Framework Program (HORIZON), under the grant agreement 101119681, Resilmesh.

References

1. Ross, R., Pilliteri, V., Graubart, R., Bodeau, D., McQuaid, R.: Developing cyber-resilient systems: A systems security engineering approach. NIST Special Publication 800-160, Volume 2 Revision 1 (1), 4–32 (2021)
2. Husák, M., Jirsík, T., Yang, S.J.: SoK: contemporary issues and challenges to enable cyber situational awareness for network security. In: Proceedings of the 15th International Conference on Availability, Reliability and Security. ARES '20. Association for Computing Machinery, New York, NY, USA (2020). https://doi.org/10.1145/3407023.3407062
3. Various.: Open Cybersecurity Alliance. Open XDR (2024). https://github.com/opencybersecurityalliance/oxa
4. Endsley, M.R.: Toward a Theory of Situation Awareness in Dynamic Systems. Human Factors 37(1), 32–64 (1995)
5. Endsley, M.R.: Theoretical underpinnings of situation awareness: A critical review. In: Proceedings of the International Conference on Analysis and Measurement of Situation Awareness, p. 24 (1995)
6. Kott, A., Wang, C., Erbacher, R.F.: Cyber Defense and Situational Awareness. Advances in Information Security, vol. 62. Springer, Cham (2014). https://doi.org/10.1007/978-3-319-11391-3
7. Brynielsson, J., Franke, U., Varga, S.: Cyber Situational Awareness Testing. In: Combatting Cybercrime and Cyberterrorism, pp. 209–233. Springer, ??? (2016). https://doi.org/10.1007/978-3-319-38930-1_12
8. Franke, U., Brynielsson, J.: Cyber situational awareness—A systematic review of the literature. Computers & Security 46, 18–31 (2014) https://doi.org/10.1016/j.cose.2014.06.008
9. IEEE Robotics and Automation Society: Ieee standard ontologies for robotics and automation. IEEE Standards, 1–60 (2015) https://doi.org/10.1109/IEEESTD.2015.7084073
10. Javorník, M., Husák, M.: Mission-centric decision support in cybersecurity via Bayesian Privilege Attack Graph. Engineering Reports 4(12), 12538 (2022) https://doi.org/10.1002/eng2.12538
11. Jakobson, G.: In: Kott, A., Wang, C., Erbacher, R.F. (eds.) Mission Resilience, pp. 297–322. Springer, Cham (2014). https://doi.org/10.1007/978-3-319-11391-3_14
12. Ede, T., Aghakhani, H., Spahn, N., Bortolameotti, R., Cova, M., Continella, A., Steen, M., Peter, A., Kruegel, C., Vigna, G.: DeepCASE: Semi-Supervised Contextual Analysis of Security Events. In: Proceedings of the IEEE Symposium on Security and Privacy (S&P) (2022). IEEE

SecAwarenessTruss: A Dynamic Training Program Based on Cyber Ranges Leveraging IT Security, Privacy, and Data Protection Culture and Awareness of Critical Information Infrastructures

Christos Gizelis, Petros Panagiotidis, Nikolaos Nikoloudakis, Constantinos Andreou, Anna Andreoudi, Athanasios Arvanitis, Nikolaos Bakalis, Evangelos Floros, Athanasios Grigoriadis, Ioannis Kapatsoulias, Konstantinos Magkos, Dimitris Merkouris, Panagiotis Mpempis, Dimitris Papanikas, George Potamos, George Pouraimis, Michail Smyrlis, George Spanoudakis, Eliana Stavrou, Stavros Stavrou, and Stelios Trikos

Abstract The SecAwarenessTruss project tackles the essential requirement of cybersecurity training within the healthcare and energy sectors and telecommunications and maritime industries. Organizations that depend on digital technologies face rising cyber threats, which demand customized training programs that address

---

C. Gizelis (✉) · P. Panagiotidis · I. Kapatsoulias · P. Mpempis · D. Papanikas
Hellenic Telecommunications Organization (OTE), Athens, Greece
e-mail: cgkizelis@ote.gr

P. Panagiotidis
e-mail: ppanagiotid@ote.gr

I. Kapatsoulias
e-mail: ikapatsoul@ote.gr

D. Papanikas
e-mail: dpapanikas@ote.gr

N. Nikoloudakis · A. Arvanitis · N. Bakalis · D. Merkouris
Public Power Corporation S.A., Athens, Greece
e-mail: n.nikoloudakis@ppcgroup.com

D. Merkouris
e-mail: d.merkouris@ppcgroupc.com

C. Andreou · S. Trikos
National CSIRT-CY Digital Security Authority, Nicosia, Cyprus
e-mail: constantinos.andreou@csirt.cy

S. Trikos
e-mail: stelios.trikos@csirt.cy

J. Soldatos and H. Abie (eds.), *Technology-Enabled Resilience: Innovations in Critical Infrastructure Protection*, https://doi.org/10.1007/978-3-032-25917-2_17

sector-specific vulnerabilities. The initiative combines regulatory requirements with GDPR and NIS2 frameworks. The cyber range platform delivers scenario-based training, which enables participants to develop incident detection and response and recovery skills through simulated cyberattacks. The training modules focus on social engineering and human factors and advanced persistent threats to develop organizational cybersecurity resilience. The implementation of generative AI and machine learning in training approaches enables personalized learning experiences which prepare cybersecurity professionals to handle current and future threats. The SecAwarenessTruss project aims to enhance critical infrastructure cybersecurity defenses through collaborative efforts that drive ongoing improvement of defense strategies.

Keywords Cyber security · Cyber range · Generic training · Critical infrastructure · Health care · Energy · Telecoms · Maritime

---

A. Andreoudi · K. Magkos
National Cybersecurity Authority, Athens, Greece
e-mail: a.andreoudi@cyber.gov.gr

K. Magkos
e-mail: k.magkos@cyber.gov.gr

E. Floros
Department of Management Science and Technology, Hellenic Mediterranean University, Agios Nikolaos, Greece

IT Department, University General Hospital of Heraklion, Heraklion, Greece

E. Floros
e-mail: efloros@pagni.gr

A. Grigoriadis · G. Pouraimis
Ministry of National Defence, Athens, Greece
e-mail: ath.grigoriadis@gmail.com

G. Pouraimis
e-mail: g.pouraimis@cd.mil.gr

G. Potamos · E. Stavrou · S. Stavrou
Open University of Cyprus, Nicosia, Cyprus
e-mail: georgios.potamos@st.ouc.ac.cy

E. Stavrou
e-mail: eliana.stavrou@ouc.ac.cy

S. Stavrou
e-mail: stavros.stavrou@ouc.ac.cy

M. Smyrlis · G. Spanoudakis
Sphynx Technology Solutions AG, Nicosia, Cyprus
e-mail: smyrlis@sphynx.ch

G. Spanoudakis
e-mail: spanoudakis@sphynx.ch

1 Introduction

The protection of critical sectors depends on cybersecurity training to combat their continuously evolving cyber threats. The growing dependence of these industries on digital technologies and interconnected systems creates an exponential rise in potential cyberattack impact. Organizations need effective cybersecurity training programs to provide employees with the necessary skills and knowledge for identifying and preventing and responding to cyber threats. The training programs need to focus on sector-specific vulnerabilities while following regulatory standards and best practices to achieve complete protection of critical infrastructure. Organizations that invest in strong cybersecurity training develop better resilience while protecting sensitive data and ensuring the uninterrupted reliability of essential services.

1.1 Cybersecurity Fundamentals

The digital era established cybersecurity as an essential discipline to protect systems and networks together with data from unauthorized access and damage and theft. Information security relies on the CIA triad which consists of confidentiality alongside integrity and availability [1]. Access to confidential data remains restricted to authorized personnel through encryption combined with access-control systems. Integrity protects data accuracy and completeness by implementing digital signatures along with checksums. The availability principle provides legitimate users with continuous resource access by implementing redundancy mechanisms and failover strategies and network architectures.

Organizations reinforce the CIA triad through supplementary security principles which they implement [2]. Systems use authentication to verify user identities and system identities through credential methods as well as biometric verification and digital certification methods. The system uses authorization to decide what actions authenticated users can perform through role-based or policy-driven access controls. The systematic auditing process records system activities while conducting analysis to detect unauthorized activities and enable forensic investigations. The implemented measures create controlled access and enable accountability while providing traceability for information systems.

Cyber threats targeting information systems span the entire threat spectrum because of existing vulnerabilities. The family of malware including viruses, worms, trojans, and ransomware causes data destruction and operational interference as well as unauthorized access to systems [3]. Phishing attacks utilize social engineering methods to distribute deceptive messages and fake websites which trick users into revealing sensitive information [4]. Networks and applications become disabled through resource exhaustion and vulnerability exploitation; denial-of-service (DoS) attacks and their distributed version (DDoS) enlarge the attack impact through coordinated host exploitation [5]. The deployment of firewalls together with intrusion detection systems alongside thorough user awareness training programs represents an effective defense strategy.

Organizations must follow structured cybersecurity and data protection protocols which are mandated by regulatory frameworks. The European Union's General Data Protection Regulation (GDPR) requires strict controls for personal data processing and strengthens data subject rights while requiring breach notification and imposing severe penalties for non-compliance [6]. The ISO/IEC 27001 standard provides criteria to build an Information Security Management System (ISMS) which guides organizations to establish and improve security controls [7]. The standards provide legal compliance and strengthen stakeholder trust while developing a robust security posture.

This work is performed in the context of SecAwarenessTruss project that aims to develop cyber range training programs for critical infrastructure operating in the domains of health care, energy, telecoms, and maritime. The SecAwarenessTruss cyber range provides scenario-based training that develops sector-specific incident detection and response and recovery skills through authentic attack pattern simulations. The training system focuses on telecommunications, energy, healthcare, and maritime sectors to address their unique vulnerabilities which protect service continuity. The training system enforces compliance with GDPR and NIS2 regulations and industry-specific standards. The platform develops organization-wide security initiative collaboration through cross-departmental exercises which creates a unified and resilient defense posture.

2 Social Engineering and Human Factors

Attackers increasingly use social engineering to bypass network defenses, manipulating individuals into revealing confidential information or taking actions that jeopardize data. According to Verizon [8], social engineering incidents are on the rise, with pretexting—often linked to business email compromise (BEC)—accounting for 40% of breaches. Phishing remains a dominant tactic, and BEC transactions average around $50,000. ENISA's Threat Landscape 2024 [9] highlights social engineering as one of the top five threats, with nearly a million phishing attacks reported in Q4 2024 alone [10]. A surge in SMS phishing has been noted, fueled by new phishing kits. Advancements like phishing-as-a-service (PhaaS) and generative artificial intelligence (GenAI) have further complicated this threat, making attacks more personalized and scalable. Combating social engineering requires a comprehensive approach focusing on education, awareness, and robust security measures.

Social engineering is manipulating human psychology to gain unauthorized access to information or systems, rather than using technical hacking methods. For example, instead of relying on malware to steal a password, an attacker might trick a user into clicking a fraudulent link and entering their credentials voluntarily. The term gained popularity in the 1990s, primarily due to hacker Kevin Mitnick, although fraudsters have used the core tactics of deception and persuasion for centuries. These methods continue to evolve with technology and changes in human behavior.

Fraudsters manipulate their victims by using emotional triggers like anxiety and urgency to create a false sense of trust. They craft convincing stories or pretexts, often based on familiar scenarios or current issues, to mislead targets. Common examples include notifications about undelivered packages or compromised bank accounts. Broader contexts, such as economic hardship or health crises, are also exploited to increase pressure on victims. Techniques like phishing, which involves fake emails from trusted sources, and baiting, which lures victims with attractive offers, are frequently used, often leading to security compromises or malware installation. The most common clues that indicate a social engineering scam are as follows:

1. The presentation of a plausible problem and urging the victim to take immediate action.
2. The promise of a reward in exchange for some action by the victim.
3. The reference to a current social situation or related events (e.g., pandemic, charity work, unemployment, product, or service offers).
4. Any attempt to activate emotions, such as curiosity, interest, joy, and anxiety.

Addressing social engineering attacks requires awareness, training, and technical defenses. Regular campaigns should educate users on recognizing phishing, pretexting, and baiting tactics. Staff should be encouraged to question unusual requests for sensitive information and verify requests received via SMS or phone. Simulated attacks, like mock phishing, can help users practice spotting suspicious messages. These campaigns should include tailored training materials and metrics to assess training effectiveness.

Technical defenses are essential for combating social engineering attacks. User identities must be verified against updated account databases for confidential information requests. Implementing phishing-resistant Multi-Factor Authentication (MFA) can safeguard against stolen credentials. Filtering email and web traffic can also block suspicious activities and prevent access to phishing sites.

Clear procedures are necessary for situations when defenses fail, such as callback procedures for financial transactions and protocols for handling sensitive data. Finally, reporting procedures should encourage transparency, avoiding any implication of blame or fear.

3 SecAwarenessTruss Cyber Range and Training Modules

The cyber range functions as a specialized training platform for cybersecurity professionals who practice against complex simulated IT environments and real-world threats [11]. The platform presents a protected operational space built from virtualized cloud-based virtual machines that replicate endpoints alongside network devices and firewalls and intrusion detection/prevention systems for safe incident response training and defensive tactic development. Advanced ranges duplicate industrial control systems including smart meters alongside SCADA networks and embedded navigation components [11, 12]. The platform allows users to configure individual elements for matching exact organizational system structures and threat patterns.

The cyber ranges allow critical infrastructure organizations such as energy, telecommunication, health care, and transportation to prepare against cyber threats through simulations that do not risk their operational systems [12]. The platform simulates ransomware and DDoS attacks which help users detect vulnerabilities while validating policies and conducting incident response tests [13]. Operational teams consisting of IT staff and executives gain practical experience through realistic training that enhances their response times and decision-making abilities as well as their cross-functional coordination. The combination of research and engineering teams uses ranges to study attacker activities and create enhanced defenses which strengthen overall security posture [12, 13].

3.1 SecAwarenessTruss Cyber Range Platform

The SecAwarenessTruss platform brings together two separate solutions into one integrated system (Fig. 1). The SPHYNX Cyber Range (SCR) delivers model-driven customization for three pilot sectors including health care, energy, and telecommunications across various technology layers (Fig. 2). The wizard-based interface of SCR enables content creators to construct and modify and maintain scenarios that specifically target sector-based threats and risks. Training sessions include standard cybersecurity principles alongside specialized exercises that focus on each pilot sector's distinct operating environment.

The Open University of Cyprus's Cyber Threat Realm (CTR) operates the MARitime Training Environment (MARTE) to fulfill maritime sector requirements (Fig. 3). MARTE enables dual onshore/offshore operation for port and sea and land simulations through its combination of specialized ICT and OT systems. The platform operates sandboxed training exercises for naval operations combined with cyber-physical scenarios.

A unified dashboard UI serves as a central hub, linking users to both platforms. The system delivers real-time visibility through data visualizations that include metrics and charts and KPIs while allowing users to access training programs and assessments along with relevant cyber-threat intelligence.

Cyber Range Modules: Generic Training

Module 1: Basic Cyber Hygiene. Basic cyber hygiene is essential for maintaining the security and integrity of digital environments [14]. This module focuses on supporting end users with the knowledge and practical skills that are necessary to protect their network systems from common cyber threats. Common skills include secure password management, recognizing phishing attempts, and implementing

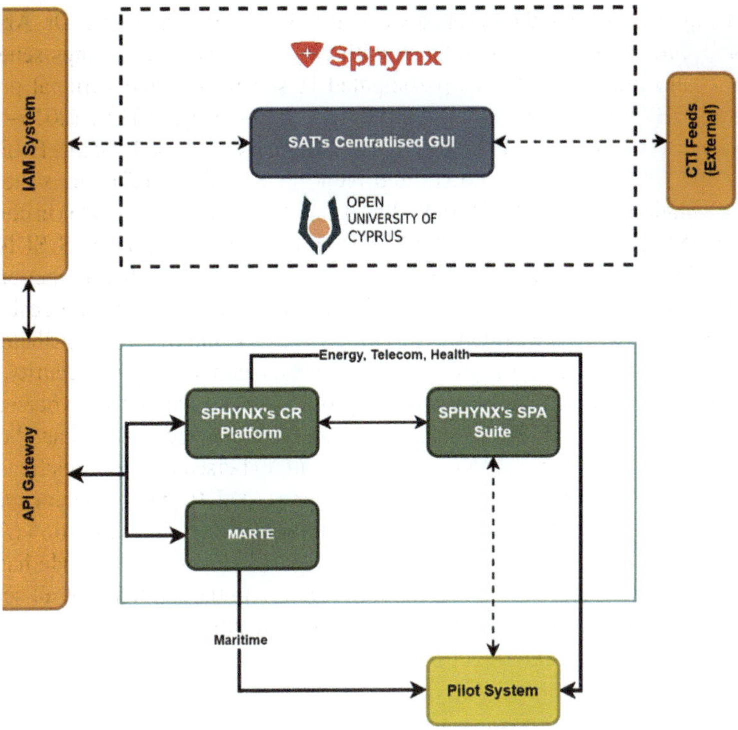

Fig. 1 SecAwarenessTruss' high-level architecture

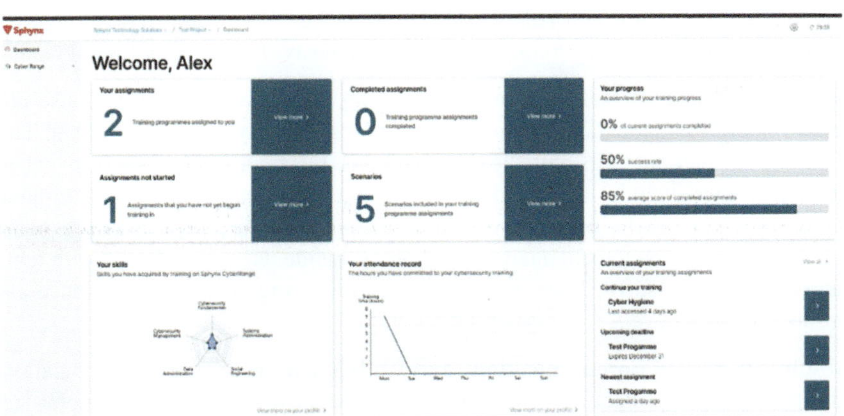

Fig. 2 Sphynx cyber range platform

Fig. 3 OUC cyber threat realm MARTE platform

regular software updates [15]. End users have learned how to identify vulnerabilities and use best practices to mitigate risks, ensuring a robust defense against cyberattacks.

Module 2: Data Protection and Privacy. Data protection and privacy are critical components of cybersecurity, especially when organizations handle sensitive information [16]. This module covers the principles of data protection, including GDPR compliance, secure data handling, and privacy by design. End users increased their understanding of how to safeguard personal and organizational data, implement encryption techniques, and respond to data breaches. The training highlights the importance of maintaining confidentiality and integrity in data management.

Module 3: Incident Response and Reporting. Effective incident response and reporting are fundamental for minimizing the impact of cyber incidents [17]. This module provides end users with the skills to detect, respond, and recover from cyberattacks. Topics include incident detection, containment strategies, and communication protocols. Participants learned to develop and implement incident response plans, conduct post-incident analysis, and report incidents in compliance with regulatory requirements. The training goal is to improve organizational resilience and ensure a rapid recovery from cyber threats.

Module 4: Secure Communications and Collaboration. Secure communications are essential for protecting sensitive information [18]. This module emphasizes securing communication channels, implementing encryption protocols, and raising a culture of cybersecurity awareness. Participants learned to recognize and mitigate risks associated with digital communication, collaborate securely across teams, and apply best practices for protecting information during collaboration. The training highlights the importance of maintaining secure and reliable communication in a digital environment.

Hands-on exercises and simulations are essential for developing practical cybersecurity skills. This module provides end users with training experiences within a controlled cyber range environment. End users engaged in realistic scenarios that replicate real-world cyber threats, allowing them to apply their knowledge and skills in a safe setting [19]. The exercises cover various topics, including threat detection, incident response, and system recovery. The training goal is to improve end users' ability to handle cyber incidents effectively and improve their overall cybersecurity posture.

Cyber Range Modules: Energy Sector Training

Modern society depends on the energy sector, but the implementation of IoT and AI and smart grids has made cyber threats more prominent. The Saudi Aramco malware attack [20], the Ukraine power-grid attack [21], the Triton/Trisis attack on a Saudi petrochemical plant [22], and the Colonial Pipeline ransomware attack [23] demonstrate that ransomware and state-sponsored APTs can disrupt operations and compromise safety and national security. Energy companies need to invest in modern detection capabilities and incident response systems while building a strong security culture for their entire workforce to combat these risks.

The digital transformation in the sector uses interconnected IT and OT systems which include advanced metering infrastructure (AMI) and wide area measurement systems (WAMS). AMI networks link smart meters from residential and commercial and industrial locations through RF, cellular, or power-line communications to an AMI Headend. The Headend aggregates meter data which becomes the input for the meter data management system (MDMS) before being used by billing and energy management platforms. Secure communication gateways together with encrypted routers protect sensitive information during its transmission. WAMS implements GPS-synchronized phasor measurement units (PMUs) across the entire grid, and phasor data concentrators (PDCs) collect synchrophasor data for real-time performance analysis which enhances grid resilience.

The organization structure together with regulatory requirements determines that training needs differ across all positions including both field technicians and senior management. To combat evolving threats, it is crucial to conduct continuous awareness campaigns for personnel. The SecAwarenessTruss cyber range modules in the energy sector conduct simulations of attacks on AMI and WAMS components through ransomware on AMI Headends as well as insider malware insertion and rogue Headend-initiated DDoS and denial-of-service on meter data and zero-day exploitation in Headend software and GPS spoofing of PMUs. The scenarios implement NIS2 guidelines along with GDPR standards and ENISA recommendations and ECSO requirements to meet both regulatory standards and protect infrastructure.

The SecAwarenessTruss cyber range platform allows energy organizations to dramatically improve their cybersecurity position. The hands-on exercises build technical abilities which produce better operational results and stronger grid performance. Organizations achieve security strength and legal compliance through adherence to NERC CIP and GDPR and ISO 27001. The combination of training measures to combat human errors minimizes incident probabilities, and ongoing professional development enhances employee commitment levels. These combined strategies develop an energy sector which demonstrates resilience against contemporary cyber threats while fulfilling demanding regulatory requirements.

Cyber Range Modules: Maritime Sector Training

The maritime sector is undergoing a significant digital transformation, integrating advanced technologies to improve vessel operations, optimize logistics, and manage critical data such as shipping routes and cargo manifests. While these innovations enhance efficiency and competitiveness, they also increase the sector's vulnerability to cyber threats [24, 25]. Maritime organizations, from shipping corporations to port authorities, now face a growing array of cyber risks, ranging from attacks on IT networks to more alarming scenarios involving shipboard systems and port infrastructure.

Prominent threats that the maritime sector is facing include automatic identification system (AIS) spoofing, ransomware, phishing, and DoS attacks. The consequences of such cyber incidents can be severe, including the compromise of sensitive data, disruption of navigation and cargo handling, and substantial financial and reputational damage. As cyber threats become more sophisticated, the maritime domain must establish robust maritime cyber situational awareness (MCSA) at strategic, operational, and technical levels.

Although several guidelines and standards exist to promote maritime cybersecurity [26–28], there is a pressing need to address challenges [29] and increase awareness and practical preparedness across the sector. This requires targeted education and training programs tailored to the specific roles and responsibilities of maritime stakeholders. Integrating cyber range-based exercises into training programs provides a hands-on, scenario-driven approach to understanding the threat landscape and building the skills necessary to prevent, detect, and respond to incidents [30].

Based on SecAwarenessTruss sector-specific risk assessment and investigations, training priorities in the maritime sector have been identified along two strategic directions: (1) Raising General Awareness: Focused on the broader maritime audience, including port administrators, shipping company personnel, ship captains, and crew, this training stream aims to enhance understanding of cyber threats and regulatory compliance obligations. (2) Developing Technical Competencies: Targeting personnel responsible for critical systems such as navigational, surveillance, and industrial IT/OT environments, this stream emphasizes technical preparedness and incident response capabilities.

The goal is to equip all levels of maritime personnel with the knowledge and skills required to foster a resilient cybersecurity culture. To guide the development of new training curricula, SecAwarenessTruss has identified specific focus areas. A core priority is building awareness of the evolving cyber threat landscape, helping personnel understand how cyber incidents can disrupt operations and impact mission-critical systems. This includes a focus on recognizing vulnerabilities in both IT and OT environments that are specific to maritime operations [31]. In parallel, training in regulatory compliance ensures that individuals across the sector are familiar with relevant cybersecurity frameworks and best practices, supporting alignment with international standards. Given the prevalence of phishing attacks, dedicated modules help personnel identify sophisticated phishing techniques that could compromise onboard and port-side systems. Additionally, specialized training is provided for those responsible for navigational and surveillance IT/OT systems, emphasizing detection and response strategies for attacks targeting these critical technologies. Finally, incident analysis and response training equips maritime staff with the necessary skills to investigate, contain, and recover from cyber incidents, thereby strengthening the sector's overall resilience and ability to maintain continuity in the face of growing cyber threats.

By delivering tailored, role-specific training curricula through SecAwarenessTruss' cyber range module, the development of practical cybersecurity skills across the maritime sector can be enabled. This tailored approach reinforces the industry's ability to proactively manage cyber risks and cultivate a strong cybersecurity culture essential for operational continuity and resilience.

Cyber Range Modules: Healthcare Sector Training

The SecAwarenessTruss cyber range modules for PAGNI (University General Hospital of Herakleion) are focused on the critical cybersecurity needs of the healthcare sector by simulating realistic attack and defense scenarios. Emphasis is placed on protecting patient data and securing Internet of Medical Things (IoMT) devices, electronic health records (EHRs), and imaging systems. The web application vulnerabilities of cross-site scripting (XSS), arbitrary code execution, and RSA signature bypass are threats to patient archives and highlight the need for secure coding, robust cryptography, and strict access controls in systems like PACS [31, 32]. The modules also address the issue of insufficient encryption and weak authentication that is a concern in healthcare IT environments [33].

Training on ransomware is key because it is one of the most critical threats to patient safety and operations. Exercises recreate network-wide infections and data encryption events which are classified as very high risk with catastrophic outcomes allowing staff to practice containment, restoration, and coordination. The scenarios include historical incidents such as WannaCry's disruption of the UK's NHS [34, 35] and emerging global trends as reported by HC3 and CISA [36]. Administrators also train on lateral movement detection within compromised Active Directory environments to strengthen internal defenses [32].

There is compliance with the evolving EU regulations throughout. The NIS2 Directive and GDPR set strict requirements for risk management, incident reporting, and resilience for hospitals [37, 38]. The SecAwarenessTruss regulatory assessment brings these requirements to life through interactive audit simulations, breach response workflows, and documentation exercises to ensure practical readiness for legal obligations [32]. This is in line with the EU's overall strategy to improve cybersecurity in essential services under NIS2 [38].

Hands-on exercises use the SPA (Security and Privacy Assurance) framework to deliver six priority attack scenarios including phishing campaigns, unauthorized access attempts, and Active Directory takeovers. These immersive modules serve both novice and experienced participants and combine phishing simulations, secure configuration reviews, and real-time incident analysis. Participants are trained in risk prioritization and containment under pressure and get practical skills that are specific to PAGNI's infrastructure [31–33].

By integrating technical defense strategies, regulatory compliance, and immersive simulations, the SecAwarenessTruss healthcare modules provide a comprehensive training ecosystem. They empower IT personnel and clinicians to anticipate, detect, and respond effectively to cyber threats and thus enhance the resilience of critical health infrastructure against current and emerging risks [36].

Cyber Range Modules: Telecommunications Sector Training

Network protection for telecommunications requires multiple layers which defend against technological weaknesses alongside human mistakes. Telecommunications operators who adopt digitalization through on-premises and cloud hybrid infrastructure face advanced threats to Active Directory systems along with internet-accessible applications and customer portal infrastructure. The SecAwarenessTruss cyber range provides organizations with a telecom security module that trains teams through authentic attack simulations. The participants carry out defensive drills against Golden Ticket persistence exploits together with simulated supply chain ransomware deployments and DDoS campaigns. Through these practical exercises, security personnel can enhance their environment defenses with established technical security measures.

The handling of extensive sensitive customer information by telecom providers requires absolute priority on data integrity and confidentiality. According to SecAwarenessTruss research on breaches, a 2024 AT&T metadata exfiltration incident exposed information of 109 million users [39]. The training modules focus their efforts on secure credential management alongside robust encryption across data lifecycles and access control architectures that block lateral movement. Through signal-to-noise simulation activities, participants learn to identify both substandard and exemplary practices.

Service availability remains essential because millions depend on continuous telecom services. The cyber range simulates extensive web of death DDoS attacks against French providers during the July 2024 Paris Olympics [40]. The training program helps participants recognize early attack indicators before teaching them to activate filtering systems and rate limits and maintain service continuity under heavy network usage. The training exercises develop the reflexes essential for quick incident response during actual cyberattacks.

Theory as a stand-alone method fails to prepare defense teams against sophisticated telecom security threats. The SecAwarenessTruss cyber range fills the knowledge gap through four-phase exercises which start with network environment setup followed by baseline review before executing complex attacks (Golden Ticket and ADFS credential theft and large-scale DDoS) followed by attack pattern analysis and prevention strategy implementation for vulnerability mitigation testing.

The entire attack life cycle from initial compromise through lateral movement to data exfiltration or service disruption enables telecom security teams to develop their confidence and proficiency. Genuine threats trigger these trained responders to deliver necessary defense capabilities which protect the essential communications infrastructure that supports modern society.

4 Advanced Cyber Range Exercises

4.1 Red Team/Blue Team Exercises: Attack and Defense Simulations

These exercises model and simulate real-world cyberattacks to test and improve the organization's defenses. The Red Team, which acts as attackers, uses tactics, techniques, and procedures (TTPs) that adversaries use to penetrate the organization's systems. Meanwhile, the Blue Team, which consists of the organization's cyber security personnel, is responsible for detecting, responding to, and mitigating simulated attacks in real-time [41].

The exercises follow an organized process to ensure realistic attack simulations and effective defense strategies. This process includes the definition of the scope, simulating the attack, detecting and responding, and analysis of the outcomes for improving the security processes.

4.2 Threat Hunting and Incident Response Challenges

The project aims to improve the skills and resilience of cybersecurity professionals to detect, respond, and mitigate cyber threats minimizing the risk of harming production systems [42]. The training programs focus on identifying vulnerable systems, simulating attack and response scenarios, including ransomware and DDoS attacks, and enhancing situational awareness [43]. The exercises are designed to replicate real-world IT environments, providing a controlled and immersive environment for cybersecurity professionals to conduct various drills and simulations.

4.3 Advanced Persistent Threat Simulation and Analysis

The project addresses zero-day threats and APTs through training scenarios that cover detection, prevention, simulation, analysis, policies, automation, and knowledge enhancement. The training programs aim to assess cyberattack detection accuracy, use tools for attack recognition and mitigation, and increase automation and cyber-resilience. The scenarios involve planning mitigation actions, re-training models for infrastructure changes, and addressing zero-day threats and APTs [44].

The project is a comprehensive initiative aimed at enhancing cybersecurity awareness, knowledge transfer, and skills development through advanced cyber range exercises. It provides realistic and immersive training environments, ensuring that participants are well prepared to defend against advanced cyber threats. The effort on situational awareness and sector-specific training further improves the efficiency of the training programs, making the project a valuable resource for critical infrastructure organizations.

5 Assessment and Evaluation

A comprehensive approach with well-defined, clear metrics and key performance indicators (KPIs) is required to assess the effectiveness of training. We use time-based metrics for the most part to gauge our training programs. They tell us whether or not we are being efficient with our time. The basic question we are trying to answer with these metrics is: Are our students completing the training in a timely manner? We have completion rates, and we evaluate them based on some expected levels we have set. Basically, if you are not completing the training on time we have designated, you are failing. That is the basic premise of time-based metrics. Then you get into the whole discussion about: Does time equal efficiency? Does time equal effectiveness? Does time equal competency? And somehow, if we cannot marry up those time-based metrics with effectiveness, efficiency, or competency metrics, then we are just left with this cloud of ambiguity.

Before training starts, establishing core competencies in cybersecurity is critical if an organization wants to develop an effective program. Assessing participants before training not only helps bridge the knowledge gap but also builds better, more personalized programs. Some organizations use standardized assessments alongside core competencies to ensure individuals have what it takes to make it through the program. These assessments also help gauge where the organization is starting from and enable more accurate trend measurements later on. The assessments themselves should not be easy, and they should cover material that is fundamental to cybersecurity and the competencies trained later on. Of course, that assumes the competencies are understood and agreed upon by the organization and the authors of the training.

The landscape of cybersecurity is changing too fast for static training programs to be beneficial. To be effective, the programs need to be continuously improved and be based on real-time feedback mechanisms. So, if we are to closely relate the SecAwarenessTruss framework, to going from a world of Objective, Metrics, to a world of Objectives, Metrics, and Feedback, which is the world of the Continuous Improvement program; this is what we are doing. We collect all what was wrong, we collect feedback from participants, we collect performance metrics, and we near the same path traveled by the hackers to collect information they use to "improve" their work.

6 Legal and Ethical Considerations

6.1 *Data Privacy and Security in Cyber Range Training*

The cyber range technology duplicates complicated enterprise and government infrastructure networks together with critical infrastructure systems in physical, virtual, and cloud-based settings which help professionals conduct offensive and defensive drills using genuine information and streaming content and authentic system vulnerabilities [45]. Higher realism levels directly result in rising privacy and security problems [45, 46]. The research investigates fundamental difficulties along with suitable approaches under existing legal and ethical standards.

6.2 *Privacy Concerns in Cyber Range Training*

Several essential privacy areas emerge during cyber range training because authentic data usage through real or pseudonymized information exposes re-identification vulnerabilities which may violate GDPR rules [46], and detailed participant activity logs for performance evaluation potentially violate individual rights unless users give consent to proper data management systems. Unclear data maintenance procedures violate data minimization and purpose limitation principles [44]. Furthermore, training platforms that span across borders require data transfer to jurisdictions with unique legal frameworks which weakens users' rights enforcement [44].

6.3 *Ethical Use of Generative AI in Cybersecurity Education*

The adoption of generative AI through large language models (LLMs) and code generators together with synthetic data tools enhances scenario generation as well as personalized tutoring and synthetic data creation and adversarial behavior simulation capabilities. The adoption of any revolutionary technology brings essential moral issues to the forefront. Cybersecurity education faces enhanced ethical challenges because it includes both student ethical conduct and national defense concerns together with data security measures and proper dual-use technology practices. Generative AI brings multiple educational benefits to cybersecurity training programs through its capabilities:

- Scenario Generation: Automated crafting of realistic attack narratives.
- Intelligent Tutoring Systems: Personalized feedback based on learner input.
- Synthetic Data Creation: Statistically valid datasets without exposing sensitive information.
- Simulating Adversarial Behavior: AI agents replicating real-world attacker tactics.

These capabilities can dramatically reduce instructional overhead, enhance the realism of training, and tailor learning experiences to individual skill levels.

6.4 Responsible Disclosure of Vulnerabilities and Security Incidents

The discovery of vulnerabilities and their subsequent reporting form the foundation of maintaining a safe digital environment [45]. When training facilities encounter new discovered flaws, their teams must address all ethical considerations together with legal aspects and social consequences [47]. The responsible disclosure process also known as coordinated vulnerability disclosure (CVD) requires private vulnerability reports to vendors for pre-publication remediation. It is governed by:

- Ethical guidelines that prioritize harm reduction.
- Legal constraints on unauthorized access and reverse engineering.
- Communication frameworks defining timelines, acknowledgments, and credit.

6.5 Best Practices for Responsible Disclosure in Training

A cyber range must establish an official disclosure procedure which includes reporting contacts along with participant protections and employ safe simulated environments that do not interact with live systems unless specifically intended for that purpose to minimize actual system exposure [48]. Teaching students about disclosure ethics requires including case studies and decision-making exercises throughout the curriculum. The training program should establish partnerships with vendors and computer emergency response teams (CERTs) to simplify and validate reporting procedures, and it should promote student involvement in bug bounty programs to utilize established legal and ethical frameworks for responsible skill development.

The ethical principle of responsible disclosure surpasses formal procedures. The practice of cyber education needs to strengthen three core values: integrity and accountability and public interest during its adoption of realistic offensive techniques. The combination of coordinated vulnerability reporting with ethical guidelines in cyber ranges develops experts who have both technical proficiency and moral principles.

7 Future Trends in Cybersecurity Training

As cyber threats continue to evolve, cybersecurity training needs to also adapt, to meet the increasing complexity of modern, digital ecosystems. Traditional awareness programs are being supplemented by innovative methods incorporating artificial intelligence (AI), machine learning, and extended reality to create immersive, responsive, and personalized learning environments [49, 50]. Emerging pedagogical models emphasize hands-on learning through simulation platforms and secure design courses, equipping learners with real-world skills to defend against sophisticated attacks [51]. These experiential approaches are key to preparing cybersecurity professionals capable of navigating increasingly intricate and unpredictable threat landscapes.

Recent advancements in generative AI, particularly GPT models, are transforming the landscape of cybersecurity training by enabling more dynamic, personalized, and effective learning experiences. GPT-powered training programs can adapt content to individual learning styles and roles, significantly enhancing engagement and knowledge retention compared to traditional methods [52]. In addition to training, GPT-4 has demonstrated the capability to generate comprehensive cybersecurity policies, especially for mitigating threats like ransomware, with results often surpassing human-created guidelines in certain contexts [53]. Generative AI was leveraged in [54] to develop phishing-tailored attack scenarios in the maritime domain. The scenarios can be embedded in awareness raising and training activities that can be delivered, e.g., over cyber ranges, aiming to enhance the workforce's cyber-resilience against phishing attacks.

Platforms such as Crimson further enhance GPTs for strategic reasoning by integrating CVE threat data with frameworks like MITRE ATT&CK [55], while CyGPT boosts network security training through knowledge graph augmentation to provide accurate, domain-specific outputs [56]. GPT models have also been employed in cyber intelligence training, helping analysts extract structured insights from historical incident reports to support threat modeling and simulation exercises [57]. A technical review by Gao [58] further highlights how GPT models, along with derivatives like SecureBERT, are contributing to improved threat detection and adaptive learning in cybersecurity applications.

Another major shift is the increasing demand for interdisciplinary and soft skills in cybersecurity education. Future cybersecurity training is expected to blend technical instruction with ethics, teamwork, and continuous learning to develop well-rounded professionals [59]. As cyber threats intersect with legal, economic, and sociopolitical domains, the inclusion of cross-disciplinary knowledge enhances situational awareness and problem-solving capabilities. Optimized curricula that integrate these broader competencies, alongside technical proficiencies, are critical in cultivating adaptable, resilient workforces ready to handle dynamic cyber environments). A core soft skill that can be empowered by generative AI is self-efficacy [60]. As demonstrated in [61], generative AI can be leveraged to empower professionals taking ownership of their learning by assisting them to create a personalized cybersecurity professional development plan. ENISA's European Cybersecurity Skills Framework was utilized to inform prompt engineering and align the investigations with cybersecurity career roles.

Looking ahead, predictive cybersecurity and real-time threat modeling will play an increasingly central role in training programs. With AI and machine learning gaining prominence in cybersecurity operations, training must evolve to include these technologies for proactive risk detection and mitigation [62]. Furthermore, platforms like CyberSec Labs emphasize hands-on exercises in cloud computing and penetration testing to bridge the gap between theoretical knowledge and industry-ready skills [63]. This signals a future where cybersecurity training is not only technologically enriched but also highly adaptive, modular, and learner centric.

Finally, GPT models can significantly enhance cybersecurity awareness programs by tailoring content to individual learning styles, increasing engagement and knowledge retention. A recent study demonstrated that GPT-enabled training modules outperformed traditional methods by delivering dynamic, role-specific content and real-time feedback [52].

8 Conclusions

The SecAwarenessTruss framework represents a transformative approach to cybersecurity education for critical infrastructure sectors. Our findings demonstrate that sector-specific cyber range training modules yield superior results compared to generic approaches, addressing the unique vulnerabilities of telecommunications, energy, healthcare, and maritime environments. The framework's innovative use of immersive, scenario-based learning effectively counters sophisticated social engineering attacks that target human vulnerabilities rather than technical weaknesses.

Key recommendations include implementing customized training content focused on sector-specific threats, integrating regulatory frameworks like GDPR and NIS2, and adopting emerging technologies such as Generative AI to enhance training effectiveness. The framework's emphasis on three core elements—CIA triad understanding, social engineering threat identification, and incident response development—provides a comprehensive foundation for cybersecurity preparedness.

The SecAwarenessTruss approach facilitates this through adaptive training scenarios, performance metrics, and feedback mechanisms that drive ongoing improvement. As cyber threats and defensive technologies evolve, training programs must continuously adapt, developing both technical competencies and essential soft skills among security personnel.

By establishing a dynamic training framework that balances technical proficiency with human factors awareness, SecAwarenessTruss contributes significantly to building cyber-resilient critical infrastructure sectors. This approach not only enhances incident response capabilities but also fosters a proactive security culture that strengthens the defense of vital services against increasingly sophisticated and interconnected cyber threats, ultimately ensuring the continuity and reliability of essential operations that underpin modern society.

Acknowledgements This project has received funding from the European Union's Digital Europe research and innovation program under grant agreement No 101128049 (SecAwarenessTruss). Views and opinions expressed are however those of the author(s) only and do not necessarily reflect those of the European Union. Neither the European Union nor the granting authority can be held responsible for them.

References

1. Samonas, S., & Coss, D. (2014). The CIA Strikes Back: Redefining Confidentiality, Integrity and Availability in Security. Journal of Information System Security, 10(3), 21–45.
2. Sandhu, R. S., & Samarati, P. (1996). Authentication, Access Control, and Intrusion Detection. In The Computer Science and Engineering Handbook (pp. 1929–1948). CRC Press.
3. Alazab, M., Venkatraman, S., Watters, P., Alazab, M., & Alazab, A. (2013). Cybercrime: The Case of Obfuscated Malware. In Cybercrime and Cloud Forensics: Applications for Investigation Processes (pp. 1–18). IGI Global.
4. Jagatic, T. N., Johnson, N. A., Jakobsson, M., & Menczer, F. (2007). Social Phishing. Communications of the ACM, 50(10), 94–100.
5. Mirkovic, J., & Reiher, P. (2004). A Taxonomy of DDoS Attack and DDoS Defense Mechanisms. ACM SIGCOMM Computer Communication Review, 34(2), 39–53.
6. Voigt, P., & Von dem Bussche, A. (2017). The EU General Data Protection Regulation (GDPR): A Practical Guide. Springer International Publishing.
7. ISO/IEC. (2013). ISO/IEC 27001:2013 Information technology—Security techniques—Information security management systems—Requirements. International Organization for Standardization.
8. Verizon (2024). Data Breach Investigations Report. Retrieved from: https://www.verizon.com/business/resources/reports/dbir/
9. ENISA (2024). Threat Landscape Report 2024. Retrieved from: https://www.enisa.europa.eu/publications/enisa-threat-landscape-2024
10. APWG (2024). Phishing Activity Trends Report 4th Quarter 2024. Retrieved from: https://apwg.org/trendsreports/
11. European Cyber Security Organisation (2020). Understanding Cyber Ranges: From Hype to Reality. Retrieved from: https://ecs-org.eu/?publications=https-ecs-org-eu-documents-publications-5fdb291cdf5e7-pdf
12. Cyber Range Project Team, NICE Community Coordinating Council (2023). The Cyber Range - A Guide. NIST. Retrieved from: https://www.nist.gov/document/cyber-range
13. European Cyber Security Organisation (2024). Cyber Range Features Checklist. Retrieved from: https://ecs-org.eu/?publications=cyber-range-features-checklist-list-of-european-providers-2024
14. Maennel, K., Mäses, S., Maennel, O. (2018). Cyber Hygiene: The Big Picture. In: Gruschka, N. (eds) Secure IT Systems. NordSec 2018. Lecture Notes in Computer Science(), vol 11252. Springer, Cham. https://doi.org/10.1007/978-3-030-03638-6_18
15. Gupta, S., Furnell, S. (2022). From Cybersecurity Hygiene to Cyber Well-Being. In: Moallem, A. (eds) HCI for Cybersecurity, Privacy and Trust. HCII 2022. Lecture Notes in Computer Science, vol 13333. Springer, Cham. https://doi.org/10.1007/978-3-031-05563-8_9

16. Alafaa, Princess, Data Privacy and Data Protection: The Right of User's and the Responsibility of Companies in the Digital World. (January 7, 2022). Available at SSRN: https://ssrn.com/abstract=4005750 or https://doi.org/10.2139/ssrn.4005750
17. Sharifah Roziah Mohd Kassim, Shujun Li and Budi Arief. Incident Response Practices Across National CSIRTs: Results from an Online Survey. OIC-CERT Journal of Cyber Security Volume 4, Issue 1 (April 2022) p 67–84. ISSN 2636-9680, eISSN 2682-9266.
18. Oliv J Patel, Dhruvil R Patel, Riddhi A Mehta. Secure Multi-Party Computation for Collaborative Data Analysis. International Research Journal of Engineering and Technology (IRJET) Volume: 11 (Mar 2024). e-ISSN: 2395-0056, p-ISSN: 2395-0072.
19. European Cyber Security Organisation (ECSO). WG5 PAPER. Understanding Cyber Ranges: From Hype to Reality. SWG 5.1 I Cyber Range Environments and Technical Exercises (March 2020).
20. Zetter, K. (2012). Saudi Aramco hack: A look at the damage done. Wired. Retrieved from https://www.wired.com/2012/08/saudi-aramco-malware/
21. Heisbourg, F. (2016). Cyberattack on Ukraine's Power Grid: Implications for Global Cybersecurity. European Council on Foreign Relations. Retrieved from https://www.ecfr.eu/article/commentary_cyberattack_on_ukraines_power_grid_implications_for_global_cybersecuri
22. Fruhlinger, J. (2018). Triton Malware: The cyber attack on a Saudi Petrochemical Plant. CSO Online. Retrieved from https://www.csoonline.com/article/3304257/triton-malware-the-cyber-attack-on-a-saudi-petrochemical-plant.html
23. Perlroth, N. (2021). Hackers behind Colonial Pipeline attack: What we know about the group behind the hack. The New York Times. Retrieved from https://www.nytimes.com/2021/05/10/technology/colonial-pipeline-hackers.html
24. Potamos, G., Theodoulou, S., Stavrou, E., Stavrou, S. (2022a), Building maritime cybersecurity capacity against ransomware attacks, Proceedings of the International Conference on Cybersecurity, Situational Awareness and Social Media: Cyber Science 2022, 20–21 June, Wale
25. BIMCO (2021), The Guidelines on Cyber Security Onboard Ships, https://www.bimco.org/about-us-and-our-members/publications/the-guidelines-on-cyber-security-onboard-ships
26. ENISA (2022), Guidelines - Cyber Risk Management for Ports, https://www.enisa.europa.eu/publications/guidelines-cyber-risk-management-for-ports
27. IMO (2022), Guidelines on Maritime Cyber Risk Management, MSC-FAL.1/Circ.3, https://wwwcdn.imo.org/localresources/en/OurWork/Security/Documents/MSC-FAL.1-Circ.3-Rev.2%20-%20Guidelines%20On%20Maritime%20Cyber%20Risk%20Management%20(Secretariat).pdf
28. Chupkemi, D. C., & Mersinas, K. (2024). Challenges in Maritime Cybersecurity Training and Compliance. Journal of Marine Science and Engineering, 12(10), 1844. https://doi.org/10.3390/jmse12101844
29. Potamos, G., Theodoulou, S., Stavrou, E., Stavrou, S. (2022b) Maritime cyber threats detection framework: Building capabilities, IFIP World Conference on Information Security Education (WISE), 2022.
30. Potamos, G., Stavrou, E., Stavrou, S. (2024) Enhancing maritime cybersecurity through operational technology sensor data fusion: A comprehensive survey and analysis, Sensors 2024, 24(11), 3458; https://doi.org/10.3390/s24113458
31. F. Cremer, B. Sheehan, M. Fortmann, A. N. Kia, M. Mullins, F. Murphy, and S. Materne, "Cyber risk and cybersecurity: A systematic review of data availability," The Geneva Papers on Risk and Insurance - Issues and Practice, vol. 47, no. 3, pp. 698–736, Jul. 2022, https://doi.org/10.1057/s41288-022-00266-6
32. M. N. Katsantonis, A. Manikas, I. Mavridis, and D. Gritzalis, "Cyber range design framework for cyber security education and training," International Journal of Information Security, vol. 22, no. 4, pp. 1005–1027, Aug. 2023, https://doi.org/10.1007/s10207-023-00680-4
33. C. S. Kruse, B. Frederick, T. Jacobson, and D. K. Monticone, "Cybersecurity in healthcare: A systematic review of modern threats and trends," Technol. Health Care, vol. 25, no. 1, pp. 1–10, 2017, https://doi.org/10.3233/THC-161263.

34. S. Ghafur, S. R. Kristensen, K. Honeyford, G. Martin, A. Darzi, and P. Aylin, "A retrospective impact analysis of the WannaCry cyberattack on the NHS," NPJ Digit. Med., vol. 2, no. 1, pp. 1–7, Oct. 2019, https://doi.org/10.1038/s41746-019-0161-6
35. M. Aljaidi et al., "NHS WannaCry Ransomware Attack: Technical Explanation of The Vulnerability, Exploitation, and Countermeasures," 2022 International Engineering Conference on Electrical, Energy, and Artificial Intelligence (EICEEAI), Zarqa, Jordan, 2022, pp. 1–6, https://doi.org/10.1109/EICEEAI56378.2022.10050485
36. Ojo, Abraham. (2025). "Ransomware trends and mitigation strategies: A comprehensive review". Global Journal of Engineering and Technology Advances, vol. 22, https://doi.org/10.30574/gjeta.2025.22.3.0038
37. Shin, Yongjoo & Kwon, Hyukjin & Jeong, Jaeyeong & Shin, Dongkyoo. (2024). A Study on Designing Cyber Training and Cyber Range to Effectively Respond to Cyber Threats. Electronics. 13. 3867. https://doi.org/10.3390/electronics13193867
38. European Union Agency for Cybersecurity (ENISA), Guidelines for the Implementation of the NIS2 Directive, 2023. [Online]. Available: https://www.enisa.europa.eu/publications/guidelines-on-implementing-the-nis2-directive
39. J. Greig, "AT&T reportedly paid ransom for deletion of stolen call logs after culprit allegedly detained," [Online]. Available: https://therecord.media/att-ransom-data-breach
40. E. Jennings-Trace, "French telecom networks experience disruption as malicious sabotage affects several operators—but Olympics remains unaffected," 2024. [Online]. Available: https://www.techradar.com/pro/french-telecom-networks-experience-disruption-as-malicious-sabotage-affects-several-operators-but-olympics-remains-unaffected
41. Granåsen, M., Andersson, D., (2016), Measuring team effectiveness in cyber-defense exercises: A cross-disciplinary case study, Cognition, Technology & Work, 18(1), 121143. https://doi.org/10.1007/s10111-015-0350-2
42. Zichen Wang. A Systematic Literature Review on Cyber Threat Hunting. arXiv:2212.05310v1 [cs.CR] (Dec 2022)
43. Adegbindin, Mariam. "Enhancing Cybersecurity Resilience Through Threat Hunting and Incident Strategies," n.d.
44. Chen, P., Desmet, L., Huygens, C. (2014). A Study on Advanced Persistent Threats. In: De Decker, B., Zúquete, A. (eds) Communications and Multimedia Security. CMS 2014. Lecture Notes in Computer Science, vol 8735. Springer, Berlin, Heidelberg. https://doi.org/10.1007/978-3-662-44885-4_5
45. ENISA – European Union Agency for Cybersecurity. (2022). Coordinated Vulnerability Disclosure Policies in the EU. https://www.enisa.europa.eu/publications/coordinated-vulnerability-disclosure-policies-in-the-eu
46. European Union. (2016). General Data Protection Regulation (GDPR).https://gdpr.eu/
47. National Institute of Standards and Technology (NIST). (2023). AI Risk Management Framework (AI RMF 1.0). https://www.nist.gov/itl/ai-risk-management-framework
48. International Organization for Standardization. (2013). ISO/IEC 27001: Information technology — Security techniques — Information security management systems — Requirements. https://www.iso.org/isoiec-27001-information-security.html
49. bin Mohammed Almoughem, Khaled Abdelaziz. "The Future of Cybersecurity Workforce Development." Ajrsp 4.45 (2023): 37–48.
50. Lampropoulos, Georgios, and Kinshuk. "Virtual reality and gamification in education: a systematic review." Educational technology research and development 72.3 (2024): 1691–1785.
51. Sharevski, Filipo, Adam Trowbridge, and Jessica Westbrook. "Novel approach for cybersecurity workforce development: A course in secure design." 2018 IEEE integrated STEM education conference (ISEC). IEEE, 2018.
52. Al-Dhamari, Nabil, and Nathan Clarke. "GPT-Enabled Cybersecurity Training: A Tailored Approach for Effective Awareness." IFIP World Conference on Information Security Education. Cham: Springer Nature Switzerland, 2024.
53. McIntosh, Timothy, et al. "Harnessing GPT-4 for generation of cybersecurity GRC policies: A focus on ransomware attack mitigation." Computers & security 134 (2023): 103424.

54. Loupasakis, M., Potamos, G., Stavrou, E. (2024). Revolutionizing Social Engineering Awareness Raising, Education and Training: Generative AI-Powered Investigations in the Maritime Domain. In: Moallem, A. (eds) HCI for Cybersecurity, Privacy and Trust. HCII 2024. Lecture Notes in Computer Science, vol 14729. Springer, Cham. https://doi.org/10.1007/978-3-031-61382-1_5
55. Jin, Jiandong, et al. "Crimson: Empowering strategic reasoning in cybersecurity through large language models." arXiv preprint arXiv:2403.00878 (2024).
56. Ou, Lu, et al. "CyGPT: Knowledge Graph-Based Enhancement Techniques for Large Language Models in Cybersecurity." 2024 IEEE 9th International Conference on Data Science in Cyberspace (DSC). IEEE, 2024.
57. Sufi, Fahim. "An innovative GPT-based open-source intelligence using historical cyber incident reports." Natural Language Processing Journal 7 (2024): 100074.
58. Gao, Mingze. "The Advance of GPTs and Language Model in Cyber Security". Highlights in Science, Engineering and Technology, vol. 57, July 2023, pp. 195–02, https://doi.org/10.54097/hset.v57i.10001.
59. Qawasmeh, Saif Al-Dean, Ali Abdullah S. AlQahtani, and Muhammad Khurram Khan. "Navigating cybersecurity training: A comprehensive review." Computers and Electrical Engineering 123 (2025): 110097.
60. Stavrou, E. and Piki, A. (2024), "Cultivating self-efficacy to empower professionals' re-up skilling in cybersecurity", Information and Computer Security, Vol. 32 No. 4, pp. 523–541. https://doi.org/10.1108/ICS-02-2024-0038
61. Kallonas, C., Piki, A., and Stavrou, E. (2024) "Empowering Professionals: A Generative AI Approach to Personalized Cybersecurity Learning," 2024 IEEE Global Engineering Education Conference (EDUCON), Kos Island, Greece, 2024, pp. 1–10, https://doi.org/10.1109/EDUCON60312.2024.10578894.
62. Aggrey, Richard, et al. "Predicting Future Cyber Threats: Analysing Trends and Predicting Future Cybersecurity Challenges."
63. Kose, Yasin, et al. "Developing Cybersecurity Workforce: Introducing CyberSec Labs for Industry Standard Cybersecurity Training." 2021 International Conference on Computational Science and Computational Intelligence (CSCI). IEEE, 2021.